CAN WE BELIEVE IT?

CAN WE BELIEVE IT?
Evidence for Christianity

GEORGE A. F. SEBER

RESOURCE *Publications* • Eugene, Oregon

CAN WE BELIEVE IT?
Evidence for Christianity

Copyright © 2015 George A. F. Seber. All rights reserved. Except for brief quotations in critical publications or reviews, no part of this book may be reproduced in any manner without prior written permission from the publisher. Write: Permissions, Wipf and Stock Publishers, 199 W. 8th Ave., Suite 3, Eugene, OR 97401.

Resource Publications
An Imprint of Wipf and Stock Publishers
199 W. 8th Ave., Suite 3
Eugene, OR 97401

www.wipfandstock.com

PAPERBACK ISBN: 978-1-4982-8919-1
HARDCOVER ISBN: 978-1-4982-8921-4

Manufactured in the U.S.A.   3/14/2016

CONTENTS

Preface   xii

1   **What Do Science and Mathematics Prove?**   1

     1.1   The Nature of Science   1
         1.1.1   Assumptions of Science   2
         1.1.2   Legacy of Science   3
         1.1.3   Materialism and Naturalism   4
         1.1.4   Communicating and Doing Science   5
     1.2   Enter Mathematics   6
         1.2.1   Numbers in Nature   6
         1.2.2   What is Mathematics about?   8
         1.2.3   Mathematics and Reality   8
     1.3   The Scientific Method   11
         1.3.1   Falsifiability   11
         1.3.2   Mathematic Has Limits   12
     1.4   Conclusion   13

2 Is There a God?   15

- 2.1 A Matter of Faith?   15
- 2.2 Nature of Our Universe   20
- 2.3 The Argument from Abstract Objects   23
- 2.4 The Cosmological and Related Arguments   26
  - 2.4.1 Evidence of a Beginning   26
  - 2.4.2 The Arguments of Aquinas   33
- 2.5 The Teleological Argument   38
  - 2.5.1 The Role of Mathematics   38
  - 2.5.2 Fundamental Laws of Physics   40
  - 2.5.3 Subatomic Particles   43
  - 2.5.4 Fine Tuning   44
  - 2.5.5 Anthropic Principle   49
  - 2.5.6 Multiverses   51
- 2.6 Argument from Design   54
  - 2.6.1 Some General Arguments   55
  - 2.6.2 The Genetic Code   57
  - 2.6.3 Epigenetics   59
  - 2.6.4 Evolutionary Developmental Biology   61
- 2.7 The Moral Argument   61
  - 2.7.1 Universal Values   62
  - 2.7.2 The Moral Law Giver   65
- 2.8 God and Religious Experiences   67
- 2.9 Evidence and Probability   68
  - 2.9.1 The Nature of Probability   68
  - 2.9.2 Probability of God   69
- 2.10 Conclusion   72

3 Is There a Spiritual Dimension?   73

- 3.1 Who are We?   73
  - 3.1.1 Nothing But ...   74
  - 3.1.2 Looking for Meaning   76
  - 3.1.3 Are We More Than Molecules?   77
- 3.2 Are We Spiritual?   78
  - 3.2.1 Neurotheology   79
  - 3.2.2 Our Changing Brain   82
- 3.3 Mind and Brain   84
  - 3.3.1 Qualia   84

	3.3.2	Consciousness	87
3.4		Dualism or Monism	90
	3.4.1	Near Death Experiences	92
	3.4.2	Mind Over Matter and OCD	94
3.5		Biblical Concepts of Body, Mind, and Spirit	97
3.6		External Evidence of a Spiritual Dimension	98
	3.6.1	Spiritual Forces	99
3.7		Conclusion	102

4 Do We Have Free Will? — 103

4.1		What Is Free Will?	103
	4.1.1	Does Randomness Exist?	104
	4.1.2	Heisenberg's Uncertainty Principle	109
	4.1.3	Enter Quantum Mechanics	110
	4.1.4	Relativity theory	116
	4.1.5	Short Comings of Materialism	116
	4.1.6	Uncertainty and Determinism	117
4.2		Determinism, Indeterminism, and Self-determinism	118
	4.2.1	Libet's Experiments	120
4.3		God's Sovereignty and Human Accountability in the Bible	124
	4.3.1	Four Answers	125
4.4		Conclusion	130

5 Is the Bible Reliable? — 131

5.1		General Principles	131
5.2		Old Testament	134
	5.2.1	Earlier Manuscripts	135
	5.2.2	Internal Evidence	136
	5.2.3	Old Testament Canon	144
	5.2.4	Archaeological Evidence for the Old Testament	146
	5.2.5	Early Chapters of Genesis	147
5.3		Old Testament Codes	150
5.4		New Testament	151
	5.4.1	Authorship and Dating	151
	5.4.2	Archaeological Evidence for the New Testament	156
	5.4.3	New Testament Manuscripts	157
	5.4.4	Reliability of the Gospel Writers	158
	5.4.5	Consistency of the Gospels	159

	5.4.6 Choosing the Canon	164
5.5	Prophecy in the Bible	166
5.6	Alleged Biblical Contradictions	167
	5.6.1 Unpleasant Old Testament Stories or Practices	171
	5.6.2 New Testament Difficulties	177
5.7	The Bible and Science	178
5.8	Scriptural Inspiration	180
5.9	Conclusion	181

6 Who is Jesus? 183

6.1	Evidence for His Existence	183
6.2	The Ethics of Jesus	185
6.3	Outstanding Features of Jesus' Life	188
	6.3.1 He Had an Extraordinary Introduction into History	188
	6.3.2 He Had Extraordinary Abilities	192
	6.3.3 He Made Extraordinary Claims	195
	6.3.4 He Had extraordinary Powers	199
	6.3.5 He Had an Extraordinary Birth	200
6.4	He Had an Extraordinary Death and Resurrection	201
	6.4.1 What Happened?	201
	6.4.2 Paul's Early Writing	203
	6.4.3 The Evidence	205
	6.4.4 Probability Arguments	212
	6.4.5 Some Alternative Views of the Resurrection	218
6.5	Is Jesus the Only Way?	218
	6.5.1 All the Major Religions are Exclusive	219
	6.5.2 New Age	221
	6.5.3 Mystery Religions	222
	6.5.4 Relativism	223
6.6	Comparing Religions	223
	6.6.1 Their Founders	224
	6.6.2 Universality	226
	6.6.3 Nature of God	226
	6.6.4 Nature of Sin	227
	6.6.5 Scriptures	227
	6.6.6 Nature of the Afterlife	228
	6.6.7 Obtaining Salvation	228

	6.6.8	Tolerance and Compulsion	230
6.7		Those That Haven't heard	230
6.8		Is God Unfair?	232
6.9		Conclusion	235

7 Do Miracles Occur? — 237

7.1		Definitions and Theories	237
7.2		The Rise and Fall of Scientific Laws	240
7.3		Miracles and Natural Law	242
7.4		Biblical Miracles	243
	7.4.1	Categories of Miracles	249
	7.4.2	God the Free Creator	249
7.5		Conclusion	250

8 Why Does God allow Suffering and Evil? — 253

8.1		Introduction	253
8.2		Positive Aspects of Pain	257
8.3		Why Did God Create Us?	257
8.4		Some Philosophical Arguments	259
	8.4.1	Best Possible World	260
	8.4.2	Further Answers	261
8.5		Free will	262
	8.5.1	We Have Some Freedom	262
	8.5.2	We are Free to Hurt Ourselves	263
	8.5.3	We are Free to Hurt Others	264
	8.5.4	We are Free to Ignore God	264
8.6		An Impartial World	265
8.7		Natural Disasters	265
	8.7.1	Human Influence in Disasters	266
	8.7.2	Everything Falls to Bits Sooner or Later	266
8.8		Animal Suffering	268
8.9		Biblical Perspective	270
	8.9.1	Genesis Chapter 1	270
	8.9.2	Genesis Chapters 2 and 3	277
	8.9.3	The World is a Spiritual Battleground	280
	8.9.4	Animal Suffering and the Bible	281
	8.9.5	The World is to Be Restored	283
	8.9.6	Suffering and Other Religions	283

8.10		Some Traditional Answers	284
	8.10.1	Suffering is God's Judgement	284
	8.10.2	Satan Can Afflict People	286
	8.10.3	Reincarnation: Suffering from Sin in Past Lives	286
	8.10.4	Suffering is All in the MInd	287
8.11		Some Old Testament Responses	287
8.12		Some Positive Aspects of Suffering	291
8.13		An Approach to Suffering	295
	8.13.1	The Death of Christ	295
	8.13.2	Conclusion	298

9 Is Christianity a Blessing or a Curse to Society? — 299

9.1		Shameful Events	299
9.2		Missionaries	301
9.3		Health Benefits of Christianity	301
9.4		Science Flourished under a Christian Umbrella	302
9.5		Christianity and Society	306
	9.5.1	Jews and Gentiles	306
	9.5.2	Husband and wife	306
	9.5.3	Parents and children	308
	9.5.4	Masters and slaves	309
	9.5.5	Other social contributions	310
9.6		Conclusion	312

10 What About Evolution? — 313

10.1		The Nature of Evolution	313
	10.1.1	Hox Genes	315
	10.1.2	Evolution of Everything	316
	10.1.3	Can We Trust Evolution to Get it Right?	317
10.2		Some Evidences for Evolution	318
	10.2.1	Vestigial Organs in Humans	321
	10.2.2	Evidence of Poor Design	321
10.3		Evidence for Human Evolution	324
	10.3.1	Pseudogenes	325
	10.3.2	Establishing Connections	325
	10.3.3	Retroviruses	328
	10.3.4	Genetic Parasites	329
10.4		In the Beginning...	329

	10.4.1	A Prebiotic Soup?	330
	10.4.2	Evolution of Proteins	331
	10.4.3	Evolution of the Cell	334
	10.4.4	Animal Fossils	335
	10.4.5	Human Evolution	340
	10.4.6	Second Law of Thermodynamics	342
10.5	Irreducibly Complex System		344
10.6	Is Evolution Compatible with the Nature of God?		347
10.7	Conclusion		349

11 How Do We Get to Know God? — 351

11.1	Introduction		351
11.2	What is God Like?		352
11.3	What Does it Mean to Know God?		354
11.4	Is God Knowable?		354
	11.4.1	The Role of God's Creation	355
	11.4.2	God's Revelation	356
	11.4.3	Life Without God	357
11.5	Reaching Out		359
	11.5.1	Contemplation	359
	11.5.2	Steps to Communication	359
11.6	Conclusion		365

References — 366

Index — 383

PREFACE

This book is written for those who have an atheistic or agnostic world view, or are just curious and have some questions they are grappling with. Christians will also find the book helpful in dealing with questions asked of them. Christian leaders such as youth pastors will be able to use this book to handle questions from young people generally and, in particular, students from high school and university.

Recently there has been a flurry of books written by atheists, sometimes referred to as the "New Atheists." In response there have been a number of books written by Christian authors such as Craig, D'Souza, Feser, Glass, Ward, and so forth, and there are numerous debates available online. It is not my intention to become involved in this discussion as I don't wish to be tied to just topics raised by the New Atheists as I believe that generally the questions they raised have been responded to adequately. However their questions are important and tend to surface frequently. Instead I have preferred to begin afresh and look again at basic questions that thinking readers, Christian or otherwise, struggle with.

Most of the Christian responses to the atheistic literature involve substantial books and are often quite specialized, so I have endeavored to try and keep arguments precise and not get too involved with philosophical discourse as a big book would defeat the purpose of my writing. Because the topic is such a big one and encompasses many disciplines I regard my book as more of a handbook where one can look up a specific topic fairly readily. I appreciate that every subtopic can be expanded considerably, but my aim is to try and provide detailed and concise arguments.

A great many books have been written on apologetics and the question is why another book. Science evolves and consequently new questions are raised that require new answers. For example, quantum mechanics has revolutionized our understanding of reality and raised serious questions about where science is heading. The topic of epigenetics throws a new slant on genetics. Chaos theory brings a new perspective on predictability. Apologetics therefore needs to be revisited from time to time. As a mathematician having spent many decades writing books about mathematical statistics, I believe that mathematical and statistical arguments can also have a brief place in the discussion.

I became a committed Christian at the age of fourteen years. Some might ask, "What did a fourteen year-old boy know about anything in those pre-computer days?" Very little. However, God is no respecter of age, gender, race, economic status, or intelligence, and the encounter that I had with God completely changed my life for good. While at university studying mathematics and physics I had many questions about my beliefs such as "What am I here for?" and "Where did the universe come from?" and "Where does evolution fit into the big picture?" While studying for a PhD in Statistics at Manchester University England I found others interested in similar questions so I continued to read around the subject. I began writing some notes on the books I read, which was the embryo of this book.

After teaching mathematics and statistics for two years at the London School of Economics to obtain some teaching experience, I returned to New Zealand (a condition of my Commonwealth Scholarship) as a lecturer in Mathematics at Auckland University with my primary interest being in the theory and application of Statistics. During that time I began giving talks about reasons for my belief in Christianity, and also wrote a number of articles in the same vein. Eventually I was appointed to the foundation Chair of Statistics and during those earlier years I was involved with running a Christian Staff Fellowship. In my retirement years I trained as a counsellor/psychotherapist and have been counseling for over 12 years as well as writing a counseling book and continuing to write statistics books. Now in my late seventies I finally fulfilled a long-term goal and have written this book. I decided to focus on a number of key questions that form the titles of the chapters. My emphasis is on evidence, as when it comes to any kind of belief system we can never be completely certain of our beliefs, but we can at least look at the balance of evidence both for and against.

As we will be dealing with science and to a much lesser extent with mathematics throughout the book, we look in the first chapter at the nature of science and mathematics, what it can achieve, and what are some of its limitations. In the second chapter we launch into looking at evidence for God's existence, which covers a number of different subject areas. Here we are generally not focussing on the type of God whose existence is in question, but rather looking at evidence that supports the existence of a particular being that created the universe. Some characteristics of such a being can be inferred from the nature of our universe such as being timeless in some way, and being the author of the morality that we find within us. As a mathematician I am very much aware of how much mathematics, and more recently statistics, underlies science. I can perhaps be forgiven for referring to both topics in that chapter.

In looking at the question of God's existence it is natural to ask about my own existence, so in Chapter 3 I consider whether there is a spiritual dimension to my life. The current approach in much of science and particularly in psychology is reductionistic where everything is reduced to "nothing but". Some problems with materialism are discussed and evidence is given that the mind is more than the brain. With my recent interest in psychology, particularly in the amazing adaptability of our brains, I have become even more aware of what distinguishes us as human beings, and whether there is more to me than just matter. It is interesting to consider how different humans might be from the other animals with regard to consciousness and self-consciousness. If there is a spiritual dimension, then this raises the question of whether there is a continued existence beyond the grave, and whether spirit beings exist or not.

Chapter 4 is very much a continuation of Chapter 3 and is concerned with whether we have free will of not, and if we do how it might be possible. The nature of randomness is considered and materialism again comes under the spotlight when we look at reality through the lens of chaos theory and quantum mechanics with its introduction of probability. The nature and roles of determinism, indeterminism, and free will are considered in detail. I digress at the end of the chapter and take a look at a problem area for Christianity, namely the clash between God's sovereignty and human accountability. Admittedly this is theological, but the question of reconciling a God who is supposed to be omniscient with human free will and responsibility is a question that has raised much debate.

Since this book is about Christianity, and the Bible is the source book of Christianity, we shall look at the Bible in Chapter 5 from a number of vantage points and initially consider the Old Testament and New Testaments separately. It is important to understand the culture of the day as to how information was passed on and to appreciate the focus of the writers in presenting historical material. Authorship is an important question and we also need to know if we have accurate copies of the original documents; external and internal evidence for authorship and accuracy is presented along with extensive manuscript evidence. As the Bible is a collection of books we consider how the books were chosen, as there exist other competing documents. Some external evidence from archaeology is given for both Testaments that support the historicity of the documents. Differences of description about the same events do occur and reasons for this are discussed in detail. Prophecy is an important aspect of the Bible. Some difficult passages and contradictions in the Bible are discussed, and the relationship of the Bible to science is briefly considered as there are some misconceptions relating to this topic. The chapter closes with comments about the inspiration of the Bible and how the Bible sees itself.

Since the New Testament describes the life and works of Jesus Christ we are now in the position in Chapter 6 to examine his life and try and understand who he was. We begin by demonstrating his historical existence from internal and external sources. He was an extraordinary man and we examine in considerable detail his claims and abilities. Evidence for his death and resurrection is given, including some probability arguments (please forgive me!). The idea that all religions are exclusive is raised and discussed. So-called mystery religions are mentioned briefly along with their relationship with Christianity. We then compare the major religions with respect to their founders, how universal they are, their understanding of God and sin, their scriptures, their understanding of the after life, their way of salvation, and their degree of tolerance and compulsion. The chapter closes with some theological issues like the existence and nature of hell, and whether God is unfair.

When talking about the life of Jesus and the Bible in general, the topic of miracles eventually comes up. We discuss in Chapter 7 what is a miracle and look briefly at some philosophical views on miracles. We revisit some of the ideas about the nature of science mentioned in earlier chapters and discuss whether or not miracles can be accommodated within our view of science today. The nature and types of biblical miracles and their role in general in the Bible are looked at.

Chapter 8 deals with the age-old problem of why there is evil and suffering in the world, a problem that has been extensively considered by philosophers over centuries. It is often the topic raised by skeptics in an attempt to disprove God's existence. After considering some philosophical arguments we begin to look more closely at some possible answers. The Bible has a lot to say about suffering, its role, and how we can cope with it. It is hoped that the reader will find the chapter is more than just a philosophical and biblical debate, but will provide some idea on how to approach suffering more positively.

There are many skeptics today who believe that religion is bad for society. In answer to this belief I shall focus in Chapter 9 on just Christianity, and not on religion in general. While acknowledging that the Christian church is regarded by some as having a mixed track record on occasion in the past, I shall give evidence that Christianity has done an enormous amount of good for society generally, both with regard to scientific progress and in promoting equality of all people.

In Chapter 10 we look at evolution both at the chemical and biological levels. As evolution is a very controversial topic not only amongst Christians but also among evolutionists, my approach is to try and give some of the evidence used to support evolution as well as considering some difficulties with the current theory of neo-Darwinism. Since evolution is supported by many Christians, I believe that how one views evolution is not essential to the Christian faith, though no doubt some may disagree with me. We look at some features of evolution and how it occurs, and digress into topics like vestigial organs, transitional forms, and arguments about poor design features. Fossil evidence is considered, and evidence for human evolution is discussed. We look at various problems that arise with chemical or prebiotic evolution such as the earth's initial conditions, the evolution of proteins and related molecules, and the cell. In a sense the chapter is not about evidence for Christianity though some would see the direction of evolution as pointing to a creator God. However, others would try to use it as evidence against Christianity. There are many questions, and it is hoped that the chapter clarifies some issues for the reader.

Chapter 11, the final chapter, is more of an epilogue and I raise the question of how we can connect with God. Clearly there are many ways that are proposed in this multicultural world, and one could write many volumes on the subject. However, I just look at one approach that I have found to be useful and hope that the suggestions there are helpful. As a counselor/psychotherapist I am aware that the counseling literature emphasizes the importance of the spiritual side of our natures irrespective of whether the counseling is religious or secular, and I personally believe that we neglect the spiritual aspect of life at our peril.

In conclusion I wold like to express my thanks to a number of people who have read some draft chapters and provided helpful comments, namely Zachary Arden, Russell Belding, Carl Cerecke, Graeme Finlay, Wilf Malcolm, Hugh Morris, and Matthew Su. I am very grateful to Amy Henrickson for allowing me to use her modified LATEX computer package.

GEORGE A. F. SEBER

Auckland, New Zealand
February 2016

CHAPTER 1

WHAT DO SCIENCE AND MATHEMATICS PROVE?

1.1 THE NATURE OF SCIENCE

I find that today there are many misconceptions about the nature of science. A common view is that science and religion (especially Christianity) are in conflict and that a thinking person would not believe in Christianity. However, the statistics tell us that a large majority of people believe in some sort of God or super being, but tend to keep science and such beliefs in separate boxes. I personally believe that science and Christianity are not only compatible but are also supplementary and tend to focus on different questions. Science can tell us what an amazing universe we live in that demands an answer to why it is so, but it cannot answer existential questions such as what is my purpose in life, assuming of course that a purpose exists. Some believe that the only valid knowledge is scientific knowledge, and that science has all the answers, or at least will eventually provide all the answers.

A student once said to me, "I don't mind what people believe as long as they don't hurt anyone because of their beliefs." I guess this reflects the attitude today; anything goes and do it if it is good for you. Truth is no longer relevant except, for example, when issues relating to God and Christianity are raised and someone says, "prove it to me." In this short chapter we consider the nature of truth in both science and mathematics.

In the next chapter we ask the question "Is there a God?" and look to science for some partial answers. It is therefore appropriate to first discuss the nature of science; its assumptions and what it can prove and what it cannot prove. Generally speaking, science endeavors to find useful models of reality to try and explain why an observation or pattern of observations that we see actually occurs.[1] For example, in chemistry we can use a model for a gas in a container where we describe the gas as consisting of particles colliding with the sides of the container thus giving rise to gas pressure. In psychology, depression has been described in some cases as a lack of the neurotransmitter serotonin in the brain. Traditionally science has proceeded along two main lines: the natural sciences that study the natural world (e.g., physics, cosmology, chemistry, geology, zoology etc.) and the social sciences that study human behavior and society (e.g., psychology, anthropology, sociobiology etc.) However, the differences between these two streams have become somewhat blurred, with some crossing over (e.g., neuroscience, genetics).

1.1.1 Assumptions of Science

Science depends on a number of assumed beliefs, namely: (1) The universe is an ordered and rational place so that doing science makes sense (why should it?); (2) The underlying "laws of nature" governing the universe tend to be mathematical and are in a sense "beautiful" (to the mathematician!), and the question is why mathematical; (3) The laws of nature are stable over time and are the same throughout the part of the universe that we can reach with our instruments (why not chaos?); and (4) Our minds are rational and we are capable of uncovering the laws of nature (why should we be able to do this?). Clearly when we do science we make these assumptions but they cannot be proved to be true. For all we know, the scientific laws may all be different tomorrow; for example the speed of light may change.

The idea of a rational cosmos is an ancient one. However, as we shall see in Section 9.4, it was Christianity that revitalized the idea by acknowledging that God made the universe according to divine reason so that the universe is therefore a logical and rational place, and we can come to understand it as we are made in the image of God. This means that in a small way we have the power to reason and comprehend the universe, and thus achieve certain goals. David Glass, physicist and philosopher, comments:

> What about atheism? How could it account for the order in the universe? Could a scientific explanation be given for it? It seems not, since the order is found in the scientific laws themselves.[2]

---

[1] As a side issue, given the introduction of quantum theory in Section 4.1.3, we shall be faced with the question of what is reality and can we know it.
[2] Glass (2012: 123).

We need order for intelligent beings to exist, but there will no doubt be some things we find difficult to understand, as any God who has the power to create the universe would have ways and thoughts higher than our ways and thoughts.[3] As we shall see later, quantum mechanics as it is currently theorized fits into this difficult category where ideas are non-intuitive, though a more intelligible theory may one day replace it.

There are those who would say that science is the source of all truth. However that is a self-defeating or self-refuting philosophical assumption as the statement itself cannot be tested by the scientific method.[4] George Ellis, eminent physicist and mathematician, when interviewed said: "The belief that all of reality can be fully comprehended in terms of physics and the equations of physics is a fantasy."[5] One of the problems is to distinguish between science and philosophy. Ellis says, "You cannot do physics or cosmology without an assumed philosophical basis", so that some skeptics who are not philosophers present "scientific" ideas that are purely philosophical without any evidential foundation. This is particularly true with cosmology. As already indicated, the so-called laws of nature are intelligible and discoverable by beings with our level of intelligence, which suggests a providential purpose for humankind. We seem to be designed to do science.[6]

1.1.2 Legacy of Science

Having looked at some of the assumptions underlying science and scientific endeavor we now consider where has science led us and what is its legacy. Science has brought with it technology and new developments in all areas of life, but its legacy has not led to a sense of being at ease. Although science has made some huge advances, we are now faced with increasing pollution, increasing rubbish, traffic congestion, overcrowding, rising costs, aggression and crime, ghettos and slums, and tensions between races, classes, and social groups. Global climate change has also been unsettling as it has raised the specter of unstable weather patterns. We are running out of some resources, especially some minerals, and the distribution of water is becoming a major problem.

There is also a rise in psychological problems, and mass media has led to indoctrination and the downgrading of language. German theologian Hans Küng says:

---

[3] Isaiah 55:8–9.
[4] Self-defeating statements, of which there are many, are ones that contain within them a contradiction to the statement itself. For example, the only thing I am sure of is that nothing is for sure. See the internet for examples.
[5] See http://ipwebdev.com/hermit/physicistgeorgeellis.html, accessed August 2015.
[6] For comments and a good reference list for this section see Collins (2008), http://www.infidels.org/library/modern/robin_collins/design.html, accessed August, 2015.

Modern psychologists have established a significant connection between the decline in religiosity and increasing disorientation, lack of standards, loss of meaning, the typical neuroses of our time.[7]

Historian Arnold Toynbee comments:

In my belief, science and technology cannot survive as substitutes for religion. ... Science has never superseded religion, and it is my expectation that it will never supersede it.[8]

Science has brought many benefits, but has not turned out to be a cure-all. John Polkinghorne, physicist and theologian, sums it up well when he writes:

science and religion are friends not foes, 'cousins under the skin', since they both participate in the great human quest for truthful understanding attainable through well-motivated belief ...The scientist and religious believer both walk by reasonable faith and not by incontrovertible sight.[9]

1.1.3 Materialism and Naturalism

Classical mechanics and scientific developments especially in neuroscience have led a number of scientists to believe in materialism (or physicalism using a more recent term), that is everything is made of matter so all phenomena (including consciousness and mental events) are the result of material properties and material interactions. For the materialist, the material world is all there is and therefore science is all that is needed, including laws yet to be discovered; materialism is discussed in more detail in the next chapter.

A related concept is naturalism where all phenomena can be explained in terms of natural causes and laws, including laws yet to be discovered. However science is limited as it can only answer certain questions. For example, personal explanations as to why we have intentions, awareness, and do certain things cannot seem to be reduced to physical explanations such as the motion of electrons in our brains. As argued by philosopher Richard Swinburne,[10] theism is about a super-personal God (being more than just personal) with intentions, and that God's existence provides a good personal explanation as to why the universe exists and why it has various features described in the next chapter. This idea is supported by theologian Keith Ward who claims that a personal explanation does not seem reducible to a scientific explanation.[11] As such, the existence of God is not a purely scientific hypothesis requiring a scientific explanation. Glass comments that a "personal explanation takes us beyond the limits of science."[12] It does not matter if the world has a finite or

---

[7] Küng (1984: 59).
[8] Toynbee (1971: 44–45).
[9] Polkinghorne (2012: 3).
[10] Swinburne (2004: 26–47, 61–66).
[11] Ward (2008a: 23).
[12] Glass (2012: 72).

infinite past, is infinite or finite in size, or is a single universe or multiverse,[13] as we still have the question of why it exists when it might not have.[14] While still considering the nature of science, we digress for a moment to look at how science is communicated.

1.1.4 Communicating and Doing Science

When it comes to discussing science, there are two factors affecting communication problems today about science, namely the explosion of science and the increase in specialization. This specialization has come about because our theories about the world are becoming so complex that researchers can only study very specialized areas of science. I am reminded of a famous story about blind men invited to feel an elephant.[15] They all gave different descriptions as perceived by their sense of touch and the part of the elephant they touched. Today scientists are furiously examining this "elephant" (our world) with their senses or their senses augmented by special instruments, and are coming up with different ideas about what they find. Some are so busy cataloguing information about one end of the elephant (the tail) that they cannot hear what their colleagues are saying at the other end (the head). I sometimes suspect that the very existence of the elephant is forgotten.

Because of of specialization we have two problems. First, we have to rely on popularizers of science to explain in less technical terms scientific developments. Unfortunately in the "watering down" facts often get confused with personal speculations, especially with news media and science writers who inevitably have their own agendas. There is also a danger of extrapolating ideas from one branch of science into another without justification. This has been done particularly with the biological theory of evolution that has sometimes been used as a complete explanation for everything rather than a claim about biology (see Section 10.1.2). Second, because of the delay in propagating new knowledge through the educational system we are generally out of date, though the internet has helped to narrow the time gap. As one who has been engaged in doing science (e.g., ecology and human blood genetics) I want to say something briefly about actually doing science.

There is one important fact about modern science that needs to be mentioned. We now realize that an experiment cannot be performed without the experimenter having an effect in some way on the experimental outcome. Subject and object cannot be separated. At the biological level it is not possible to study a living organism without interfering with its environment. At the measurement level the theory of relativity tells us that even so-called fixed quantities like time, matter, and size all depend on the framework of the observer. As mentioned in Section 4.1.3, where quantum mechanics and related

---

[13]These topics are considered in the next chapter.
[14]Holder (2013: 83).
[15]This ancient story has several versions; see "elephant in a dark house" on the internet.

matters are discussed in more detail, at the atomic level quantum theory tells us that we cannot learn anything about an atom unless we interfere with it using at least a quantum of energy.

In doing science we need to distinguish between the facts and the scientific theories developed to explain them, as sometimes these two ideas are confused. Roughly speaking, facts are what we can observe whereas a theory endeavors to explain what has been observed and why it is so. For example, we may wish to determine a certain theory of migration throughout the Pacific islands that explains certain facts such as language items or items found in burial sites obtained from the various islands. We would like to develop some ideas of closeness or similarity to see if migration connections can be established. If the use of certain words or artifacts are more "similar" for two islands than for others we can perhaps suspect a migrational connection between the two islands. I have chosen this somewhat unusual example to not only demonstrate the difference between fact and theory, but also to show the variety of scientific theories we can have and to note that mathematics (statistics) is used even here in the definition of "closeness." We can set up a measure of "distance" between two islands based on how similar certain items are on both islands.

As mathematics plays an important role in scientific theory and people tend to believe that mathematics is truth, we shall look more closely at the role mathematics plays in science. I realize that many readers may not be too excited about mathematics so that they may wish to skip some of the following sections.

1.2 ENTER MATHEMATICS

We shall look at two aspects of the role of mathematics, first in nature and then in science. As a mathematician and one who has been hooked on mathematics from my early teens, I often wonder where my ideas come from. Mathematics seems to have a life of its own and it has its own language, as with other subjects.

1.2.1 Numbers in Nature

Science writer Philip Ball[16] in a three volume set demonstrates in many ways (e.g., hexagonal crystals, water and sand flows, branches in trees, formation of cracks, to name a few) that structure and mathematics underly nature because of physics and chemistry. With our present state of knowledge it does not seem to be a question of a particular pattern arising from evolution and having an adaptive survival value. On the theme of mathematics, there is the so-called Fibonacci sequence F_n of numbers 1, 1, 2, 3, 5, 8, 13 , 21, 34, 55, 89,... that has the property that any number in the sequence is the sum

---

[16]Ball (2009).

of the two previous numbers. The starting pair of values in the sequence is either 1,1 or 0,1. The numbers are also related to the so-called "golden ratio" or "golden proportion"[17]

$$\phi = \frac{1}{2}(1 + \sqrt{5}) = 1.6180339887\ldots,$$

which satisfies the equation $\phi^2 = \phi + 1$. Here F_n, the nth number in the sequence is

$$F_n = \frac{\phi^n - (-\phi)^{-n}}{\sqrt{5}}.$$

This ratio was defined by Euclid more than two thousand years ago because of its crucial role in the construction of the pentagram, a figure to which magical properties had been attributed. It is asserted that the builders of the pyramids and the Parthenon used it. When n is sufficiently large (e.g., 13 or more) we can effectively ignore ϕ^{-n}, the inverse ϕ^n, as it gets smaller and smaller as n gets bigger and bigger so that the ratio of consecutive numbers F_n/F_{n-1} is close to ϕ and the ratio is essentially constant, namely approximately 1.6180. One of the dangers in looking at sequences of numbers or numbers in nature is that it is very easy to see patterns that may not really exist or have significance. However, it is interesting that both the Fibonacci sequence and the golden ratio do turn up in all sorts of places. Some quote that in the human body most of our body parts follow the numbers one, two, three, and five; for example three segments in each limb, five fingers on each hand, and three bones in each finger.

There are examples in nature such as the lily with three petals, buttercup and wild rose with five petals, delphiniums with eight petals, cineraria and ragwort with thirteen petals, aster and chicory with twenty-one, plantain and pytethrum with thirty four, michelmas daisies and the asteraceae family with 55 and 89 petals respectively, and many more plants showing numbers from the sequence in their leaves, growth patterns, branches, and in some cases spiral patterns (called golden spirals). For example, pineapples have a particular double set of spirals—one going in a clockwise direction and one in the opposite direction. When these spirals are counted, the two sets are found to be adjacent Fibonacci numbers. The same is true for tapered pine cones. In the case of branches it is suggested that this design provides the best physical accommodation for the number of branches, while maximizing sun exposure.

The golden ratio appears in animals and humans, for example in our finger digits, the lengths of our forearm and hand, measurements on our four front teeth, eye width and distance between eyes, and heart rhythms. With the ratio, the proportion of the smaller part to the greater is the same as the proportion of the greater to the whole so that the dividing point is in some

---

[17] Livio (2002).

sense in "equilibrium." The eye senses the balance of this proportion and it occurs very much in art and fashion, as well as in nature.[18] Mathematics occurs in many places in nature and it will be discussed further in Chapter 3 when we look at chaos theory. We now look at the general role of mathematics in science.

1.2.2 What is Mathematics about?

Mathematics has come to play a fundamental role in our scientific theories, especially those relating to some of our most fundamental theories in physics that underlie all physical, chemical, biological, neurological or cerebral processes. Cosmologist Paul Davies sums it up well when he says,

> there is no logical reason why nature should have a mathematical subtext in the first place. And even if it does, there is no obvious reason why humans should be capable of comprehending it.[19]

With regard to the role of mathematics I like the following quote from John Polkinghorne:

> Some have suggested that humans happen to have a taste for mathematics and so they mould their accounts of physics into forms that gratify this preference. Previous discussions of the difficulty of theoretical discovery, and the way in which the universe resists our prior expectation encourage the contrary realistic view that these mathematical patterns are read out of, and not read into, the structure of the world.[20]

For me he is essentially saying two things. First, the mathematics is unexpected and unpredictable, and second, we don't impose our mathematics on the world but the world imposes the mathematics on us. We uncover the underlying structure and not impose it; the structure is discovered and not created. For example, Einstein's general theory of relativity, though counterintuitive and not necessarily in final form (as theories can change), is remarkably precise in its predictions and its agreement with the physical world. To see how mathematics interacts with science I wish to digress for the moment and look more closely at mathematics and its relationship with reality.

1.2.3 Mathematics and Reality

The reader may be surprised to find that mathematics is not really about real things. Mathematics talks about straight lines, triangles, planes, circles and spheres, but these are defined mathematically as perfect structures that may not actually exist, or if they did we would have no way of determining

---

[18]There are many examples given on the internet; for example http://www.maths.surrey.ac.uk/hosted-sites/R.Knott/Fibonacci/fib.html, accessed August 2015
[19]Davies (2006: 5).
[20]I have lost the reference for this; for further discussion see Polkinghorne (2011).

if they were perfect as we don't know how to make perfect measurements. For example, if I drew a straight line with a ruler on a piece of paper and looked at the result under a strong microscope, we would find that the line is actually full of bumps and hollows and not perfectly straight. But how do we know that the ruler is perfectly straight? We can collect a whole family of apparently straight edges by looking at the edges of books, tables, and pencils etc. and come up with the abstract idea of a perfectly straight edge or line that incorporates all these concrete versions of it as special cases. In the end we can only define what a perfectly straight line is mathematically, and this enables us to plot the line on a graph.[21]

We can then go a step further and consider an abstract triangle that has perfectly straight lines as sides. If we know the exact lengths of the sides of the triangle and the exact angles at each corner we are able to derive various mathematical formulae about the triangle such as its area. The same applies to rectangles and other mathematical objects. Fortunately our measuring devices are accurate enough for us to be able to build things and use the abstract mathematical formulae as very good approximations to deal with real situations. Furthermore, modern science has given us the laser that has the advantage that over short distances light travels in virtually a perfectly straight line so that we can use it for determining distances very accurately (but not exactly). Over very large (cosmic) distances, however, light can be substantially bent by gravity.

We see then that mathematics can often be used to give us a very good approximation to reality that helps us when we carry out certain activities. If the underlying assumptions are exactly true, then the mathematical formulae or "laws" hold exactly; otherwise our formulae are usually good enough. However, since mathematics is about an idealized world, the application of abstract mathematics to the real world can be messy and not always straightforward. Philosopher Nancy Cartwright says that we need to

> distinguish between the tidy and simple mathematical equations of abstract theory, and the intricate and messy descriptions in either words or formulae, which express our knowledge of what happens in real systems made of real materials, like helium-neon lasers or turbo-jet engines.

She adds: "Nature is not governed by simple quantitative equations of the kind we write in our fundamental theories."[22] The laws are not false, just that their nature is more complicated than has been assumed in the past.

Sometimes formulae arise through the interaction of experimentation and mathematics. If an experimenter collected a series of different sized circular objects and plotted the length of the circumference (using for example a piece of string) versus the diameter length for each object, the experimenter would find that the points on the graph were close to being all on a straight line thus indicating that the circumference of a circle appears to be roughly a constant

---

[21] The equation is $y = ax + b$.
[22] Cartwright (1983: 128–129).

times its diameter.[23] If we now assume a *perfect* circle, that is one with a known centre and having a fixed (exact) radius r from the centre to the circumference, we can then use mathematics (calculus) to prove that $\ell = \pi d$, where ℓ is the length of the circumference and $d = 2r$ is the diameter of the circle. Here π is approximately 3.14159. We can similarly prove mathematically that the area of a perfect circle is $a = \pi r^2$. Although we cannot make a perfect circle, or know if we have one, we can make one perfect enough for the formula to still hold to a high degree of accuracy and we can use it to make circular objects like a ring.

The Pendulum Experiment

I want to consider just one more example that invokes some laws of mechanics rather than just those using calculus. Suppose we have a pendulum that consists of a small metal ball on the end of a thin very light piece of wire of length L suspended from a support. The ball is given a slight push so that it swings back and forth through a small angle (θ, say), taking time T (the period of oscillation) to complete the full swing. If different length wires are used with the angle θ approximately the same each time, and T is plotted against $\sqrt{L}$ (the square root of L) for each wire, we find that the points lie approximately on a straight line indicating that the period of oscillation is closely proportional to the square root of the length of wire. If we now idealize the experiment assuming that the ball is small enough so that the air resistance is negligible, the wire is weightless and inextensible, θ is small enough, and there is no friction at the support, then using calculus and Newton's laws of motion we can prove mathematically that $T = 2\pi\sqrt{L/g}$, where g is gravitational acceleration.

The accuracy of this result with regard to a real-life pendulum will depend on how closely our pendulum conforms to the idealized pendulum. For a perfect pendulum in a vacuum the oscillation would continue indefinitely, but in practice it would slow down due to friction and air resistance, and eventually stop. We may also be interested in changing the angle θ or the weight of the ball to see what happens. Generally the results are not affected by the weight of the ball (though a bigger ball will have more air resistance) or by the angle θ provided it is small enough. However the formula needs modifying if the angle θ is much larger.[24] It turns out that the error in using our simple formula is less than one percent when θ is less than 22°. What this example tells us is that formulae may be useful up to a point, but may have to be modified to allow for other possibilities such as a bigger angle of oscillation. This idea is important below when we consider how science progresses.

---

[23]This result has a long history
(see http://www-history.mcs.st-and.ac.uk/HistTopics/Pi_through_the_ages.html), accessed August, 2015.
[24]This is because the mathematical approximation $\sin\theta \approx \theta$ required for the formula no longer holds.

1.3 THE SCIENTIFIC METHOD

Having considered something of the role of mathematics in science we now consider what we mean by the scientific method. Returning to the previous pendulum example, if we set the angle of the pendulum to $\theta = 45°$ we would find that our original formula would give us wrong results so that our formula would need modification. Science is like that. We develop a theory that we accept while it fits experimental observations. If we come across an experiment or an observation that does not fit the theory we have to modify the theory. We see then that a scientific law is always on probation; subject to good behavior. The law is accepted provisionally until a new scientific experiment indicates that it is either wrong or needs some modification. A scientific law therefore cannot be proved, only disproved or modified for the time being. Sometimes only part of a theory is at fault and the problem is to determine which part as theories are usually networked with other theories. This process of falsification is the way science advances. Some laws can be derived mathematically, as we saw above with the pendulum experiment, but they depend on assumptions that need to be verified experimentally, while others are empirical laws based solely on observations. This topic is discussed further in Section 7.2.

A good example of scientific progress relates to Newton's laws of mechanics and his inverse square law of universal gravitation. Kepler earlier obtained empirically a law that says that the square of the time of revolution of a planet is proportional to the cube of its mean distance from the sun, but Newton's laws can now be used to derive it mathematically. Some irregularities were observed in the orbit of the planet Uranus, and Newton's theory not only explained the irregularities being due to the presence of another planet, but also predicted exactly where this planet (Neptune) should be. Unfortunately Newton's model was not sufficiently accurate to explain some slight perturbations in the behavior of the planet Mercury's orbit that could not be explained by another planet being closer to the sun. One contrary observation was enough to require a new theory, and this was provided by Einstein's theory of general relativity in 1915. It explained the small discrepancy in Mercury's orbit, technically called the perihelion precession of Mercury. Newton's laws, however, are still surprisingly accurate so that the adjustment was small.

1.3.1 Falsifiability

An important concept concerning the scientific method is the notion of falsifiability due to Karl Popper who was primarily a logician.[25] There are of course some statements that cannot be tested to see if they are false. For example, if I claimed to have carried out a certain action with no witnesses present last

---

[25] Popper (1963: 33–39) called this falsifiability. For an excerpt see http://www.stephenjaygould.org/ctrl/popper_falsification.html, accessed August, 2015.

week, then my claim is unfalsifiable. What I am really referring to here is a scientific theory; it must be able to be tested so that if it is false it must be possible to make an observation, a prediction, or carry out an experiment that will demonstrate its falsehood, as with Newton's theory in the previous paragraph. This can sometimes be very hard to do, as for example when part of a theory may be correct. Sometimes a theory can be so general that it is incapable of being disproved; one can always dream up an explanation for what is observed. We see this with the theory of evolution taken out of its biological context. What I have found is that whenever some new scientific fact is discovered about animal life or us, the evolutionist will generally have an evolutionary explanation that often cannot be disproved; it usually cannot be proved either.

Another example of an untestable hypothesis that is perfectly consistent with empirical observations is the hypothesis of solipsism, namely that all of reality is the product of our minds. We just cannot test its validity as all the evidence is consistent with it and no evidence can contradict it. Solipsism is therefore not strictly a scientific hypothesis, but rather a philosophical viewpoint.

One the problems in trying to test a scientific theory is that it is often very hard to maintain the required ideal conditions for the theory to hold because of background "noise" or contamination combined with the complexity of what is going on. It is also impossible to conclude that a theory is universal as it can only be based on finite experience.

1.3.2 Mathematic Has Limits

Although mathematics underlies some of the foundational aspects of science (the reader might want to skip this interlude), I want to point out that even mathematics has it limitations with regard to proof; for instance the existence of undecidable problems. (Take a deep breath!) Turing, in 1936, showed that there exists a problem for which it is impossible to construct a single algorithm that always leads to a correct yes-or-no answer.[26] Gödel had previously gone a step further in 1930 and proved the following two theorems:

(1) If formal set theory is consistent then there exist theorems that can neither be proved or disproved.

(2) There is no procedure that will prove set theory consistent.[27]

Do you want a translation? Roughly speaking what is meant is that it is not possible to find a complete and consistent set of axioms for all of mathematics. There will always be statements about the natural numbers that are true, but are unprovable within the system, and such a system cannot demonstrate its own consistency. Mathematics is incomplete and not all powerful. As John Polkinghorne says:

---

[26] For other "undecidable problems" see the topic on the internet.
[27] Stewart (1996: 266).

Kurt Gödel has taught us that even pure mathematics involves an act of intellectual daring, as we commit ourselves to a belief in the unprovable consistency of the axiomatic system under consideration.[28]

Enuff!

1.4 CONCLUSION

Although science has made many inroads into our ignorance, it has also produced many unanswered questions; changing the metaphor, the higher we climb the greater the unchartered territory we see before us. The problem is not science itself but rather the philosophical interpretations of science that give rise to the idea that science and religion are in conflict. On the other hand there is the danger of using God to fill the gaps in our knowledge as gaps have a habit of being filled, although other gaps keep opening up. It should be mentioned that some atheists are guilty of "science of the gaps" arguments rather than "God of the gaps"; if there is something that doesn't make sense or there is a gap in a scientific explanation, then the response is that naturalistic science will one day fill that gap. This is true to a certain extent, but as I have commented above, there are some questions that science cannot answer.

Science seems to have unjustifiably assumed a God-like status in that it supposedly provides an answer to everything. Yet we have seen above that science and mathematics have limitations, and we need to bear this in mind when we launch into some science and mathematics in the chapters that follow. After this short introductory chapter we begin the process of looking at evidence for Christianity and we start with the question of whether there is evidence for the existence of God in the next chapter.

---

[28] Polkinghorne (1998a: 15).

CHAPTER 2

IS THERE A GOD?

2.1 A MATTER OF FAITH?

In the previous chapter we saw that science and mathematics have limitations and cannot provide "proof" of whether God exists or not as neither of these disciplines can "prove" anything about the real world. Mathematics is about an idealized world that endeavors to approximate the real world, while science is about accepting the status quo for the time being until an experiment or observation proves otherwise. The question, "Is there a God?" has teased philosophers down through the centuries, and there is a rich heritage of answers going back to Augustine, Aquinas, and others. A follow-up question might be, "What sort of God or gods do you have in mind whose existence you are trying to prove?" The term "God" is an ambiguous term as it means different things to different people. However, the sort of God I have in mind for the moment is at the very least a being somehow responsible for the existence of our universe; other characteristics will surface later. Philosopher Edward Feser proposes one view, namely: "when we speak of God as being powerful, intelligent, good, and so forth, we are not describing features that exist in a distinct way in God Himself", which we would see as distinct in our human experience, "But *in God* they exist as one: God's power *is* His intellect, which *is* His Goodness, and so forth, Being Itself."[1]

The sort of God we are considering could be regarded as being well beyond our comprehension and understanding, so that some would say that it is impossible to logically argue from what we know to prove God's existence or non-existence. The Bible itself says that spiritual things are spiritually discerned and not through the wisdom of men.[2] It would be like asking a man blind from birth to explain color; it is beyond his comprehension. For this reason we are really forced to talk about God in terms of metaphor and allegory. The Jews took this view in their thinking and their writings, and we see it in the Bible. That is why the parables were so popular. Some like the metaphor for God used by the 12-step AA program and a number of other addiction programs, namely "a higher power", which implies something higher than oneself that is powerful.

If we use God in a sentence we could argue that we have an undefined term and we are trying to reduce God to some sort of object. Can we use our paltry experience to argue about something outside of our experience? When people say they don't believe in God I am curious to know what sort of God they *don't* believe in as they may have only a caricature of God that I don't believe in either. Some skeptics have a habit of putting up straw men to knock down. All people have beliefs, frequently without foundations, and so often will attack the beliefs of others or the persons themselves without questioning their own beliefs or opening them up to scrutiny.

With regard to faith, Feser[3] maintains that belief in God rests "firmly and squarely" on reason rather than faith, and that the traditional arguments for God's existence going back to the early philosophers like Augustine and Aquinas are still relevant today. He castigates the so-called "new atheists" like Harris, Dennett, Dawkins, and Hitchens for their lack of knowledge and understanding of this rich religious tradition. He argues against so-called fideism, which is the view that religion rests on faith alone. The persuasive force of some arguments will depend on how much a person is prepared to accept any premises on which the arguments are based. Also, some arguments may at best make the case for God more probable.[4]

Sometimes we are more concerned with proving that one probability is greater than another than trying to prove that a particular probability is one (that is certainty) or finding the actual value of the probability; also there is the question of whether adding another piece of evidence increases the initial probability or not. Some of these ideas are used in Section 2.9 below. Küng makes the comment:"Belief in God is continually threatened

---

[1] Feser (2008: 109, his italics).
[2] 1 Corinthians 2:6–14.
[3] Feser (2008: 4–7).
[4] For a helpful discussion on the role and type of various probabilistic arguments see Swinburne (2004: chapter 1).

and—under pressure of doubts—must constantly be realized, upheld, lived, regained in a new decision."[5] The same comment applies to atheism and in fact to any belief system, as we have to live with uncertainty and accept that we do not know for sure.

As well as logical problems in discussing God, we cannot put God in a test tube either and say, "Here is a bit of God."[6] As I inferred above, the existence of God is not a scientific hypothesis. Clearly "test tube" proof is out of the question and we need to resort to "circumstantial" evidence as one might meet in a court of law.[7] Glass points out that to believe *that* God exists is very different from a belief *in* God, as the latter involves trust in God.[8]

In this book we will look at both scientific and historical evidence. Theologian and historian Michael Licona[9] notes that both approaches have similar problems. What scientists see in their telescopes is already thousands or more years out of date, because light takes time to travel, while geology requires a significant amount of guesswork. Physics refers to entities like quarks and strings that are not directly accessible, and evolutionists have no way of directly verifying if one major kind of life-form evolved from another. Similarly history does not have direct access to the past but must rely on documents and artifacts from the past.

Some skeptics are good at putting up alternative possible explanations (even if unlikely) of how something came to be and then expect theists to disprove them when often this is not possible. Logically such explanations cannot be ruled out so that so with this tactic one is only limited by one's imagination. For example, a person judged to be sane is convicted of murder and offers in defense that an alien appeared and forced him or her to commit the murder, or took the person's gun and shot the victim. This is of course possible and, although logically impossible to prove or disprove, it is not likely to be accepted in a court of law. It is highly unlikely that it happened that way, and it all boils down to the likelihood or probability that the defense offered gives the correct explanation for what happened. In the end it is a matter of balancing evidence. However, one might ask how important is evidence for one's beliefs, as some require more evidence than others.

If God does exist, then God surely would want to us to know that he exists and provide a way for us to know this. If the Bible's teaching in Genesis that we are made in the image of God is true, then we can know about God through that spiritual part of us that is related to the image. Plantinga[10]

---

[5] Küng (1984: 78).
[6] cf. 1 Kings 8:27.
[7] This question is considered in Chapter 8 on suffering and evil.
[8] Glass (2012: 49).
[9] Licona (2010: 103).
[10] Plantinga (2000).

effectively argues this. Christians are encouraged to give a defense of their faith as it is defensible.[11]

Which Belief?

Given the evidence or lack of it, one must then take a step of faith towards one of the following broad categories of belief (with some variations): atheism (believing God does not exist), theism (belief in God), agnosticism (cannot know), or a version of pantheism (e.g., everything is mind or spirit). A view related to atheism is naturalism that says that nature is all there is and there is no supernatural realm and/or spiritual intervention in the world. A problem with this approach is that because we can only rely on our perceptions and on the abstract structure of scientific theory (especially quantum theory), we in fact know very little about the material world, the human mind, and the actual nature of reality.

The philosopher Kant argued that we have no basis for assuming that our perception of reality actually resembles reality itself; all we have are experiences, so that science does not give us a complete description of reality. Now it is certainly the role of science to look for material explanations, even for something we might think of as being immaterial like the mind, but it is naive to think that materialism can therefore explain everything, and that materialism is all there is. The theist who is a scientist does not deny the validity of scientific reasoning, but also allows the possibility of other kinds of knowledge. Galileo once said: "Science tells you how the heavens go, and the Bible tells you how to go to heaven." I like the comment made by John Barrow, eminent theoretical physicist and mathematician, "A universe simple enough to be understood is too simple to produce a mind capable of understanding it."[12] Another view related to atheism or agnosticism that generally disbelieves in the supernatural and relies on mankind alone, reason alone, and nature alone (i.e., nothing outside the system), is secular humanism, though its adherents are perhaps more willing to work with theists towards a better and brighter future. In my view, humanism resting on naturalism lacks the ability to deal adequately with human nature and the psychological realities of life, and it doesn't deal with the whole person.

A mixture of theism and naturalism is seen in deism in which God is described as creating the world but leaving it running by means of unchanging laws without any supernatural interference. On this basis miracles don't exist. Deism was a a view more common in earlier centuries, but is apparently held by some cosmologists and philosophers today. On the other hand we have metaphysical naturalists who are either materialists in which everything reduces to matter, or are pantheists in which everything reduces to mind or spirit. Both views deny a supernatural intervention. Pantheism (from "pan" meaning "all") claims that all is God and God is all. There are various

---

[11]See, for example, 1 Peter 3:15.
[12]Barrow (1990: 342–343).

versions of pantheism and one in particular called "panentheism" (meaning "all in God") that is also related to so-called "process theology", believes for example that God is changing and finite. However, it can be argued that if there is a God who created the universe, then God must somehow be in all the universe sustaining its existence and willing its properties, but is not dependent on it. The world makes sense because God is behind it and it is what makes science possible. Also, in Jeremiah 23:24 God speaks about no one being able to hide from God as God fills heaven and earth, i.e., is everywhere. If God created the universe, then God would transcend space and time.

The atheist who says he or she has no faith is illogical as the final step to holding any worldview has to be by faith as there is no certainty. God-worship can be replaced by Science-worship! We often hear the phrase "blind faith" bandied around by skeptics, yet everyone has a religious or nonreligious belief system whose truth may not be proved logically, so that everyone has faith. This faith does not mean blind trust in the absence of evidence as people like Dawkins would argue.[13] Faith as simply an idea is not irrational as such, as faith always implies faith in something or someone. As noted by Frank Turek, everyone is a fundamentalist with fundamental beliefs about the nature of the world and how we should live. He says:

> Atheists, for example, believe that there is no God; that life arose from non-life without any intelligent intervention; that there is no afterlife; and that science is the supreme if not exclusive source of all truth. Those fundamental beliefs usually result in moral fundamentals.[14]

We need to add to this the belief that human reason is reliable. By definition, faith is blind or at least only partially sighted. Yet if you think about it, we exercise faith in almost everything we do whether it be in people or in things (e.g., our car), as life is full of uncertainty. In the end it is a question of what is more rational, theism or the other alternatives.

Turek and Geisler[15] argue that a person needs more faith to be an atheist than a theist. Religious faith is not in opposition to reason, but rather it can help us to answer those fundamental questions about life and existence that are outside of reason alone by providing access to revelation. The Bible tells us that faith is a requirement to please God who rewards those that wholeheartedly seek Him.[16] D'Souza sums it up well with the words:

> The Christian has faith even though he is not sure, while the unbeliever refuses to believe because he is not sure. But they agree in being unsure. The skeptical habit of mind is as natural to Christianity as it is to unbelief.[17]

---

[13] See McGrath (2005).
[14] http://crossexamined.org/crossexamined-solution/, accessed August 20-15.
[15] Turek and Geisler (2004).
[16] Hebrews 3:6.
[17] D'Souza (2007: 195).

We have to make our decision and in doing so we have to weigh up the consequences if we are wrong. If it seems there is a five percent chance that Christianity is right, is it worth the risk of rejecting it if eternity is at stake? C. S. Lewis expressed this idea somewhat differently when he said: "Christianity is a statement which, if false, is of *no* importance, and, if true, of infinite importance."[18] This idea goes back to so-called "Pascal's wager."

2.2 NATURE OF OUR UNIVERSE

Before looking at some arguments for God's existence I want to first look at some deep issues about the nature of our universe to give a brief indication of where scientific ideas seem to be heading. This section is somewhat speculative and is an attempt to try and convey some ideas concerning some current thinking about our universe; thinking which is still changing. A reader may wish to move to the arguments relating to God's existence beginning in the next section (or else take a deep breath and read on).

How big is our universe? Is it finite or infinite? We don't know as it depends on our model of the universe. Space is expanding at an increasing rate under two opposing forces—the initial momentum versus the pull of gravity (which depends on the density of the universe). This process is complicated by the existence of so-called "dark energy" apparently making up about 68.3 percent of the mass-energy of the universe and by "dark matter" taking up about 26.8 percent; the remaining 4.9 percent is the ordinary matter we can observe. We currently don't know what dark energy is though it is related to the so-called cosmological constant, a constant in the equations of general relativity that describes vacuum energy. It is a hypothetical form of energy that permeates space and exerts a negative pressure or "anti-gravity" that repels visible matter and causes the accelerated expansion. On the other hand, dark matter is a form of subatomic particle that does not reflect or emit electromagnetic radiation and causes a positive gravity pressure that attracts visible matter. Both are inferred rather than observed from observations of gravitational interactions between astronomical objects.

Currently measurements suggest that the universe is "flat" (whose large scale geometry is the usual Euclidean geometry) so that it could expand forever. It is not quite as simple as this as gravity from the sun and stars warps space and time locally. Time varies with mass, acceleration, and gravity, but we shall not go into details. We note that the universe could also be finite, especially if it had a beginning, and only be infinite in the infinite future. Because it is apparently expanding at an accelerating speed greater than the speed of light, we can only see part of it, and as time goes on we will see less of it because the expansion will "outrun" light. This does not contradict the idea that objects cannot go faster than the speed of light as the expansion

---

[18]C. S. Lewis (2001d: 101).

is not to do with the objects in space but is due to the expansion of space itself. It's like marks on a balloon that further separate from each other as the balloon is pumped up; the space between the marks expands.

How many dimensions has our universe? General relativity talks about a space-time continuum involving four interrelated dimensions—three spatial dimensions and a time dimension. It seems that having three spatial dimensions is critical for our existence. For example, Barrow and Tipler[19] argue that the stability of atoms and the inverse square law that gives stability to the solar system over a long time period depend on there being three dimensions. Roughly speaking, relativity theory tells us that time can only exist when there is space, and gravity from a large stellar body bends the space-time geometry, slowing time and bending light. Einstein defined a set of field equations that explained the way that gravity behaved in response to matter in space-time and were used to represent the geometry of space-time and establish the theory of general relativity.

String Theory

Another theory, so-called *String Theory* and its offshoot Membrane Theory (M-Theory), sometimes called the "theory of everything," is a controversial developing theory that endeavors to reconcile general relativity (gravity and the cosmos) with quantum physics (nuclear particle theory), namely quantum gravity that combines the physics of the very large with the physics of the very small. At some energy levels these two theories haven't agreed in the past. The quantum gravity theory attempts to unify the four forces in the universe— the electromagnetic force, the strong nuclear force, the weak nuclear force, and gravity— together into one unified theory. String theory provides the idea that the fundamental building blocks of the universe are very small pieces of vibrating string with the different atomic particles corresponding to different vibrational states of the string. It is not clear at present whether the idea is true or not, though it has been around for some time. There are various string theories requiring 26 space-time dimensions, or 10 dimensions for superstrings, and 11 for M-Theory, where the additional dimensions are "compactified" or hidden in some way.[20] In the latter case we have the concept of a brane, a higher dimensional string, which has been used recently to suggest a new type of cyclic universe.

Our understanding of the universe is becoming more and more esoteric and our physics more and more counter-intuitive. Given what has happened in the past, our ideas about the nature of the universe will no doubt continue to change. The problem is that we don't know what matter is, and Ward asks, "Is it quarks, or superstrings, or dark energy, or the result of quantum fluctuations in a vacuum?"[21] Most of matter is empty space and, as discussed

---

[19] Barrow and Tipler (1986: 261).
[20] See, for example, Greene (2003), Jones and Robbins (2009), and Holder (2013).
[21] Ward (2008a: 14).

in Section 4.1.3, we don't know for certain whether matter is "real" or not, or just what we perceive. As Overman[22] notes, a theory of everything will simply amount to equations describing the laws of physics that are self-consistent but are not self-sufficient. They do not explain why something physical exists and where the laws come from in the first place. A typical answer to the latter question is that the laws just are, which is not very helpful. It can be argued that the mind that produces the equations cannot be completely described by them. A theory of everything would point even more strongly to a creator.

Without getting tangled up with string, it is possible to conceive of a God in a higher dimensional space. This might help us to begin to understand how God is able to do some of the things that God might do, for example seeing all that is happening at a particular instant, seeing the end from the beginning, and being very close to us but distant in another sense (theological terms used are immanent and transcendent.)[23] It would seem to me that the past and the future possibly co-exist (in 4-dimensional space), otherwise we have the strange situation where, for every second, we move from a state of non-existence to another state of non-existence through a point in time that is continually changing! However, all this is pure speculation on my part. The important thing for me is that God exists apart from space, yet is present and causally active at every point in space.

How God interacts with time is a mystery to me as God has been temporally involved with the world when he[24] created it; this question is considered further in the next chapter. It is interesting that in the Bible God calls himself YHWH (pronounced Yahweh), meaning "I am whom I am," the ever-present one. The problem is that we don't really understand the nature of time, and how it is defined depends on what theory of physics we are considering. Philosopher and theologian William Lane Craig, in referring to a lecture by the well-known physicist C. Rovelli said:

> Time as it plays a role in physics is an operationally defined quantity varying from theory to theory: in the Special Theory of Relativity it is a quantity defined via clock synchronization by light signals, in classical cosmology it is a parameter assigned to spatial hyper-surfaces of homogeneity, and in quantum cosmology it is a quantity constructed out of the curvature variables of three-geometries.[25]

I have made this quote, not to bewilder you (and confuse me) with science, but to simply point out that time is an elusive quantity once we shift from our simple view of time as a succession of events, say in counting in our minds, to the strange world of quantum physics.

---

[22] Overman (2009: 41).
[23] cf. Isaiah 57:15 and Acts 17:28.
[24] In this and later chapters I shall refer to God as male even though God is neither male nor female; my apologies to the reader. It's a pity that English does not have a suitable word.
[25] Craig (1998a).

With regard to a so-called theory of everything, Hawking and Mlodinow make the following comment:

> ... if there really was a complete unified theory, it would also presumably determine our actions—so the theory itself would determine the outcome of our search for it. Might it not equally well determine that we draw the wrong conclusion?[26]

As an attempt at a partial answer, Hawking and Mlodinow appeal to the idea of evolution and natural selection in which people who draw the right conclusions survive better so that as we evolve we get closer to the right conclusion. However, they question whether it is still the case now. With regard to evolution it is interesting that the abstract mathematics that underlies our physics seems to have little survival value!

One further question. Does the universe have a centre? The Cosmological Principle might suggest that we are at the centre. Theory says there is no centre, but every point is a centre. To see this, consider the analogy of a two-dimensional world with time as the third dimension, where the two spatial dimensions are the latitude and longitude on the surface of a sphere (or balloon). The centre of the sphere corresponds to time zero and, as the sphere expands with time, distances between points on the sphere expand uniformly in all directions on the surface. There is no centre on the surface of a sphere. Once the sphere is very big the surface is essentially (locally) flat. To a two-dimensional object like an ant walking on the surface, the surface would appear infinite, although it is actually finite, as the ant would never reach the end (though it could eventually return to the beginning of its journey). For us in three dimensions looking at the sphere, we are outside its surface but we can get very close to its surface because of the extra spatial dimension. In terms of theological language we can therefore be both "immanent" (close by) yet "transcendent" (in a different dimension) with respect to the ant. Unfortunately the analogy breaks down when we go up an extra spatial dimension as we cannot imagine the three-dimensional surface of a four-dimensional sphere, but must turn to abstract mathematics. Hawking and Mlodinow comment that in combining quantum mechanics and general relativity,

> space and time together might form a finite, four-dimensional space without singularities or boundaries, like the surface of the earth but with more dimensions.[27]

We now consider some arguments for the existence of God.

2.3 THE ARGUMENT FROM ABSTRACT OBJECTS

I have used the word "abstract" rather loosely up till now, but care is needed if it is introduced into a philosophical argument. Usually an abstract object

---

[26]Hawking and Mlodinow (2005: 24).
[27]Hawking and Mlodinow (2005: 216; see also p. 155).

is thought of as one that is not "concrete", that is it does not exist in space and time. Okay, it is not a perfect definition as there is the glaring counter example: if God exists who is outside of space and time, yet is a personal agent and can interact with the world, then we have a non-concrete being that is not abstract. Clearly an abstract object is one that is not involved with any cause-effect relation; it doesn't cause anything. We come back to this question later.

As a mathematician I have always been intrigued with the question of why do abstract objects exist and where do they hang out! Mathematics is a good example of abstract ideas. We talk about numbers, sets, propositions, equations, and higher dimensional spaces of more than three dimensions. Although motivated by the real world, these sorts of ideas and objects are not part of the physical world as we know it. For example, we know about five oranges and five apples, but what about the number five by itself? Certainly number 5 does not cause anything.

As mentioned in the first chapter, straight lines, planes, circles and spheres are defined mathematically as "perfect" objects, which generally may not physically exist, or if they did we would have no way of determining if they were perfect as we cannot make perfect measurements. Given certain assumptions and these abstract objects, we can then use mathematics to derive certain formulae about the objects. Our abstract idea of a triangle is irrespective of its shape, size, and color, and although it has many physical representations its "essence" or "form" as Plato would describe it, does not exist in the material world. Nor does it depend on human minds for its existence, although it exists in human intellect.

Going deeper, there are very large numbers that are beyond our physical and mental experience. Although not actually not a number, infinity—and there are various kinds —is a good example of something treated like a number. David Hilbert, one of our great mathematicians, stated,

> The infinite is nowhere to be found in reality. It neither exists in nature nor provides a legitimate basis for rational thought ...The role that remains for the infinite to play is solely that of an idea ....[28]

However, in mathematics and in particular in the modern topic of nonstandard analysis, infinities (along with infinitesimals) can be used to play a similar role to that of ordinary numbers.

We see then that although mathematics is a model for reality, it is not reality as we imagine but has a separate existence in the abstract: as already noted, infinity is a mathematical idea and not something that actually exists in real life. Similarly there is no single point corresponding to time zero, as a point is a mathematical construct of zero "length." The question is where did the mathematics come from and how is it retained in our brains. It could be argued that whenever we think of an abstract object like a triangle for

---

[28] Hilbert (1964: 151).

example, we actually visualize a concrete version of it, but it seems that we can hold the concept of an abstract triangle in our minds without visualizing a particular triangle.

To accept mathematics, in fact science itself, means the acceptance of abstract concepts and objects. Of course abstract ideas are not confined to just mathematics, as every aspect of life has its own abstraction. Every object such as a table, chair, dog, and human, and especially ideas like beauty and justice, have an abstract form representing the essence of a family of objects or ideas. We can think of an abstract family of chairs or an abstract family of beautiful objects. This is true even if we tend to think of such objects in our imagination using some sort of physical representation. The question then is where do abstract ideas come from as they cannot come from matter unless the abilities of self-organization and self-awareness are latent in molecules. As Michael Ruse, the non-Christian Darwinist philosopher said: "Why should a bunch of atoms have thinking ability? ... No one, certainly not the Darwinian as such, seems to have any answer to this."[29] Evolutionists who are atheists must accept that molecules must somehow have this inherent property and that abstract ideas evolved along with the ability to reason, otherwise there is a discontinuity from non-thinking to thinking. Without some inherent property or possibly additional outside input, we are faced with a well-known argument put forward by the atheist J. B. S. Haldane which goes like this:

> It seems to me immensely unlikely that mind is a mere by-product of matter. For if my mental processes are determined wholly by the motions of atoms in my brain, I have no reason to suppose that my beliefs are true. They may be sound chemically, but that does not make them sound logically. And hence I have no reason for supposing my brain to be composed of atoms![30]

This is an argument against naturalism as, if appearances are right, there must be more than just natural processes at work in my brain; there must be some sort of "reason" that is outside of natural processes.

If we say that our minds are the accidental result of a purely material universe and have somehow emerged from material, then we need to be able to distinguish between reason and order on the one hand, and chance and disorder on the other to actually make the comparison. We may try to get round this by saying that the whole is greater that the sum of its parts, that is a collection of molecules has some higher level of reality than individual molecules; there are different layers of reality and thinking is an emergent property. For example, in living collections such as an ant colony or beehive, it seems that the group has a "mind" of its own even though an individual member has few neurons to rub together. However, we are still arguing for something beyond the parts. Such an argument and a discussion about naturalism is considered in greater detail in Chapter 3.

---

[29] Ruse (2001: 73).
[30] Haldane (1927: 209).

Some abstract ideas like infinity are certainly created by the human intellect, and we ask why we can do that, but other abstract ideas like a triangle and justice simply exist. What is their source? Since abstract ideas are external to our material existence, it can be argued that God provides the best explanation for the existence of abstract entities. For example, if we perhaps think of abstract ideas in terms of God's "thoughts", then he is in causal relation with them, and also with us. Therefore there should be no problem as to how it is that we could know something about them. The Platonist would argue that abstract ideas are mind independent, which conflicts with the idea that they depend on the mind of God, or even on us.[31] Further comments on abstract ideas are made in Sections 3.1.3 and 3.4.2 in reference to the mind being more that the brain.

2.4 THE COSMOLOGICAL AND RELATED ARGUMENTS

There are a number of philosophical arguments that endeavor to prove the existence of God, and I consider some of these below as they are part of a rich history of ideas. It is with some trepidation that I do this as I am not a philosopher but I wish to give something of the flavor of the arguments so as to give the reader a brief idea as to their nature. It is not my intention to get too involved in philosophical discussion and logic, important though it may be, as my focus is ultimately on evidence. Before launching into that discussion I shall first look at the evidence supporting the idea that the universe has a beginning.

2.4.1 Evidence of a Beginning

When we look at theories about the origin of our universe we are in very deep and somewhat unchartered waters so that some of the current models about the universe's origin must be tentative, as theories are always open to change and refinement.[32] New information continues to arise, such as the recent discovery by NASA's Hubble Space Telescope using a special technique of one of the farthest, faintest, and smallest galaxies ever seen. It is estimated to be over 13 billion light-years away.[33] The cosmological debates are endless (e.g., about the nature and magnitude of the cosmological constant) as many ideas are not currently testable, as seen in past issues of the Scientific American. However, the most widely accepted viewpoint from cosmology, which is regarded as well supported but still debated, is that the universe had a finite beginning. There is strong scientific evidence for this. This theory says that the universe began with a "big bang" (explosive expansion) when a tiny

---

[31] For further philosophical discussion on this topic see Craig (2012).
[32] For a readable history of cosmology see Hawking and Mlodinow (2005).
[33] http://www.sciencedaily.com/releases/2014/10/141016140851.htm; accessed August, 2015.

particle of almost infinite density at an enormous temperature exploded, and physical space, time, and matter as we know it all came into being.[34] This does not mean that there was empty space and the universe began in it. There was initially nothing, not even space or time.

Support for the idea of a big bang comes from the fact that the universe is expanding outwards at a very high speed and if the motion of the galaxies is traced backward in time they all merge together at some time in the past. The Big Bang model is based on Einstein's general theory of relativity and the so-called Cosmological Principle that states that the matter in the universe is homogeneous and isotropic, that is the matter is distributed uniformly everywhere and in every direction when averaged over very large scales. Einstein could not provide an explanation of his theory that would not require a beginning or a Beginner and later wrote of his desire "to know how God created the universe."[35]

A further question then for current research is how did chaos and disorder that occurred immediately after the Big Bang become transformed into an ordered "smooth" universe, where at any time one part of the universe looks much the same as any other part. Distant stars and galaxies are distributed much the same everywhere and resemble the local ones in terms of their compositions and motions. As far out as we can detect, the laws of physics appear to be universal. Where did the order and uniformity come from?

The Big Bang model postulates that about 13.82 billion years ago (according to the recent Planck Mission dating of the universe) the part of the universe we can observe now was then very tiny.[36] Technically it began as a mathematical "singularity" (like a point of infinite density) and expanded outwards from a hot very dense state into becoming the cooler universe we can see today. We cannot explain scientifically how the universe started off as it is believed that the laws of physics were actually produced by the Big Bang. Also, the singularity didn't appear in space, but instead space began inside of the singularity. Prior to the singularity, nothing existed, not space, time, matter, or energy — nothing.

We note that a vacuum or even "empty" space is not nothing. To ask what was there before the Big Bang does not make sense, as time as we know it did not exist then; there was no "before." The universe came out of nothing. There were no particles, no fluctuation, no laws, no principles, no potentialities, and nothing quantum. Atheists will tell us that the universe created itself, which in my mind requires greater credulity than believing God created the universe. If a universe can appear out of nothing why not other

---

[34] With a particle of this size Einstein's theory breaks down; we need a quantum theory for this that we don't have yet.
[35] Einstein (1954: 84).
[36] The actual age of the universe is not my concern at present; my point is that it appears to be finite.

objects. Why can't the thief tell the judge that the money he has had just appeared!

Some have put forward the argument that things such as subatomic particles can pop briefly in and out of a quantum vacuum—a so-called vacuum fluctuation. However a quantum vacuum is not absolutely nothing, but is a source of fluctuating energy.[37] Also the particles are generally virtual particles, that is theoretical particles whose existence is not clear, being possibly just theoretical constructs. In any case we still have to account for the existence of the quantum vacuum. Another approach used by some is to simply redefine "nothing." However, we cannot really define nothing in the first place except as the absence of something (the so-called "empty" set in mathematics). We cannot name "nothing" as something as it is like personifying "nobody."

Evidence for the Big Bang is given by the presence of remnants of the expanded original matter detected by microwave detectors as a uniform glow across the sky.[38] Because the universe was once very dense, it should have been very hot, as matter heats up when compressed and cools down when expanded. The intensity of the radiation diminishes as the universe expands so that the tail-end of the radiation of the hot matter would leave a faint glow. Further evidence for the Big Bang was given in 1992 by the finding of so-called "galaxy seeds," namely tiny temperature "ripples" that enable matter to collect into galaxies. We note that the microwave background has the same temperature in every direction we look when averaged out over the very small variations. This would mean that the initial state of the universe would need to have almost the same temperature everywhere, and the initial rate of expansion would have to be chosen very precisely to avoid a consequent collapse. As Davies points out,

> Our universe has picked a happy compromise: it expands slowly enough to permit galaxies, stars and planets to form, but not so slowly as to risk rapid collapse.[39]

Since an explosion cannot be expected to produce a smooth organized expansion, Davies asks,

> How has the universe contrived its explosive genesis with such exactitude that there is no distinguishable difference across the sky, even between regions that have never been in causal contact?

A second line of evidence of the Big Bang is the origin of light elements. The heavy elements like carbon and iron, for example, are created in the interior of stars, and then through supernovae are exploded out into space. About ten billion years were needed for this to happen so that the stars could

---

[37] For a good discussion of this point by philosopher David Albert see http://www.nytimes.com/2012/03/25/books/review/a-universe-from-nothing-by-lawrence-m-krauss.html?_r=0, accessed August, 2015.
[38] Microwaves have a longer wavelength than visible light.
[39] Davies (2006: 61–62).

evolve along with the chemical elements needed for the generation of life. On the other hand, the very light elements like deuterium and helium cannot be synthesized inside a star as you need a much hotter source to create them. The Big Bang provided immense temperatures of millions of degrees that could produce these elements.

A third line of evidence for a finite beginning comes from the laws of thermodynamics. We begin with the first law, often called the Law of Conservation of Energy, which states that although energy can change its form it cannot be created or destroyed so that the total amount of energy in the universe (or any isolated part of it) is constant. This relationship between energy and matter, which are regarded as interchangeable, is described by Einstein's famous equation $E = mc^2$, where E is energy, m is mass, and c is the speed of light (all in appropriate units). We now look at the second law in more detail.

Second Law of Thermodynamics

Before discussing the second law I need to introduce some definitions in order to be clear about the technical applicability of the law. A closed system is defined to be one in which no mass can transfer across the boundary of the system, though energy may be exchanged across the boundary. An isolated system is one that has no such exchanges whatever.

The Second Law is a mathematical one that states that the entropy of an isolated system not in equilibrium will tend to increase over time, approaching its maximum value at equilibrium. Any change in an isolated system will be to maximize entropy. What then is this creature called entropy? Briefly, the entropy of a system is a measure of the probability that the system has its particular arrangement of mass and energy within it. High entropy corresponds to a high probability that the arrangement exists, so that as a random arrangement is highly probable (that is more likely to occur) it would be characterized by a high entropy, while something that is highly structured and not random will have low entropy as it would be less likely to occur. For example, a watch has low entropy as it is highly structured and unlikely to occur by chance, but if I smash the watch it becomes less structured and more "random" and the entropy goes up. The entropy increases as a system becomes less ordered and more random, and decreases as the system becomes more ordered and structured.

My personal interpretation of the Second Law is that "everything falls to bits sooner or later", that is tends to become more random. For example, in the supermarket you may have seen a pile of canned goods neatly displayed in a stack. If I was to knock it over I would increase the entropy of those cans because they then become randomly scattered rather than forming a structure. If the cans were left long enough they would disintegrate until you had a shapeless stinking (random) mess. In that case there would then be an even further decrease in structure with a consequent further increase in randomness.

An important aspect of the Second Law is that differences in energy levels tend to disappear and energy becomes less available. This is because differences in energy levels are less likely (lower entropy) than no differences (higher entropy) as things tend to level out. For example, heat cannot flow from a colder to a hotter body; the energy flow has only one direction and is irreversible. If I add hot water to cold I end up with lukewarm water; the hot water doesn't get hotter and the cold water doesn't get colder. When combining hot and cold we lose the energy we could have extracted from using the temperature difference. Similarly a difference in air pressure produces energy in the form of wind power to drive turbines and generate electricity.

A similar example is provided by a dam. The water at the top of the dam is stationary, but it has so-called potential energy because of its height. As it descends through the dam the potential energy is turned into kinetic energy, or motion, and this is used to turn turbines to create electrical energy that gets used by consumers. Any moving water at the bottom of the dam will eventually lose most of its energy in one way or another as it flows away. The net result is that the original potential energy available for use is in the end largely dissipated (evened out) and there is less useful energy available as the total energy is constant by the first law. This means in practice that the amount of useful energy available in the world is steadily decreasing so that the universe is running down—the so-called "heat death"— when everything ends up at constant temperature. This is estimated to take a long time; for example our sun is slowly burning up as it exhausts its hydrogen necessary to fuel its nuclear fusion reaction, and in about 5 billion years will expand to destroy our solar system and become a red giant.[40] Eventually stars will also burn out and we will end up with a dark, cold, and ruined universe.

We can however cheer up as at present we still have available energy to do work and have not run out yet. The universe therefore must have a finite age otherwise if it had an infinite past our energy supply would have run out by now.[41] This running down of the universe provides evidence of a finite beginning, otherwise the sun would have burnt out by now. Also the fact that the universe is expanding also indicates that there was a beginning when the expansion began, and as it expands it runs out of energy. The key point to be made is that the Second Law implies a beginning with low entropy that requires a very highly ordered beginning, and this would have required unbelievable accuracy with regard to entropy to kick start the universe.[42]

I should mention, for completeness, that there has been an addition to the second law called the law of maximum entropy production. This states that

---

[40] Another possibility is that there might be a so-called "big crunch" when expansion stops and the universe collapses in, or else we somehow destroy ourselves (see, for example, the Doomsday argument on the internet).
[41] This reduction of available energy agrees with my favorite saying: "As I get older and older I get better and better at taking longer and longer to do less and less!"
[42] The origin of the initial low entropy is still under debate.

a system will select the path or assembly of paths out of otherwise available paths that minimizes the potential or equivalently maximizes the entropy at the fastest rate, given the constraints.[43] Sometimes this will initially lead to choosing a path that leads to order, as ordered flow in the end can lead to producing entropy (randomness) faster than disordered flow. This means that there may be parts of a system that become more ordered even though the system overall is more disordered. Hmm, well!

It should be noted that the earth is not an isolated system as it receives energy from the sun, and each one of us is not a closed system as we take in food. However, it is also argued from the evidence that the second law appears to apply to open systems as well as closed systems and, since the input of energy to the earth is comparatively small, the earth is approximately a closed system. This means that bits of a system can become more ordered at the expense of other bits becoming more random; it is the total amount of randomness that matters. The universe as a whole can be regarded as an isolated system.

Other Models of the Universe

A we have seen above, according to the Big Bang the universe began as a subatomic particle of essentially infinite density, and is now expanding outwards. The Second Law rules out the idea of a cosmic rebound where the universe continually collapses (contracts) and then rebounds (expands) forever, as useable energy would be lost with each cycle and the amount of disorder would increase with each cycle, implying a finite beginning. It transpires that this oscillating model apparently contradicts the laws of physics because of the initial singularity, as determined by general relativity. One of the problems of relativity is that the theory breaks down when applied to the initial subatomic particle, which is in the realm of quantum theory, so that the theory needs some modifications. Consequently other models have been proposed, but they have various problems as well as a lack of physical evidence. All the viable models involve having a beginning, though the beginning might not be a particular point in time but rather have a "fuzzy" beginning. All the models require so-called "fine tuning", a concept discussed below.

In looking at inflationary models where the universe is still expanding, the question has arisen as to whether the model can be extended to the infinite past, thus avoiding the problem of an initial singularity. Using mathematics, the answer according to cosmologists Avande Borde and Alexander Vilenkin[44] is no, as long as some reasonable physical conditions apply to the models (including multiverses and cyclic models). This work was then extended by Borde et al.[45] to allow for some quantum effects and uses more general assumptions so that the need for a beginning also applies to other cosmological models as well as inflationary ones. Further results are given by Audrey Mithani

---

[43] Swenson (2000).
[44] Borde and Vilenkin (1994).
[45] Borde, Guth, and Vilenkin (2003).

and Alexander Vilenkin[46] with regard to some oscillating and static models. Vilenkin concludes:

> It is said that an argument is what convinces reasonable men and a proof is what it takes to convince even an unreasonable man. With the proof now in place, cosmologists can no longer hide behind the possibility of a past-eternal universe. There is no escape, they have to face the problem of a cosmic beginning.[47]

However there will always be alternative models proposed such as that recently by physicists Ahmed Ali and Saurya Das[48] based on some changes to the underlying mathematical assumptions (in this case the nature of geodesics), that gets rid of the big-bang singularity and suggests the possibility of an infinite age of our universe. Their result is the consequence of the assumptions they make,[49] and other assumptions might lead to different results. No doubt further models will be forthcoming and what is needed is scientific evidence to back new models. Such models can arise because of the impossibility of knowing what happened during the first 10^{-43} seconds.[50] Up till that time the classical ideas about gravity and space-time do not hold, and we wait for a quantum theory of gravity for that initial period in the universe's history.

Holder[51] describes how we can arrive at variety of models by classifying the models according to two parameters, the curvature of four-dimensional space-time, which can be negative, positive, or zero (i.e. "flat" as described above), and the cosmological constant that can also be negative, positive, or zero, giving us nine possible paired combinations. Attempts to explain why the cosmological constant is so small (a current value is about 10^{-24}) have continually failed. As there is no scientific evidence to support the idea that hydrogen atoms are coming into existence to keep the universe from running down, the steady-state theory of the famous astronomer Fred Hoyle is currently not supported. Hoyle does however acknowledge a Creator of life.[52] While we are looking at alternative models, mention can be made of a model proposed by Stephen Hawking that endeavors to get rid of the singularity in the mathematics while still having a sort of a beginning. He uses a device called "imaginary time" to achieve this and open the door to other models by removing the problem of the physical laws breaking down at a singularity and removing the need for a boundary or edge. However, when

---

[46] Mithani and Vilenkin (2011).
[47] Vilenkin (2006: 176).
[48] Ali and Das (2015).
[49] For some comments on the model by Hugh Ross see Ross (1998) and http://www.reasons.org/articles/have-quantum-physicists-disproven-the-big-bang, which was accessed August, 2015.
[50] This is known as Planck time, the time taken for light to travel the distance of one Planck length, namely $\sqrt{hG/2\pi c^5}$, where G is the gravitational constant, c is the speed of light in a vacuum and h is Planck's constant.
[51] Holder (2013: 22).
[52] Hoyle (1983: 24, 27, 150).

complex numbers[53] are replaced by real numbers, the singularity reappears, so the question of a beginning still applies. My point in boring you with such imaginary details is to show the lengths that some scientists may go to avoid letting God have a foot in the door.

Looking at the subatomic particles we are made of (or more generally "strings," if the string theory mentioned above is correct) we see that we are basically vibration, and vibration needs duration and hence time to exist. We therefore need the concept of time to exist in our present physical form. However time is a strange concept when we think about it as Einstein has shown that it is relative to our position in the universe. The faster we travel through space the slower we travel in time. I wonder how this anomaly with time relates to the concept of "eternity" mentioned in the Bible?

2.4.2 The Arguments of Aquinas

Aquinas gave a number of arguments for the existence of God and these include his famous Five Ways given in one of his five volume work *Summa Theologica* written 1265–1274, namely: (1) from effects to a First Cause; (2) from motion to an Unmoved Mover; (3) from contingent being to a Necessary Being; (4) from degrees of perfection to a Most Perfect Being; and (5) from design in nature to a Designer of nature.[54] For the philosophically minded reader (who wants to stay awake at night!) we shall look very briefly at just some of these. As I mentioned above, I am not a philosopher and my aim is simply to highlight some of the ideas. But before doing so it may be helpful for the reader to be aware of Aristotle's four causes that are often mentioned in philosophical discussion.

Aristotle's Four Causes

Using a child's rubber ball as an example,[55] Aristotle's four causes are: (1) the *material cause* (the underlying stuff an object is made out of, namely rubber); (2) the *formal cause* (the structure, pattern or form of the object, namely spherical, solid, and bouncy, in other words its "essence"); (3) the *efficient cause* (what brings a thing into being, or in philosophical jargon actualizes a potentiality in a thing, namely the manufacturing process that produced the ball); and (4) the *final cause* (the end goal or purpose of the thing, namely to provide amusement for a child). The material and efficient causes underly materialism and deal with the question "how?" while the formal and final causes have a particular role to play in considering arguments about God's existence and deal with the question "why?"

---

[53] An "imaginary" or so-called complex number takes the form $3 + 5i$, where $i = \sqrt{-1}$ does not exist as a real object except in the mind of a mathematician as a useful extension of arithmetic.
[54] Aquinas (1920), cf. Geisler (1999: 725).
[55] Feser (2008: 62–63).

The final cause is the cause that determines the other causes, or the "cause of all causes" according to Aquinas. It underlies the "potentiality", the final form, and the efficient cause of an object; there is a reason or goal (final cause) for the object's existence. To realize the goal, the object would have to have an appropriate form to fulfill that goal, and an appropriate material structure such that its form is realizable. Therefore in this rubber ball example we see that the final cause underlies the other three causes. The materialist, however, would say that there are no formal or final causes and that any appearance of final causality is illusory, as everything is fundamentally meaningless. However the language used by materialists and atheists is far from meaningless as we read what certain human organs are used for, how animals have adapted their behavior, why living forms behave in a certain way, that genes have selfish ends and purposes, and there are phrases used like "information laden." Teleological (purpose-laden) language is heavily used for example in discussions about evolution.

Cosmological Argument

The first argument we shall consider is the cosmological argument. This has a long history and is associated with a long line of early famous philosophers beginning with Plato, Aristotle, and Thomas Aquinas, and more recently with the defenders of religious belief like William Lane Craig, Alvin Plantinga, Richard Swinburne, and others. There are several variations of the argument and it is not my intention to get involved with some of the niceties of philosophical discourse but simply make some broad comments. (I am sure the reader does not want to wade through the many turgid arguments both for and against that would take a whole book on their own so I shall just give a taste of the sorts of arguments available.)

The argument as it is often understood says that every effect in the world requires a cause, and nothing *in the world* is the cause of its own existence. If you think about it, everything that happens follows from something prior to it so that working our way backwards to the beginning there must have been a very first cause or uncaused cause we can call God that gave rise to our universe. (Otherwise we have, as someone has said, nobody plus nothing equals everything!) There had to be a beginning to the sequence of causes for the sequence to exist. Feser[56] notes, however, that this understanding of the argument is a straw man and points out that what the argument should say is

> every *actualization of a potential* has a cause, or *whatever is composite* has a cause, or *whatever has a feature only by participation* has a cause, or *whatever is contingent* has a cause.

God is outside any regress back into the past by noting that God is pure actuality devoid of any potentiality, is simple rather than composite, is being itself, and is absolutely necessary rather than contingent. Clearly we can

---

[56] http://edwardfeser.blogspot.co.nz/2014/07/clarke-on-stock-caricature-of-first.html, accessed August, 2015.

get into a deep philosophical discussion here in trying to sort out what the above quote means, and Feser's article should be consulted for further clarification. However, I want to sidestep this and point out that a first cause argument, as properly understood, has evidential support from the current cosmological theory described above that the universe and space-time had a finite beginning. This means that if God brought the world into being, there wasn't a continuum of time already existing into which God dropped the world. Rather, time was created along with space, with time being our mode of existence. This does not mean that God existed prior to the Big Bang, but rather God was causally prior and not temporally prior to the Big Bang; God was outside of time. To ask the question who made God is therefore not appropriate as it suggests a time prior to the existence of God.

It should be noted that arguments used by Aquinas did not require the universe to have a finite past; there may be other space/times as well as this one. Even if we regress backwards in time from a cause to a prior cause and to one before that, and so on, so that we have an infinite sequence of causes, this sequence must depend on something outside the sequence to exist. As God is outside the sequence, the rules of causation do not apply to God. Digressing for the moment, there is a well known variation of the cosmological argument called the Kālam cosmological argument.

Kālam Cosmological Argument

Part of this argument deals with the idea that the world cannot have an infinite past. Infinity is a very difficult concept to get our minds around. For example, all the numbers $1, 2, 3, \ldots$ form an infinite set, but if you take away another infinite set consisting of all the odd numbers $1, 3, 5, \ldots$ you are still left with an infinite set of even numbers $2, 4, 6 \ldots$. To say that the world has an infinite past means that we have now reached the end of an infinite set, which is not possible. It would be like starting somewhere(?) at infinity and counting backwards to 1, then reversing the process. The problem has been expressed by the following rather brief argument.[57]

1. It is impossible to traverse an infinite sequence by successive addition.

2. The temporal series of past events in our universe has been formed by successive addition (e.g., $1, 1 + 1 = 2, 2 + 1 = 3, 3 + 1 = 4$ and so on).

3. Therefore the series cannot be infinite.

There are other arguments that I won't go into,[58] but will settle for a couple of metaphors that might make the argument clearer. Having a past infinite series would be like reaching the top of a ladder with an infinite number of rungs or like jumping out of a bottomless pit! Time therefore must have had

---

[57]See also Feser (2008: 93).
[58]See the internet under http://www.philosophyofreligion.info/theistic-proofs, accessed August, 2015.

a beginning since if it didn't we would not have reached the present time (confused?). This means that the universe had a finite beginning, which leads to the following steps:

1. Whatever begins to exist has a cause.

2. The universe began to exist, that is, have a beginning.

3. Therefore it follows that the universe has a cause.

Science is about finding causes and, except for some scientists, takes the principle of causality for granted in the world. However, there is no reason not to extend causality from objects in the world to the world itself. There is also the question of what keeps the world going now that it exists (e.g., what makes physical regularities hold true day after day?), and this is irrespective of whether the world had a beginning or not. The final steps in the argument go like this:

1. The universe has a cause.

2. If the universe has a cause, then an uncaused, personal Creator of the universe exists, who without the universe is beginningless, changeless, immaterial, timeless, spaceless and enormously powerful.

3. Therefore God exists

We see that God must be uncaused and therefore eternal and changeless, being the creator of time. Since God created space, God must transcend space and therefore be immaterial and not physical in nature. These arguments have led more recently to a great deal of debate, and a helpful accessible reference about the debates is given on the internet.[59] With regard to the question "Who caused God?" the answer is that God did not begin to exist as God is eternal, so we cannot go to the next step in the argument and say that God has a cause.

Further Arguments of Aquinas

A second argument for God's existence that stems from Aquinas that we briefly consider next is that about the unmoved Mover. He argues that as change exists, there must be something that causes the change, and something that causes that change, and so on. Progressing back in time to the beginning there must be an unchanging changer, namely God. Such a being would have to be eternal and not come in and out of existence as that would imply a change.[60] The third argument is the argument from contingency, which can be summarized as follows:[61]

---

[59] For example, http://en.wikipedia.org/wiki/Kalam_cosmological_argument, accessed August, 2015.
[60] For the interested reader, such an argument is discussed in detail by Feser (2008: 91–102).
[61] http://www.philosophyofreligion.info/theistic-proofs/the-cosmological-argument/the-argument-from-contingency/, accessed August, 2015.

1. Everything that exists contingently (that is does not need to exist) has a reason for its existence.

2. The universe exists contingently.

3. Therefore the universe has a reason for its existence.

4. If the universe has a reason for its existence then that reason is God.

5. Therefore God exists.

At the risk of putting the reader to sleep I want to say a bit more about contingency. Everything that exists is either contingent (dependent on something else for its existence) or necessary (independent of anything else). When we say that the universe is contingent we mean that it need not have existed and it could have been different from what it is. In order to find out if it could have been different we need to know how it actually is through scientific investigation. In looking at this question, philosopher Mortimer Adler[62] focuses on two of the four causes of Aristotle mentioned above. We have the cause of coming into existence and an efficient cause of continuing existence. For example, my mother was the cause of my existence but now that she has passed on she is not the cause of my continuing existence.

Adler also distinguishes between superficial contingency and radical contingency. Something with superficial contingency is contingent but is transformed into something else when it ceases to exist, while something with a radical contingency becomes nothing when it ceases to exist; it has the potential for nonexistence. Every object in the world appears to be superficially contingent. If you burnt this book out of sheer frustration, the book would cease to exist as a book, but it would be transformed into a charred mass. Yes, it would still exist but in a new state. Lawyer Dean Overman continues Adler's approach with the statement:

> Although the components of the universe are only superficially contingent, the universe as a whole is radically contingent, because the universe is only one among many possible universes.[63]

These other universes don't have to actually exist; they only have to be logically possible. As our universe could be other than what it is with different laws, it has the potential for nonexistence (annihilation). It is therefore radically contingent and not a necessary universe. This means that our universe requires a cause for its continuing existence as it could cease to exist. If we define God as a necessary rather that a contingent being, then God cannot be part of the universe. Overman then says that :

> God is the necessary cause of the continuing existence of the universe and all of its components, even if the universe did not have a beginning.

---

[62] Adler (1980).
[63] Overman (2009: 27–29).

This means that God is not simply a cause that began the universe and then let it run on its own, but is rather a cause that is continually preserving the universe. Further, if the universe is contingent, we don't have an explanation from natural laws as to why the universe exists. The current laws of physics don't apply with the initial event of creation (time zero or just after) so that another (personal?) explanation is needed. The universe cannot explain its own existence. All the above arguments tend to be variations on a similar theme. We therefore now ask the question, "If God exists, what can we say about the nature of such a God?"

The Nature of God

What is God like? It can be argued that God is pure existence and pure "actuality", and therefore has no potentiality as everything that changes has potentiality. Hence God cannot change (the Unmoved Mover described above), and has no potential to actualize through change.[64] Can there be more than one such God? If so then there would be more than one purely actual being with no unrealized potentialities, but each of them must have some feature that the others lacked to distinguish between them. However, lacking some feature means having an unrealized potentiality, which is a contradiction. Therefore there can be only one purely actual being. For example, you cannot have one such being more powerful than the others as that being has an unrealized potential for increasing individual power. Such arguments however do not necessarily tell us much about the nature of God. This is considered in later chapters.

2.5 THE TELEOLOGICAL ARGUMENT

We will now look at some of the the special features of our world that have led to our existence and to life being possible on our planet, and begin with the special role that mathematics plays in our understanding of the universe. Some fundamental laws of physics are then examined, which leads us to the existence of certain fundamental constants. These constants are found to be closely fine-tuned for the existence of life and in fact the existence of our universe.

2.5.1 The Role of Mathematics

When we consider the fundamental mathematical equations underlying cosmology and quantum theory we notice two features. First, the equations, of which some are given below in Table 1, are surprising simple and elegant (to the mathematician; the reader might think otherwise!) and second, they may

---

[64]See Psalms 102:25–26; Malachi 3:6; and James 1:17.

depend on so-called universal constants, along with many other fundamental physical constants. From a relatively small number of basic equations it is possible to derive a large number of other equations. There are, of course, other empirically based laws that we use to describe experiments and biological phenomena, but here we talking about the fabric of the universe. Abstract mathematics postulated the existence of antiparticles (from Dirac's equation)[65] and black holes (from Einstein's equation of general relativity) before they were discovered. The following quote from Eugene Wigner, recipient of a Nobel prize in Physics, that sums up many a reaction to the equations says:

> The enormous usefulness of mathematics is something bordering on the mysterious .... There is no rational explanation for it .... The miracle of the appropriateness of the language of mathematics for the formulation of the laws of physics is a wonderful gift which we neither understand nor deserve....[66]

Even Einstein was amazed at the comprehensibility of the world. However, unlike Wigner and Einstein, but more in keeping with Newton, many physicists today see the beauty and elegance of these remarkable equations not as a mystery but as evidence for the existence of a designer/creator, or God who created the laws.[67] The successfulness of mathematics raises similar questions to those about fine tuning discussed below.

Beauty

The concept of beauty has even guided mathematics in the development of new theory, though we have to be careful that we don't equate beauty with truth.[68] For example Steven Weinberg, a Nobel prize winner in Physics and a convinced atheist, points out, "mathematical structures that confessedly are developed by mathematicians because they seek a sort of beauty are often found later to be extraordinarily valuable by the physicist." He also says: "Physicists generally find the ability of mathematicians to anticipate the mathematics needed in the theories of physics quite uncanny," and "sometimes nature seems more beautiful than strictly necessary."[69]

While there is a certain "beauty" or even "simplicity" in the mathematics of physics we should not forget that the universe is full of beauty at all levels from the snow flake and the butterfly to the sunset and the galaxy that appears to be an objective reality, rather than an artifact of our minds. This beauty, which includes proportion and symmetry and amazing mechanisms (e.g., the trapdoor spider, bird navigation, and shelters constructed by marine dwelling

---

[65] See, for example, http://en.wikipedia.org/wiki/Dirac_equation, accessed August, 2015.
[66] Wigner (1960: 2).
[67] Jeremiah 33:25.
[68] For a discussion on this point with regard to the role of beauty in mathematics by science writer Phillip Ball see http://aeon.co/magazine/philosophy/beauty-is-truth-theres-a-false-equation/, accessed August, 2015.
[69] Weinberg (1994: 153, 157, and 250).

Foraminifera[70] points to an intelligent designer that values form and beauty. I have already mentioned the Fibonacci sequence of numbers in Chapter 1. If God is the greatest being there is, then it is not unreasonable to assume that such a being has perfect aesthetic sensibility and has fine-tuned the world for simplicity, beauty, and elegance.[71]

2.5.2 Fundamental Laws of Physics

Having referred above to the "simplicity" and "beauty" of mathematics in the laws of physics I now want to look more closely at some of these laws given in Table 1 below. The first thing we notice is that the equations are quite short and therefore, in a sense, simple in form (even if incomprehensible to the reader!).

Table 1: Fundamental Laws of Physics

Mechanics (Hamilton's equations) $\dot{p} = -\frac{\partial H}{\partial q}$ and $\dot{q} = -\frac{\partial H}{\partial p}$

Electrodynamics (Maxwell's equations):

$$F^{\mu\nu} = \partial^\mu A^\nu - \partial^\nu A^\mu \text{ and } \partial_\mu F^{\mu\nu} = j^\nu$$

$$\nabla \cdot E = 4\pi\rho, \ \nabla \times E = -\frac{1}{c}\frac{\partial B}{\partial t}, \ \nabla \cdot B = 0, \text{ and } \nabla \times B = \frac{4\pi}{c}J + \frac{1}{c}\frac{\partial E}{\partial t}.$$

Statistical Mechanics (Boltzmann's equations):

$$S = -k\sum_i P_i \log P_i \quad \text{and} \quad \frac{dS}{dt} \geq 0$$

Quantum Mechanics (Schrödinger's equation):

$$i\hbar \frac{\partial \Psi}{\partial t} = \hat{H}\Psi, \quad \hbar = \frac{h}{2\pi} \quad \text{and} \quad \Delta x \Delta p \geq \frac{\hbar}{2}$$

General Relativity (Einstein's equation): $G_{\mu\nu} = -8\pi G T_{\mu\nu}$

There are other mathematical equations and I will choose one to illustrate how we may need other conditions associated with the use of the equations, namely the initial or boundary conditions together with the values of certain fundamental constants that are not determined by the laws themselves.[72]

---

[70] Hardy (1965).
[71] For a theology of beauty see Dubay (1999).
[72] I am grateful to engineer Walter Bradley for ideas from his article "The Designed Just So Universe" on the internet.

Consider the following mathematical equation derived from Newton's laws of motion and gravitation describing the situation where an object is thrown down from the top of a building of height h with velocity v. (Non-mathematical readers may wish to take a coffee break at this stage or jump to the next paragraph!) If $h(t)$ is the height of the object at the time t after it was thrown, then

$$h(t) = h - \frac{1}{2}\frac{Gm}{r^2}t^2 - vt,$$

where G is a universal gravitational constant, and m and r are the mass and radius of the earth. This is an example where we assume perfect conditions such as no or negligible air resistance and that all measurements are made so accurately that any measurement errors are negligible. Suppose we want to find the height of the building by simply dropping the object (i.e., the initial velocity is zero or $v = 0$) and timing its descent (i.e., the initial time is zero and it takes time t_0, say, to descend to zero height). When the object hits the ground the height $h(t)$ of the object is now zero and, using the above equation,

$$0 = h - \frac{1}{2}\frac{Gm}{r^2}t_0^2 \quad \text{or} \quad h = \frac{1}{2}\frac{Gm}{r^2}t_0^2,$$

which gives us our answer. We can make three observations from our initial equation:

(1.) We note the mathematical form that "nature" (or our idealization of nature) takes.

(2.) We need to know the value of the universal constant G.

(3.) The "boundary" or initial conditions must be specified (here the initial value of v_0). If we were on the moon we would have different values for m and r; G would be the same as it is a universal constant.

We see then that all that can be controlled are the boundary conditions; the rest is out of our hands. This is true when any engineer designs something such as a car. The boundary conditions determine the shape, the component parts, and the actual working of the the car. What makes such a design difficult is that all the parts tend to be interdependent in the way they work. For example if one engine part is only slightly out then the engine may seize up.

This leads me to my first point. Any boundary conditions would have been set right from the beginning when the universe began. For example, if the initial velocity was too high, the expanding matter would not form planets, stars and galaxies, while if it was too low the universe would soon collapse under the force of gravity; the two have to be balanced. Various figures suggest that the precision of the big bang to create the universe is, as mentioned below, one part in 10^{123}, which is extremely precise.[73]

---

[73] Penrose (2010).

My second point is that there are a number of fundamental constants, some of which are given in Table 2 below like G above, as well as dimensionless and other physical constants, that have to be very precise.[74] The coupling constants have to be obtained experimentally, and some of the values quoted in the literature vary, depending on the definitions used (e.g., α_g is based on the ratio of two particles, with different particles possibly used) and the energy levels involved. Note that the constants are quite different in magnitude.[75]

TABLE 2: Physical Constants

Fine structure and coupling constants (dimensionless)

Gravitation - dimensionless $\alpha_g = 5.9 \times 10^{-39}$

Weak coupling $\alpha_w = 1/29.6$

Fine structure constant $\alpha = 1/137$

Strong coupling $\alpha_s = 0.1187$

Some general constants

Boltzmann constant $k = 1.38 \times 10^{-23}\ JK^{-1}$

Planck's constant $h = 6.63 \times 10^{-34}\ Js^{-1}$

Speed of light $c = 3.00 \times 10^8\ ms^{-1}$

Gravitational constant $G = 6.673 \times 10^{-11}\ m^3 kg^{-1} s^{-2}$

Mass of elementary particles

Pion (π^0) rest mass (energy) $m_\pi = 0.238 \times 10^{-24}\ kg\ (135 MeV)$

Neutron rest mass (energy) $m_n = 1.675 \times 10^{-27}\ kg\ (939.6 MeV)$

Electron rest mass (energy) $m_e = 9.11 \times 10^{-31}\ kg\ (0.511 MeV)$

Proton rest mass (energy) $m_p = 1.673 \times 10^{-27}\ kg\ (938.3 MeV)$

Unit charge $e = 1.6 \times 10^{-19} coul$

We note that the mathematical form and boundary conditions alone are not sufficient to guarantee having a universe that is a suitable habitat. A whole range of other conditions including appropriate values of the universal constants are required as well for the existence of the universe and of life. These conditions would have been set right from the initial Big Bang. As some writers say, our universe appears to have been be fine-tuned for existence and for the building blocks of life.

---

[74]Various constants are listed in http://physics.nist.gov/cuu/Constants/, accessed August, 2015.

[75]Though this depends in some instances on the units of measurement used.

2.5.3 Subatomic Particles

Before talking about fine tuning, I want to digress for a moment and look at the amazing world of subatomic particles. Although matter looks solid to us it is almost entirely empty space with the particles of matter extremely small and some lasting less than a billionth of a second.[76] There are over 200 subatomic particles, all of which have a necessary part to play. The reader is no doubt familiar with the fact that atoms are made of nuclei and electrons, and nuclei are made of protons and neutrons. However it may not be so well-known known that protons and neutrons are each made of three quarks of two varieties.[77] The lighter particles such as electrons, neutrinos, and muons are collectively called leptons, while the heavier particles that are nuclear particles and products of their interactions such as protons, neutrons, and pions are collectively called hadrons. As well as the muon and tau particles that are negatively charged we have a very common particle called the neutrino that has no charge. We then get combinations of the neutrino with the electron, tau, or muon, as for example the muon-neutron, to give further particles with no charge. In particular, fermions refer to the class of particles with one type of spin, while bosons (both elementary and composite) refer to those with another type.

The menagerie increases as there are also antiparticles, composite particles (e.g., baryons, mesons, and hadrons mentioned above), and hypothetical particles postulated by theory (e.g., tachyons, gravitons, and the Higgs boson that has now been discovered, which through its associated field provides mass to particles). Particles can also have different spins, and symmetries play a fundamental role at both the subatomic and atomic levels. For example atoms are highly symmetric and most of the molecules and all crystalline materials have some elements of symmetry. These symmetries can be explained by an abstract branch of mathematics called group theory. Why is this so?

One of the mysteries of our world is why there is any matter around at all. Every particle has an antiparticle that has the same mass as the particle but with opposite charge and quantum spin, and antiparticles combine together to form antimatter as ordinary particles combine to form matter. Both particles and antiparticles are equally affected by gravity as gravity depends on mass. Mixing particles and antiparticles or matter and antimatter together can lead to the annihilation of both with a release of energy according to the formula $E = mc^2$, where E is the energy, m is the combined mass of the particles, and c is the velocity of light. This means that when the world came into being there was a preponderance of matter over antimatter rather than there being equal amounts, the reason for which, at the time of writing, is an unsolved problem of physics. It has also been found that vacuum space is far more

---

[76] It is interesting that the Bible talks about the things that are seen being made from things that are unseen (Hebrews 11:3).
[77] There six varieties or "flavors" of quarks altogether.

complicated than anyone had previously imagined. Elementary particles can spontaneously pop out in space and disappear again in such a very short time that they cannot be measured directly, but do have an observable effect. Clearly the research in this area is still ongoing.

2.5.4 Fine Tuning

Returning to the constants, let us now look first at the so-called fine-structure constants listed in Table 2 above. I have listed some values to give just a rough idea as to their relative magnitudes. These constants are related to the four fundamental physical forces: gravity, electromagnetism, and the strong and weak nuclear forces. The strong force holds atoms and composite subatomic particles together while the weak force is involved in radioactive decay. It is proposed that all four forces were unified in some way up to 10^{-43} seconds (called one Planck time) after the big bang so that during this brief time the laws of physics were not operating and the particles of quantum physics did not exist because of the extremely high temperature.

The first thing we find about the fundamental constants is that they have just the right values, with very little room to maneuver if one was trying to set these up to allow the formation of life in a universe through some laws of nature. As Hawking has noted,

> The laws of science, as we know them at present, contain many fundamental numbers, like the size of the electric charge of the electron and the ratio of the masses of the proton and the electron.... The remarkable fact is that the values of these numbers seem to have been very finely adjusted to make possible the development of life.[78]

Studies have shown that the relationships of these constants are absolutely critical. A slight variation of the ratios of the constants can cause disaster at both the atomic level and the stellar level so that the resulting universe would be unsuitable for life of any imaginable type, and perhaps not even exist. A few examples of these requirements are outlined briefly as follows:[79]

(1.) The early density of the universe was just right. Stephen Hawking has suggested that

> if the density of the universe one second after the Big Bang had been greater by one part in a thousand billion, the universe would have re-collapsed after ten years. On the other hand, if the density of the universe at that time had been less by the same amount, the universe would have been essentially empty since it was about ten years old.[80]

---

[78] Hawking (1988: 125)
[79] For a list see www.godandscience.org/apologetics/designun.html, accessed August, 2915. Further references are Barrow and Tipler (1986), Rees (2000), Collins (2009), and Barnes (2012).
[80] Hawking (1988: 121–125).

Davies said that the matching of the exploding thrust of the Big Bang with the inward gravitational attraction had to be accurate to one part in 10^{60}, which is an incredible accuracy.[81]

(2.) According to mathematician Roger Penrose,[82] the original phase-space volume would need to be accurate to one part in $10^{10^{123}}$ to set the universe going. He arrives at this figure on the basis of the second law of thermodynamics as the universe must initially be highly ordered since it is becoming more disordered as the entropy increases. Even though a very much larger probability given by one part in $10^{10^{60}}$ would give us our universe and life, it is still negligible. Rodney Holder, astrophysicist and theologian, concludes:

> The implication is that our universe is vastly more special than required merely in order for us to be here. It is much, much more special than a universe randomly selected from the subset of universes conducive to life. This is a very serious challenge for the multiverse idea but totally consistent with design.[83]

The multiverse theory is discussed below.

(3.) Life depends upon a number of principles operating in the quantum realm. For example, the Pauli Exclusion Principle, which states that no more than one particle of a particular kind and spin is permitted in a single quantum state, plays a key role in guaranteeing the stability of matter.

(4.) The frequency distribution of electromagnetic radiation produced by the sun is also critical, as it needs to be tuned to the energies of chemical bonds on earth. It is also matched to the unique window of light transmission by water, water being an essential ingredient of living tissue and eyes in particular. The mass of an electron and Planck's constant are also involved in the matching.

(5.) The strong force and the electromagnetic force are precisely balanced; these determine the quantum energy levels for nuclei. The ratio of these two forces cannot vary by more than 1% without drastically affecting the needed abundance of carbon, a building block of life. A 2% increase in the ratio would mean that the universe would have no hydrogen, no long-lived stars that burn hydrogen, and there would be no water—an essential for life. There is also a close relationship between the strong force and the relative masses of the elementary particles.

(6.) The ratio of the weak force to the electromagnetic force is also finely balanced. If the ratio was slightly larger, neutrons would decay more

---

[81] Davies (1984: 184)
[82] Penrose (1989: 339–345) and http://www.ws5.com/Penrose/, accessed August, 2015.
[83] Holder (2013: 145).

rapidly thus reducing production of deuterons, helium, and elements with heavier nuclei. If the ratio was slightly weaker, the big-bang would burn up most of the hydrogen, turning it into helium leaving little or no hydrogen and many heavier elements instead. This would mean no long-term stars and no hydrogen containing compounds such as water. We also need a right mix of helium and hydrogen, which is obtained by balancing the weak force and gravity.

(7.) The force of gravity cannot be a lot greater. Also, the ratio of the electromagnetic force to the gravity force is critical (accurate to 1 in 10^{40}). It is related to the radiation required for the operation of necessary chemical reactions that are essential for life.

(8.) The energy density of empty space (currently related to the cosmological constant, though still controversial) has a precise value. The cosmological constant is the result of the almost perfect cancellation of a very large number of comparatively very large physical constants. For example, a change in the strength of the gravitational or nuclear force by as little as one part in 10^{100} could entirely ruin the cancellation, making space expand or contract furiously.

(9.) One particularly famous case of fine tuning involves the production of carbon and oxygen in the stars through nuclear reactions that requires a delicately arranged process where hydrogen burns to form helium, which then ignites to form carbon and oxygen. The energy levels for these two are just right for life and famous astronomer Fred Hoyle, who is not noted for theistic beliefs, describes this as design and suggested that

> there is an enormous intelligence abroad in the universe...a super intellect has monkeyed with physics, as well as with chemistry and biology, and that there are no blind forces worth speaking about.[84]

(10.) The relative masses of the elementary particles also need to have precise values. I give just one example. Hawking has noted that the difference in the mass of the neutron and the mass of the proton must be approximately equal to twice the mass of the electron, which is the case.

There are many other requirements that are satisfied.[85] Philosopher Robin Collins[86] describes in detail how (5)–(10) above give particularly strong evidence of fine tuning. Philosopher John Leslie notes that:

> A force strength or a particle mass often seems to need to be more or less exactly what it is, not just for one reason but for two or three or five. Yet obviously it could not be tuned in first one way and then another, to satisfy several conflicting requirements.[87]

---

[84] Hoyle (1981: 12).
[85] See Leslie (1989) and his article http://www.leaderu.com/truth/3truth12.html, accessed August, 2015.
[86] Collins (2003).
[87] Leslie (1989: 64).

Prominent New Zealand physicist Jeff Tallon in a New Zealand newspaper article sums up the fine tuning in a straightforward manner:

> The field of cosmology tells us that the universe is exquisitely finely balanced. Its density, back at the first moments of the "big bang", was critically balanced to better than one part in one billion billion billion billion. A fraction more dense and it all would have collapsed again. A fraction less dense and it all would have evaporated — no galaxies, no stars, no planets, no mother Earth. All the known forces of nature are tightly balanced relative to each other. A little this way and protons do not form. A little that way and neutrons don't form. Tweak another way and no particles at all. Tweak another way and everything is hydrogen only. Now if the universe were truly random this would not happen.

When we look at our planet earth we find other striking coincidences. For example:

(1.) If we look at the collection of all possible universes, only an extremely small number are life permitting, while the rest are life prohibiting.[88] A number of authors cite very many different properties of the universe that are being kept in perfect balance so that the universe hangs together and we are able to live on earth.[89] (I do well to make the meat, potatoes, and vegetables all come out at the same time properly cooked when I am cooking for dinner!) A few of these properties that are related to earth follow.

(2.) If our solar system was too close to the center of our galaxy or star cluster, earth would be devastated by cosmic radiation. There is also a huge "black hole" at the centre. A black hole is a mathematically defined region of space-time exhibiting such a strong gravitational pull that no particle or electromagnetic radiation can escape from it. It is black because the intense gravity prevents light from escaping.

(3.) Earth is just the right size to support the right kind of atmosphere and atmosphere thickness that we need. If it was smaller like Mercury, it couldn't support an atmosphere, while if larger like Jupiter it would have a poisonous atmosphere of hydrogen.

(4.) Our earth is placed at just the right distance from the sun and the other planets in our solar system. Any closer to the sun we'd burn up, and any further distance away we'd freeze. With a nearly circular orbit we enjoy a comparatively narrow range of temperatures compared with the moon and the other planets. The earth's 24 hour rotation ensures even heating.

---

[88] For a discussion about the possibility of life elsewhere in the universe see Strobel (2004: chapter 7).
[89] For a helpful reading list about fine tuning see Barnes (2013).

(5.) If our magnetic field was much weaker, we would be devastated by cosmic radiation, while if it was much stronger, we would be devastated by severe electromagnetic storms.

(6.) The moon is the right size and distance from us to keep the oceans circulating and not stagnating. It also stabilizes the tilt of the earth's axis, the tilt being responsible for our seasons.

(7.) Water, one of the strangest substances known to science, plays a key role as life is mostly water and we need it to live. For example it has a high specific heat, which means that chemical reactions within the human body will be kept stable. It also has surface tension so it can move upwards nourishing plants. In contrast to everything else, it freezes from the top down so that fresh water and sea creatures can survive underneath. If ice didn't float we would freeze up. Because ice is a poor conductor of heat, it does not absorb heat energy from the water beneath the layer of ice, which prevents the water freezing. In addition to its solvency water has other important properties. For example with salts in solution, it conducts electricity.

(8.) Carbon, with its unusual chemistry that allows it to bond to itself as well as to other elements, can lead to the creation of highly complex molecules that are stable over prevailing terrestrial temperatures and are capable of conveying genetic DNA information. Silicon falls far short of carbon when it comes to properties for some kind of life. Furthermore, carbon dioxide has the property, unique among gases, of having at ordinary temperatures about the same concentration of molecules per unit volume in water as in air so that it is everywhere available for photosynthesis. It should be noted that the universe has to be sufficiently old for the natural evolution of the intrastellar synthesis of carbon. Several billion years are needed for the production of carbon and other heavy elements inside dying red giant stars.

(9.) The appearance of photosynthesis, an extremely complicated process, right at the start of life on earth was essential for all future life. This created the oxygen in our atmosphere.

We find then that the universe is fine tuned so that life and humans can exist. This suggests humans have a special place in the cosmos and it can be argued that we are the result of a creator who has special concern for us. We are not just a random speck on a tiny part of the universe, but there is order and structure rather than chaos; the earth was designed to be inhabited.[90] In Psalm 19:1 we read that the heavens declare God's glory and the sky reveals God's handiwork. Paul also states that God's power and deity are revealed

---

[90] Isaiah 45:18.

in God's creation so that we have no excuse.[91] It is all there for us to see, though some, however, refuse to see it!

It was apparently the fine tuning of the universe that helped lead Antony Flew, famous philosophy professor and former leading atheist, to come to believe in God at the age of 81. He said in a telephone interview with ABC News (12 Sept. 2004) that a "super-intelligence is the only good explanation for the origin of life and the complexity of nature." He felt that the fine-tuning of the universe at every level was too perfect to be due to chance. The astronomer Lee Smolin wrote that "Luck will certainly not do here. We need some rational explanation of how something this unlikely turned out to be the case."[92] In response to this idea, there is an argument sometimes presented that says that that if the universe was designed to be fine-tuned for life, it should be the best one possible, and that some argue that it is not because of evil and suffering. The existence of the latter is usually regarded as an argument against God's existence. This topic will be discussed in Chapter 8.

2.5.5 Anthropic Principle

In an endeavor to explain why the universe is fine-tuned, some writers have introduced the idea of the "Anthropic Principle", which is stated in various ways and has been proposed in roughly two forms—the weak and the strong versions. The weak version says that the nature of the physical universe must be compatible with the conscious life that observes it, that is, what we can expect to observe must be restricted by the conditions necessary for our presence as observers. If our universe wasn't hospitable to our life we wouldn't be here to ask why! However, I cannot see that this argument makes it any more likely that this type of universe would come into existence. There is the illustration[93] of you facing a large firing squad of, say, 100 trained marksmen because of trumped up charges while you are traveling in another country: they all fire and all miss. What do you conclude? Perhaps you might say that there is nothing more to be said as if they hadn't missed you wouldn't be here to be surprised about it. However, this does not detract from the fact that the event was exceptional and you would suspect that the whole thing was a set up and deliberately planned.

The strong principle (SAP) asserts that the universe must be such that it has within it those properties that allow life to develop at some stage of its history. A second version of the SAP, derived from the findings of quantum mechanics, has been called the Participatory Anthropic Principle (PAP) by physicist John Wheeler. This version holds that observers are necessary to bring the universe into being. The PAP follows from the standard Copenhagen interpretation of quantum mechanics, in which some type of living

---

[91] Romans 1:20.
[92] Smolin (1997: 24).
[93] Craig (2010: 116).

consciousness is required to make events "real" (see Section 4.1.3). According to this interpretation, developed by physicist Neils Bohr (1885–1962), there is no such thing as a concrete quantum reality until a living observer exists to "collapse" the appropriate quantum wave function (got that?).[94] Without this act of observation, reality seems to be held in a paralyzing state of indecision. But such an approach has problems as Holder notes:

> It is, however, highly paradoxical to assert both that non-conscious matter pre-exists and causes my existence, and that I cause its existence. And surely the result of a quantum experiment, such as photons passing through two slits on a screen, is decided by the imprint on the photographic plate at the time the experiment happens, rather than by me three months later when I take the plate out of the cupboard and look at it.[95]

It could be argued that God is the one who makes the first observation and brings the world into existence. We look more fully at quantum theory in Section 4.1.3.

Some argue that the universe was compelled in some sense to have conscious life eventually emerge (the strong anthropic principle), which is a form of biological determinism, and why this should happen just by natural processes evades me! They believe that life has adapted through evolution to the world and not the world has adapted to (or was designed for?) life. It has been suggested that it must be possible to observe at least some universe, and hence the laws and constants of any such universe must accommodate that possibility. As I have already indicated, this does not adequately deal with all the fine tuning we observe, and the infinitesimal probability of it happening by chance. Further, some multiverse theories (and there are several) are not falsifiable so that they fit into the category of non-testable hypotheses.

There seems to be a strong agreement among cosmologists that our world is fine-tuned for the existence of intelligent, carbon-based life, or rather "fine-tuned for the building blocks and environments that life requires."[96] The question is why. A related question to ask is: "Can life have evolved elsewhere?" After all, our own galaxy the Milky Way contains some hundred thousand million stars and there are another hundred thousand million galaxies, any of which could spawn life. However, assuming there is only one universe and it has occurred only once, the probability of a getting the finely-tuned constants and associated quantities happening under a chance event is vanishingly small. Furthermore, the additional pattern of information required for life to occur would seem to be independent of the initial event thus adding to the unlikelihood of it all occurring by chance. Even allowing for other galaxies, the probability of life under the above assumption of one universe is negligible. Robert Jastrow, the famous agnostic astronomer, observed

---

[94] For further details see Section 4.1.3.
[95] Holder (2013: 78).
[96] Davies (2003: 118).

that the anthropic principle "is the most theistic result ever to come out of science."[97]

As already mentioned, some theorists have gone so far as to argue that life is necessary to make the universe itself real. However, I do not wish to get drawn into the extensive philosophical debate about the pros and cons of fine tuning and these principles but will briefly comment on some alternatives to the design idea that have been suggested. For example, suggestions to explain the fine tuning such as alien design, alien big bang, our universe is an alien computer simulation, and our physical world is an abstract mathematical structure[98] have been proposed and one is only limited by one's imagination!

2.5.6 Multiverses

I want to mention one theory in particular that seems to have more support, the so-called Multiverse or Many Worlds theory that assumes the existence of many universes with the same natural laws but different physical constants, some of which are hospitable to intelligent life. It is argued that if all possible values occur in a large enough collection of universes, then viable ones for life will surely be found somewhere. Because we are intelligent beings, we are by definition in a hospitable one. Worlds that are not fine-tuned, however, will have no observers in them so they cannot be observed. Without the right conditions we would not have found ourselves to be here now, but perhaps somewhere else, at some other appropriate time; if life was impossible, no one would know it! Astrophysicist Luke Barnes puts it nicely when he writes, "If observers observe anything, they will observe conditions that permit the existence of observers."[99] He says this is of course tautological, but certainly not trivial.

One idea promoted is that the fine tuning is due to the sampling idea of "selection bias" as, in the long term, only survivors can observe and report their location in time and space.[100] Barnes gives an example of the Malmquist bias where in any survey of the distant universe we will only observe objects that are bright enough to be detected by our telescopes. He gives the example of quasars, which are very bright, but the brightness is not due to selection bias. He says the (best) answer is: because quasars are powered by gravitational energy released by matter falling into a super-massive black hole, and not because we otherwise wouldn't see them. Selection bias may have a part to play as to why we are here, but I don't believe you can explain a highly improbable event by simply pointing to us being on the scene to observe it. In any case a multiverse can still have a beginning.

---

[97] Jastrow (1984: 22).
[98] Tegmark (2008).
[99] Barnes (2012).
[100] See, for example, Bostrom (2002).

There is another problem. All current multiverse models need a mechanism for generating universes, and this itself would need an underlying complex system of laws overseeing which universe got which set of laws. If these underlying laws were slightly different, the generator might not be able to to produce a life-sustaining universe. As philosopher Robin Collins notes,

> The multiverse generator itself, whether of the inflationary variety or some other type, seems to need to be 'well-designed' in order to produce life-sustaining universes.[101]

We are then back to a first cause! Also the laws governing the generation of multiverses would themselves have to be fine-tuned to give rise to even just one universe that allows life to develop. The fine-tuning extends not just to the physical constants and the physical laws, but also to the initial distribution of mass-energy in the universe. The problem is that the range of values that would allow life to be supported is extremely small compared with the range of non-life values.[102] Furthermore, our universe is not fine-tuned for just simply observers, this being the main kind of fine-tuning used to explain the multiverse hypothesis, but also for observers that can interact with each other.

Multiverse models have been discussed under the umbrella of Hubble volumes and an infinite universe, pocket universes, cyclic models (John Wheeler), black holes producing new universes (Lee Smolin) with "mutated" constants, parallel universes (Hugh Everitt), quantum alternatives (Max Tegmark), inflation models, and string theory possibilities that allow for different values of the constants not predicted by the theory. The question is why should there be a collection of universes that are life permitting.[103] The problem is how to test such theories. Davies offers a variety of nonscientific arguments against the multiverse theories. He admits there are some regions of the universe that lie beyond the reach of our telescopes, but says: "somewhere on the slippery slope between that and the idea that there are an infinite number of universes, credibility reaches a limit."[104] He goes on to say:

> Extreme multiverse explanations are therefore reminiscent of theological discussions. Indeed, invoking an infinity of unseen universes to explain the unusual features of the one we do see is just as ad hoc as invoking an unseen Creator. The multiverse theory may be dressed up in scientific language, but in essence it requires the same leap of faith.

George Ellis, eminent mathematician and cosmologist, who mentions seven questionable arguments supporting multiverses, comments:

---

[101] http://home.messiah.edu/ rcollins/, accessed August, 2015
[102] For examples of fine tuning see Collins at http://www.home.messiah.edu/ rcollins/Fine-tuning/Stenger-fallacy.pdf, accessed August, 2105. Barnes (2012) provides a refutation of Stenger (2011) who endeavors to give a natural explanation for fine tuning.
[103] For a recent discussion see Vilenkin (2006).
[104] Davies (12 April 2003). "A Brief History of the Multiverse." New York Times.

Nothing is wrong with scientifically based philosophical speculation, which is what multiverse proposals are. But we should name it for what it is.[105]

However this subject will no doubt be the focus of further mathematics and a search for observational evidence. For me, a belief in multiverses is no more credible than a belief in "heaven" and "hell" (see Section 6.8), as the latter could be regarded as alternative universes operating out of space and time.

There are further problems with the multiverse approach. First, there is no evidence for any of these other possible worlds, nor can there perhaps be any such evidence in the future because these alternative domains are believed to be utterly beyond human observational powers, even in principle. Second, this approach begs the question, since it assumes the prior and unexplained existence of the multiverses themselves. Where did they come from? Third, if one of the possible universes was created by God, then why could we not exist in it. Keith Ward[106] argues that if God actually exists as a necessary being, God must exist everywhere and always, so that there would be no possible universe that could exist without God. Finally, the use of an infinite number of unobservable worlds to explain the existence of our own world is an unprecedented violation of Ockham's Razor, which states that the simplest explanation in any set of natural circumstances is probably the correct one.

It should be noted that even if we have an infinite number of universes to choose from, ours might not be one of them. For example, the even numbers form an infinite set, but it does not contain the number one as it is an odd number.[107]

Some rethinking might be needed if we found a completely different form of life elsewhere in the universe. This does not seem likely as the laws of physics appear to be universal. It is not a question of life appearing on just our planet, but rather the whole universe may have had to be formed in a particular way to contain any life at all. Also the existence of a multiverse doesn't explain the laws of nature that we have and why this particular multiverse exists and not some other. It could be argued that multiverses do not rule out God as God may have somehow used a "chance" process to obtain the universe that we have.

None of the discussion so far explains why we not only observe the universe, but we also understand it! Some would argue that this is also a self-selection effect, but this makes the probabilities get even smaller in that we now need two events to occur—a multiverse and we understand it. If one is going to try and use probabilistic arguments, then why cannot the order we see simply be

---

[105] Ellis (2011).
[106] Ward (2008a 73–75).
[107] It should be mentioned that some theists support the multiverse idea and for a general discussion of the topic see Holder (2012: chapter 12).

confined to just our solar system rather than to the whole universe, as local order is more likely from a probability point of view.

Biblical Viewpoint

Summing up, creation by God seems to be a much simpler explanation for the existence of our universe rather than as a a chance event or as a chance selection of one out of a large number of multiverses. I believe that a Creator provides the best explanation for why the universe exists instead of nothing. The Bible talks about God creating the world out of nothing, which is in complete contrast to other pagan and Greek myths of creation. For example, in the Babylonian creation myth *Enuma Elish*, the heavens and the earth were made out of dividing up the corpse of an evil goddess destroyed in battle, but in the very first verse in the Bible (Genesis 1:1) we read that in the beginning, that is at the commencement of time, God created the heavens and the earth (it is not clear from the Hebrew if it meant created out of nothing or made from existing material). However, the Hebrew word *Bara* almost invariably means create out of nothing,[108] while the New Testament uses the Greek word *Ktisis*.[109] We also have Romans 4:17 that talks about God calling into existence things that do not exist, and Hebrews 11:3 makes a similar comment. In contrast to Christian and Hebrew thought, where the material world was pronounced to be good by God the maker, other creation stories take a negative view of the material world.

2.6 ARGUMENT FROM DESIGN

The appearance of design is very strong in living things. The eminent biologist and atheist Richard Dawkins says, "The illusion of purpose is so powerful that biologists themselves use the assumption of good design as a working tool",[110] and: "Biology is the study of complicated things that give the appearance of having been designed for a purpose."[111] Davies comments that: "The impression of Design is overwhelming."[112] According to materialists, the appearance of purpose in nature is an illusion. They would argue that a particular feature like feathers, for example, is simply the product of purposeless and meaningless causal processes. Such a feature appears to have a purpose because the creatures that first developed them through random mutations just happened to have a competitive advantage over those who did not, so the result looks like purposeful design. Any purpose we see is not actually in the physical world but is projected onto the feature as a "mind-dependent" activity that

---

[108] For example, Genesis 1:27, 2:4, 5:1–2; Deuteronomy 4:32; and Isaiah 42:5, 45:12.
[109] For example, Romans 1:20; Ephesians 3:9; Colossians 1:16; and Revelation 4:11, 10:6.
[110] Dawkins (1995: 98).
[111] Dawkins (1996: 1).
[112] Davies (1987: 203)

needs to somehow be explained in terms of meaningless processes in the brain. The question of mind and brain is taken up further in Section 3.3.

2.6.1 Some General Arguments

The design argument is based on the idea that, as the world appears to be designed, the world must have a designer. This idea was popularized by Paley, for example, who appealed to the fact that the world appears to be very complicated. He argued along with other "intelligent design" supporters that it is all a question of probabilities and gave the example of someone finding a watch. The finder could conclude that the watch just happened or, because of its complexity, it was designed by a watchmaker. The universe is like a watch, being extremely complex and orderly! Although this approach has merit, it has been attacked by others from a number of points of view. For example, some argue that the world could have happened by chance by appealing to evolution and natural selection. They claim that Paley was invoking the so-called "God-of-the-gaps" idea, as further science may eventually provide natural explanations for the complexity. For example, philosopher and astrophysicist Victor Stenger in an online article entitled "Is the universe fine-tuned for us?" said:

> The fine-tuning argument and other recent intelligent design arguments are modern versions of God of the gaps reasoning, where a God is deemed necessary whenever science has not fully explained some phenomenon.[113]

This Science-of-the-gaps argument, which is always used when we don't know anything, is not unreasonable as science does fill in gaps, but it does not negate the existence of God. A more powerful argument was put forward by Aquinas called his Fifth Way (essentially using Aristotle's fourth cause) which is described very briefly as follows.[114]

The material in the universe interacts according to regular patterns of cause and effect that from our perspective enables us to construct working laws about the universe. It is assumed that the world is a rational place. There are also other biological entities in creation that have no consciousness but act in a regular fashion to maintain their continued existence. Trees, for example, will send roots to seek out water (often through cracks in rocks), spread their branches to collect light for photosynthetic leaves, and then use various methods to spread their seeds. They act with a sense of direction in terms of self-preservation, yet are apparently unguided by their own consciousness. All these physical and biological processes are directed toward a certain end or final cause, though the physical laws themselves are material and efficient causes, but not final causes. Even humans, who are conscious beings, are made up of unconscious and unintelligent components that show final causality. It

---
[113] www.colorado.edu/philosophy/vstenger/Cosmo/FineTune.pdf, accessed August, 2015; see also Stenger (2011).
[114] See, for example, Feser (2008: 114–118) and (2009) for further details.

is not possible for anything to be directed towards an end unless the end exists in a mind that directs it. This means there is a Supreme Intelligence that we can call God who is outside our universe doing the directing toward the final causes, and must therefore be the ultimate cause. This argument does not invoke the complexity of the world and the probabilities of getting a fine-tuned universe, or the origin of the universe, nor is it upset by appealing to evolution. In fact it can be argued that evolution itself has a sense of direction, thus supporting the idea of a final cause (see Chapter 10). We note that this created and sustained system has a kind of independence that means it can be studied independently without referring to God.

Revisiting Paley's watch, we can analyze the materials of the watch and see that it operates under the same physical laws governing molecules as other materials. If I smash the watch the same physical laws still operate, but the watch will not keep time any more. There are two key properties about a watch. The first is the set of boundary conditions or, in the words of Michael Polanyi, scientist and philosopher,[115] its shape or morphology. The second is the function of the watch as a timepiece. We cannot describe these two properties in terms of molecules. We need a higher law or purpose to describe both of these. The same is true of a human being and this idea is taken up further in Chapter 3.

There are two particular arguments sometimes brought up against the design argument.[116] The first proposed by philosopher David Hume[117] says that if there is order in the universe and organized complexity in nature, then we would need to account for the even greater organized complexity in the mind of the designer, namely "who designed the designer?" If the probability of life emerging on earth is extremely small, then it would be argued that the probability such a designer exists is even smaller. The second argument claims that science through Darwinism (and its variations) explains away the need for a designer as the organized complexity in nature has arisen through evolution.

Simplicity of God

With regard to the first argument, which is philosophical and not scientific, we need to consider what the nature of God might be like. Organized complexity arises in a system of physical parts that interact together to provide function, as for example in the human brain. Aquinas, in his Summa Theologica (Part 1 Article 7), argued that God is not physical (even though the Bible uses anthropomorphic metaphors like eyes and arms) and does not have parts so although complex in one sense God is essentially "simple." The same would hold with nonphysical parts. God's mind is not divided up into separately existing ideas; the mind comes first with the ideas an integral part

---

[115] Polanyi (1967).
[116] See Glass (2012: chapter 6).
[117] Hume (1779).

of the mind rather than there being a collection of ideas adding up to a mind. God's properties like goodness, omniscience, and justice are not separate properties of God but are God's being: God is goodness, omniscience, and justice. Furthermore, this multiplicity of attributes is an illusion as they are all really the same thing. All of these characteristics are not separate but are simply different ways of looking at the same thing.[118]

Aquinas provided a number of other arguments for the simplicity of God. For example, every composite has a cause, as items in the composite cannot unite unless something causes them to unite. However God is uncaused, being the first cause, and is therefore not composite. We therefore cannot say that something has greater organized complexity than God. Also, since God is the first cause, we saw above that it doesn't make sense to ask the question who created God, which is the same as asking who designed the designer. However, I am digressing here away from evidence.[119]

Design and Neo-Darwinism

In the case of the second argument that Neo-Darwinism provides an explanation for the existence of design, it does not deal with the fine tuning of the universe and therefore falls short of being an explanation for cosmic design. However, some fine tuning is essential for life and for evolution to even occur on planet earth, and as we saw above the probability of this happening is extremely small. There is also the open question of how life began from unguided natural processes. However, it is conceded that evolution can at least partially explain some evidence of design, but does not negate evidence of a designer. One argument raised against a designer is the existence of so-called poor design features. These are discussed in Section 10.2 along with a discussion about vestigial organs.

An interesting issue has been raised about just one aspect of the complexity of life, namely the so-called irreducibly complex biochemical systems. This is discussed in Section 10.5 along with other biological systems that are extremely complex. In the following section we consider the genetic code because it is not just a complex system, but it also has a very special information content.

2.6.2 The Genetic Code

The subject of genetics has made enormous progress since the discovery of the DNA code in the twentieth century. For example, about 35 papers were published by The Encyclopedia of DNA Elements (ENCODE) on September 6-7, 2012, in the journals *Nature*, *Science*, *Genome Biology*, and *Genome*

---

[118]For a helpful discussion see "The Complexity of a Simple God: Aquinas Response to Richard Dawkins" on the internet.
[119]In any case, even if God's existence has a low probability this does not automatically rule out the possibility that God designed the universe.

Research. Huge amounts of genetic data are appearing and, as some questions are answered, more questions arise! DNA molecules are responsible for passing on to us physical characteristics from our parents. Our DNA guides our development from the size of a full stop at the end of a sentence to adulthood! The molecule takes the form of a double helix like a long ladder twisted into a spiral with its sides made up of sugar and phosphate molecules with its "rungs" consisting of four bases adenine, thymine, guanine, and cytosine. Each base and its sugar and phosphate attachments form a molecule called a nucleotide. These bases are like letters of a genetic alphabet and combine in sequences to form words, sentences, and even paragraphs like a language; referred to as the DNA code. Each cell contains sections of DNA code forming genes that are on chromosomes, and these provide instructions to guide the functioning of the cell.

The amount of information in our DNA is enormous and extremely precise. For example, the largest human chromosome, number 1, has approximately 220 million base pairs. The "letters" in a DNA molecule must follow an exact order otherwise garbled and sometimes fatal instructions are given to cells. This is what a mutation does to instructions. Apart from just a few exceptions, the genetic code is universal suggesting a single origin rather than multiple origins. Other aspects of genetics are discussed further in Chapter 10.

A key idea is that the message given by the genetic code is not inherent in the DNA molecules. For example, the letters PAIN in English mean bread in French. The word GIFT in English means poison in German, but is meaningless in French. Meaning depends on the language and on the grammar. What distinguishes a language is that certain random groupings of letters have come to symbolize meanings according to a given symbol convention. Nothing distinguishes the sequence a-n-d from n-a-d or n-d-a for a person who doesn't know any English. Within the English language, however, the sequence a-n-d is very specific, and carries a particular meaning. Another example is the two sentences "You do love me" and "Do you love me?" have the same words but give a totally different message.

We see then that the DNA message involves a particular ordering of its nucleotides and has an associated language that is not inherent in the molecules themselves. For instance, the nucleotide sequence in the bacterium E. coli looks very much like a random sequence, yet that sequence is capable of biological function. We therefore have to explain the origin of the information in the genetic message. It is an example of a final cause. We cannot say that the information arose just from material forces as the material base has nothing to do with the message, being independent of the message and having no affect on the message. The message transcends chemistry and physics. Some theories assume that self-organizing properties within the chemicals themselves created the information in the first DNA. This would be like me writing on a blackboard in chalk that I like oranges and saying that the chalk caused me to write the message.

We have plenty of examples in everyday life where the message is independent of the medium. For example we can send the same message by morse code, semaphore, mail, e-mail, phone, or even by smoke signal! The message in fact transcends chemistry and physics. I hasten here to point out that any analogy with crystals is inappropriate as crystals incorporate atoms, molecules, or ions arranged in an orderly, repeating pattern extending in all three spatial dimensions, and are specified but simple. A crystal fails to qualify as living because it lacks complexity, while a chain of random nucleotides fails to qualify because it lacks specificity. We see then that nonliving things are either unspecified and random (e.g., lumps of granite and random nucleotides) or specified and simple (e.g., snowflakes and crystals). Some have suggested that the chemicals themselves or external self-organizing forces created the information in the first DNA molecule. However information scientists tell us that you cannot have a code without a code-maker. We ask the question: What is information? It is not the medium that conveys it as we saw in DNA, nor is it matter or energy. It is nonmaterial. Feser points out that the common use of words like "encoding," "information," "instructions," and "blueprint" used to describe DNA "all involve directness of something toward an end beyond itself, and thus to final causality."[120] The existence of the DNA code provides very strong evidence of an intelligent Designer.

In addition to the DNA code, DNA itself has an amazing structure. Charles "Chuck" Misler, engineer and author, describes it particularly well as follows:

> Suppose you were asked to take two long strands of fisherman's monofilament line—125 miles long—and then form it into a double helix structure and neatly fold and pack this line so it would fit into a basketball. Furthermore, you would need to ensure that the double helix could be unzipped and duplicated along the length of this line, and the duplicate copy removed, and the master returned, *all without tangling the line.*[121]

The process of duplicating is amazing as the unwinding of the DNA involves speeds estimated as approximately 8000 rpm without tangling the DNA strand. There is a lot we don't know about cell division as the cells not only divide, but they somehow begin to specialize into various organs, bone, muscle, skin etc.

2.6.3 Epigenetics

A discussion of DNA would not be complete without mentioning a new developing topic called epigenetics, where the Greek suffix *epi-* means over, outside of, or around. As an extension of genetics, it provides a second tier of heritable information including anything affecting the genome that is not encoded

---

[120] Feser (2008: 129).
[121] Misler (1999: 322–323).

in the DNA itself, and refers to changes in the regulation of the way genes express themselves; genes can be switched on or off, or their expression can be increased or decreased. For example, when lactose is available it can remove an inhibiting protein that allows *E. coli* to turn on an entire set of genes to metabolize a particular class of sugars, namely beta-galactosides. In general, DNA can respond to signals from outside the cell so that as well as some DNA making proteins, other parts originally referred to as "junk DNA," are responsible for the control mechanisms. (In humans only about 2% of the DNA encodes protein.) Technically, the switching is done through DNA methylation, histone acetylation, and microRNAs, and can be triggered by the environment, as noted above with *E. coli*. There is another more interesting example where there is some evidence that exercise can be beneficial with regard to switching on some helpful processes.[122]

What is revolutionary in genetic thinking is that these switches can be inherited and passed down through multiple generations. For example, if the mother is overweight during pregnancy, it can effect weight control mechanisms in her child, leading to obesity or even diabetes when the child is older. Also identical twins who have very different life styles but identical DNA can turn out to be quite different, as their genes are expressed very differently. This means that the consequences of our actions that affect how our genes are expressed before we have a child can be passed on to the child, and perhaps the way we bring up a child can affect the expression of the child's genes.[123]

We see then that some DNA can determine how we look (i.e. our form), while other DNA can determine how we behave (i.e. our function). As with DNA itself, the epigenetic "language" is over and above the proteins synthesized as the latter must be arranged into higher-level systems of proteins and structure. There is an additional layer of information that is not provided just by the DNA itself. Meyer discusses in detail the various sources of information in embryonic cells such as cytoskeletal arrays, membrane targets, ion channels, and the so-called sugar code stored in the arrangement of sugar molecules on the exterior surface of the cell membrane.[124] These sugar molecules can be combined in many more ways than the amino acids making up the proteins, thus providing a larger vocabulary for the "language" or code. He also says:

> If DNA isn't wholly responsible for the way an embryo develops—for body-plan morphogenesis—then DNA sequences can mutate indefinitely and still not produce a new body plan, regardless of the amount of time and the number of mutational trials available to the evolutionary process.[125]

We now look further at embryonic development.

---

[122] For example, Denham et al. (2014).
[123] It gives sense to a puzzling Bible verse that talks about the sins of fathers being visited on their children to the third and fourth generation (Exodus 34:6–7).
[124] Meyer (2013: chapter 14).
[125] Meyer (2013: 281).

2.6.4 Evolutionary Developmental Biology

We now introduce the field of evolutionary developmental biology (popularized as "evo-devo"), where gene switches are involved in embryonic development. In his book Sean Carroll,[126] molecular biologist, explains that there is a small number of "tool kit" gene complexes or clusters that determine body pattern. In a review of this book evolutionary biologist Jerry Coyne comments:

> The book's centerpiece is the unexpected discovery that the genes that control the body plans of all bilateral animals, including worms, insects, frogs and humans, are largely identical. These are the 'homeobox' (Hox) genes, whose products bind to the DNA of other genes, triggering a cascade of processes that ultimately yield eyes, limbs, hearts and other complex structures.
> The evolutionary conservatism of these genes across long-diverged species is staggering.[127]

Carroll further says: "All of the genes for building large complex animal bodies long predated the appearance of those bodies in the Cambrian Explosion."[128] What he is saying is that there are common genes across the living world that were there at the beginning and it is the way they are expressed that determines development and speciation.

A further question to consider is what about epigenetic mutations with regard to embryonic development. Can they provide the necessary changes along with natural selection to generate new forms of life? Meyer argues that they do not for several reasons.[129] First, the structures where we have epigenetic information such as cytoskeletal arrays and membrane patterns are not so vulnerable to alteration by many typical sources of mutation. Second, when alterations do occur that can change cell structures, they are likely to have "harmful or catastrophic consequences" such as death of the embryo or sterile offspring. As they are generally harmful they are eliminated by natural selection. He refers to experiments that demonstrate such processes.

2.7 THE MORAL ARGUMENT

We now move away from looking at evidence for God's existence in the external world and we now look for evidence in our inner world. A major argument for the existence of God centers around the apparent existence of morality in our world today and in our personal lives. This topic will now be discussed in detail.

---

[126] Carroll (2005a).
[127] Coyne (2005).
[128] Carroll (2005a: 139).
[129] Meyer (2013: 285).

2.7.1 Universal Values

I have always been intrigued by the evidence that certain emotions including anger, disgust, fear, joy, sadness, and surprise have universal facial expressions (though sometimes modified by learning and culture).[130] For example, an angry face is recognizable in any culture. This universality also extends into the moral realm where we see the so-called "golden rule" (or ethic of reciprocity) of treating others as you would like others to treat you operate across cultures and religions. Although what is regarded as right and wrong may vary from culture to culture, and some cultures are more extreme than others, basic moral principles do exist even though there may be some disagreements about their implementation.[131] There are many things in society, for example the protection of marriage and the family, that suggest that there is a universal sense of right and wrong. Actions like rape, torture, and child abuse are not just socially unacceptable behavior; they are simply unacceptable. Murdering one's own people, lying, cheating, and stealing are generally regarded as wrong.

Morality seems both natural and universal; we need to do good and shun evil, though some people in all cultures act in extreme ways, and there are some cultures that carry out practices that most other cultures would disagree with (much like individuals). In any case, if we believe that there is no objective morality, then we have no right to criticize such things as slavery, cannibalism, child brides, female genital mutational, and honor killing, if practiced by certain cultures.

There is also the problem that a moral relativist, that is one who maintains that all morality is relative, has no objective standard to measure moral improvement by, though some norms might be better for a particular society than others, but they will depend on the society. They might argue for tolerance, but there is no reason why we should enforce tolerance on others! Even relativists will be upset if their favorite values are ignored or even insulted; for example a drunk driver crashes into their new car! They may also have strong feelings about human rights.[132] We might add that if morality is relative, perhaps social justice should be as well.

Some have argued that evolution has led to morality, but evolution cannot explain why certain behaviors are good or even why moral obligations exist irrespective of our belief about them. Why do we have "oughts?" Why should I set aside my own well-being when it conflicts with another person's well-being? Perhaps having others suffer more to make my life better is more logical from an evolutionary viewpoint. Why help others who have physical or mental problems when, after all, it is the survival of the fittest that counts?

---

[130] Ekman (1980, 1993, 1999).
[131] See, for example, Brown (1991) and Wilson (1993).
[132] It is very hard, perhaps impossible, for an atheist to be consistent about morality. We all have difficulty living up to even our own ethics.

Why show altruism toward those outside one's in-group, especially if no one knows about the altruism exercised. Glass summarizes it well:

> If there is no God and if we are the accidental result of an unplanned evolutionary process, what could possibly make it a fact that the well-being of the human race is morally significant? In the long run we are all dead. We shall pass into oblivion, and no one will regret or even note our passing.[133]

However if God is behind evolution it could be argued that God engineered the evolutionary process to produce the morality that exists, and that the human race is morally significant. Either way we have support for God's existence.

The question of human rights raises a problem as we can ask what right do people have to impose their beliefs on another culture with regard to say oppression and gender inequality. Also there are situations where having a particular right such as freedom of speech may not be in the best evolutionary interest of a society. Law expert Michael Perry[134] raises the problem of trying to find a secular ground for human rights, and indicates that a religious ground seems to fit rights better. Can we justify human rights in a world without God when deep down we believe that human rights exist? We have a passion for justice, as part of being human, but the world carries on being unjust.

Guilt is another indicator of ultimate right and wrong. It is a painful feeling that occurs when we think we have violated a moral rule—an "ought." We have a conscience! Some may find conscience a "spoilsport," but its role is to help keep us on the straight and narrow and not allow us to be sabotaged by our feelings and inclinations. Parents find that children, without being taught, complain that something isn't "fair." Who taught them about fair? The concept of fairness is about an internal awareness that there's a certain way that things ought to be. It's not limited to three-year-olds who are unhappy that their older siblings get to stay up later. We see the same thing on "Save the Whales" bumper stickers. Why should we save the whales? Because we ought to take care of the world. Why should we take care of the world? Because we just should, that's why. It's the right thing to do. There's that sense of "ought" again.

Clearly men and women are moral beings concerned about making right and wrong decisions. Also the question is not just about the existence of morality but the existence of absolute morality. Some atheists can be very moral people and some Christians very immoral, but that is beside the point as we have to look at the overall picture. As Glass comments:

> The point is not that God is necessary as a motivation for us to be good, but that without God there is no basis for any action to be good.[135]

---

[133] Glass (2012: 215).
[134] Perry (2007).
[135] Glass (2012: 217).

Marc Hauser,[136] extensive writer and researcher in many fields, argues that humans have an innate moral faculty (a kind of intuition), which he ascribes to evolution, whereby any action is to benefit the propagation of certain genes. However this does not account for a number of human acts that show strong altruism and provide no genetic advantage. D'Souza[137] gives some examples such as giving blood, Mother Teresa giving her life to looking after the sick and dying in Mumbai (Calcutta), showing altruism toward strangers who are never seen again, one person giving his life for another, and the Christian idea of loving your enemies. Evolution cannot explain why a particular behavior is good, so morality is still unexplained. However, if God exists then it is possible that atheism cannot provide a basis for ethics. There is then the question of whose ethics.

Morality is also about ignoring self-interest and doing something that ought to be done, out of duty. If however God is left out of the equation, there is no reason to believe that any morality evolved by mankind is objective. After all, if there is no God, then what is special about human beings? They are just accidental by-products of nature who evolved relatively recently on an infinitesimal speck called the planet earth. Animals kill each other, and according to materialists we are just animals, so why shouldn't we do the same? Why should we have moral obligations? However, we are different from the animals in many respects. For example, in spite of our kinship with chimpanzees (discussed in Section 10.3.2), it has been suggested that chimps don't have morality.[138] However Frans de Waal, primatologist and ethologist, has more recently devised experiments showing some form of moral behavior in chimps such as reciprocity (fairness), empathy (compassion and consolation), and prosocial tendencies.[139] He has also demonstrated some of these aspects with elephants and dogs. In a world supposedly governed by the ruthless survival of the fittest, why do we see acts of goodness in both animals and humans, a problem that plagued Charles Darwin in developing his theory of evolution through natural selection.

In another approach, philosopher Jay Budziszewski [140] explains the rational foundation of what we all really know to be right and wrong and provides strong arguments for the natural law tradition (natural morality), with an emphasis on religious belief. Natural law is using reason to determine what is right and wrong, which according to the apostle Paul is written on human hearts.[141] The famous philosopher Immanuel Kant wrote,

[136] Hauser (2006).
[137] D'Souza (2007: 234–235).
[138] de Waal (1996: 209).
[139] de Waal (2005, 2009).
[140] Budziszewski (2011).
[141] Romans 2:15.

> Two things fill the mind with ever new and increasing admiration and awe, the more often and steadily we reflect upon them: the starry heavens above me and the moral law within me.[142]

If materialism is all there is, then there is no objective basis for morality. Why should we live "good" lives, if objective moral good and evil do not exist. It would not matter what sort of person you become! If we deny final causes, nothing has any goal or purpose including reason itself, so why pursue the good. Philosopher Nicholas Wolterstorff[143] sets out to see if human rights and justice can be grounded in secularism. He concludes that no one has succeeded in developing a secular account of human dignity and "it seems unlikely that it can be done." He presents a theistic grounding based on the idea that we are made in the image of God[144] and that God loves us, thus giving us worth.

2.7.2 The Moral Law Giver

We have seen above that there appears to be a universal moral law. C. S. Lewis[145] in chapter 1 (book 1) of his book argued that this idea is supported by a number of facts: We all assume that moral disagreements make sense, that moral criticisms make sense, and that it is important to keep our promises and for nations to keep treaties. Furthermore, when we break this moral law we generally feel uncomfortable and come up with all sorts of excuses for our behavior. We believe we ought to behave in a certain way, but we don't! Lewis, in his chapter 2, then deals with a number of objections to a universal moral law. First, it is not just a "herd instinct." Suppose you heard a cry for help from someone in difficulty (e.g., drowning). The herd instinct would prompt you to help, but your survival instinct would tell you it is too risky. Lewis argues that there is a third thing that tells you to follow the impulse to help rather than the impulse to run away; it cannot be either of them. In fact this third thing usually goes against what you want to do in spite of the risks and the stirring up the herd instinct, so it cannot be the herd instinct. The stronger impulse doesn't always win. Also there are no instincts or impulses that always agree with the rule of right behavior. Even love and patriotism are sometimes considered wrong.

A second objection is that the moral law is a social convention. In answer to this Lewis notes that not everything we learn from our parents or from society is a social convention (like mathematics for example). As we have mentioned above, similar basic moral laws are found in virtually all societies, both past and present, thus contradicting the idea that they are just socially constructed. Moral progress is often made when reformers actually go against

---

[142]Kant (1788).
[143]Wolterstorff (2008: chapters 15 and 16).
[144]Genesis 1:27–28.
[145]Lewis (2001a).

society (e.g., slavery). A third objection that the moral law is just a law of nature is ruled out by the fact that the latter simply describes what happens rather than what ought to be.

In Chapters 4 and 5 of Book 1 Lewis considers the question of what lies behind the moral law that we find inside of us. He argues that a universal moral law requires a Universal Moral Law Giver that gives moral commands, as expected by a lawgiver (we feel we are being got at!), and is intensely interested in us carrying out right conduct. Deep down we support the idea of right conduct and feel better when we comply. Lewis goes on to argue that this Law Giver must be absolutely good so that our moral effort is not futile in the long run, and the source of all good must be absolutely good. Therefore there must be an absolutely good Moral Law Giver we can call God so that we have an absolute moral law. Otherwise if there is no God then objective moral values do not exist.

Some would argue that all morality is relative and that "We can be absolutely sure that there are no absolutes," which is an absolute statement! We can come across similar self-refuting statements like the previous one. For example, a person who says there is no such thing as truth is therefore not making a true statement by that criterion. I have heard it said that only science gives us the truth, yet science cannot verify that statement scientifically as it is philosophical one.

Returning to the moral question, some may argue that God cannot be good as the world is imperfect and seems so unjust and cruel. This again raises the question of where do the ideas of perfection and injustice come from.[146] Lewis argues as follows:

> If the whole universe has no meaning, we should never have found out it has no meaning: just as if there were no light in the universe and therefore no creatures with eyes, we should never know it was dark.[147]

If we believe in an afterlife and God's justice, then when we can consistently make moral choices contrary to our self-interest and even undertake acts of extreme self-sacrifice we know that such decisions are not in vain, but have eternal significance. We are therefore accountable for all our actions knowing that good will prevail.

In concluding this section I want to refer briefly to a dilemma[148] sometimes put forward by atheists that asks whether something is good because God wills it or God wills something because it is good. If God wills it it is arbitrary, while if it is good (irrespective of God) then good exists apart from God and God is not necessary for goodness. However, we don't need to engage either of these alternatives as God wills something as He is good; goodness is what God

---

[146] The questions of suffering and evil are discussed in Chapter 8.
[147] Lewis (2001a: 39).
[148] Referred to as Euthyphro's dilemma.

is, and God's nature provides the standard.[149] God wills something because He is good, and something is good because God wills it.

2.8 GOD AND RELIGIOUS EXPERIENCES

Many people claim to have had religious experiences that seem to them to be experiences of God. From the so-called principle of credulity and in the absence of any positive counter-evidence, apparent perceptions ought to be taken at their face value. The kind of counter-evidence against such a belief would be evidence against the existence of God. As pointed out by Swinburne,[150] if God exists we might expect God to not only take an interest in the human race, provide some sort of revelation, and perhaps even answer some prayers, but God might also interact with some appropriate individuals who are capable of thinking about God and worshipping God. However evidence of such contact and public manifestations would be minimal, otherwise our free will (discussed later in Chapter 4 and Section 8.5) to choose between good and evil would become compromised.

The argument from religious experience claims that such experiences have often occurred and adds accumulatively to the evidence for God's existence. Of course so-called religious experiences are extremely varied and subjective, and according to Swinburne there are five kinds of such experiences described by the following examples: (1) a natural public object (e.g., a starry night) arousing feelings of awe about God's handiwork, (2) a person observing and being affected by an unusual public object or phenomenon (e.g., people speaking in tongues, observing some kind of healing miracle, or being moved by religious music), (3) a private religious experience that has describable sensations such as a dream or vision, or even hearing a voice, all seeming to be real, (4) a private experience that has sensations analogous to normal sensations, but only analogous (e.g., a mystical but physical experience such as a feeling like warmth or experiencing light), and (5) a private experience without physical sensations (e.g., a conviction that God is telling a person something they should do or, using the right brain hemisphere, to visualize God acting in some way). Swinburne debates whether such experiences are of God and his chapter provides details, especially in relationship to the principle of credulity.

It is not my intention to delve further into this topic but simply set it out for the reader. All I can say is that I have had some of the above experiences, and the fruits of Christian conversion (a major religious experience) have led to changed lives not only externally with regard to their behavior, but internally as well as they now see things very differently. People I would

---

[149]See Section 2.6.1 where it talks about the philosophical idea of God being a "simple" being.
[150]Swinburne (2004: chapter 13).

never have believed to change have undergone a complete transformation, and this is replicated all over the world. Such experiences are not proof of God's existence, but I don't believe they can be readily discounted as purely psychological phenomena. If it is argued that such experiences are simply due to natural laws the question then is, in terms of say evolution, what adaptive advantage does such an experience have as some atheists maintain that religion is bad for society (see Chapter 9). You cannot have it both ways! Others have tried to attribute such changes to brainwashing. Admittedly this may sometimes take place in mass Christian evangelistic meetings with some people. However there is much more to that as for example when conversion takes place in the privacy of one's room after much thought and perhaps internal struggle in trying to hand over personal control of one's life to God.

2.9 EVIDENCE AND PROBABILITY

A number of writers on both sides of the theistic divide sometimes throw around probabilities to support their arguments. If probabilities are going to be used, their computation must follow the rules of probability theory. Sometimes events are treated as though they are independent so that event A is assumed to be independent of whether event B occurs or not, when in fact the occurrence of A depends on the occurrence of B, for example. In this case conditional probabilities need to be used such as the probability of A occurring given B has occurred, and when used they have to be combined using so-called Bayes' theorem. Statistician David Bartholomew[151] focused on this in a paper presented to the Royal Statistical Society that described how probability can be applied to subjects such as theology and its relationship to physics, evolution, and philosophy. It is a difficult topic to pin down, as was indicated by the extensive and varied comments that followed the paper. Probabilistic arguments have not been as prominent as they might be, perhaps for at least three reasons.

First, they are often very difficult to formulate and somewhat subjective, as a number of assumptions are usually involved. Second, the mathematics can be too complicated for the average reader. Third, as I have already noted, some get it wrong and may confuse conditional and unconditional arguments. However, probabilities do have a part to play in assessing evidence and they need to be considered where feasible. I shall now give an application after some introductory comments about probability.

2.9.1 The Nature of Probability

Probabilities come in three flavors; theoretical (classical), empirical (frequentist), and subjective. The theoretical approach can be described by reference

---

[151] Bartholomew (1998).

to a box that contains 5 white balls and 7 red balls that are all identical except for their color. If a ball is chosen at random out of the box, what is the probability it is is red? Since each ball has the same chance of being chosen, there are seven chances out of 12 that a red ball is selected, so that from the mathematical concept of equally likely events the probability is 7/12 of a red ball being selected.

The empirical approach deals with an unknown probability that we endeavor to estimate. For example, suppose we have a machine that produces light bulbs, but occasionally produces a faulty one. Then the probability p of producing a faulty bulb is unknown, but it can be estimated by checking a large number of bulbs and calculating the proportion that are faulty. The larger the number of bulbs checked, the closer will the (empirical) estimate be to p. Another example of an empirical probability is the statement on the News that there is a 25 percent chance of rain tomorrow. Here we have a prediction based on past weather patterns so, although empirical, it will not be as clear cut as the previous example and will probably have a subjective element.

This leads to the third type of probability, or subjective probability, that describes our degree of belief that a certain event will take place. These are personal and will vary from person to person, and may be based on a guess, a hunch, some mixed numerical information, or opinions of others. For example, a sports follower may say that their team has an eighty percent chance of winning, or on a more conservative note say that the team is more like to win than lose so that the probability of winning is suggested to be greater than 0.5. Such probabilities tend to get thrown around rather casually without any proper empirical justification such as you are more likely to be struck by lighting than be involved in a certain kind of accident.

2.9.2 Probability of God

In this book all three types of probability will arise occasionally. Relevant to this chapter are subjective probabilities like the probability that God exists. We begin with a guess for a given probability and then see how this changes when we include further evidence. For example, let G be the statement that God exists and created the universe (the God hypothesis) and let $\sim G$ be the contrary statement that G does not exist. Define $Pr(G)$ to be our guess for the probability that the God hypothesis is true (usually referred to as a *prior* probability) and let $Pr(\sim G)$ be the probability that it is not true so that by the probability rules $Pr(\sim G) = 1 - Pr(G)$. If E_1 is the physical evidence about our universe mentioned for example in this chapter, then what we are interested in is $Pr(G \mid E_1)$, the probability that the God hypothesis is true given (that is conditional on) the evidence. Here $Pr(G \mid E_1)$ is usually referred to as the *posterior* probability. Provided we can give numbers to the conditional probabilities $Pr(E_1 \mid G)$ and $Pr(E_1 \mid \sim G)$ that the evidence exists given that the God hypothesis is true and the evidence exists given the God hypothesis is not true, respectively, then we have the formula (known as Bayes' rule)

$$Pr(G \mid E_1) = \frac{Pr(E_1 \mid G)Pr(G)}{Pr(E_1 \mid G)Pr(G) + Pr(E_1 \mid \sim G)Pr(\sim G)}$$
$$= \frac{a}{a+b} \quad \text{say.}$$

Here $b = Pr(E_1 \mid \sim G)Pr(\sim G)$ so that if a person is an atheist $Pr(\sim G)$ will be close to one and b will be closely equal to $Pr(E_1 \mid \sim G)$, the probability of getting the evidence given that God does not exist. Now it is accepted by non-theists generally (except perhaps those who take a multiverse approach to reality) that the probability of getting a world such as ours by chance that is livable is vanishingly small so that this probability and therefore b will be extremely small and small compared to a. This means from the above equation that $Pr(G \mid E_1)$ will be just less than one. To make this more concrete, assume that the prior belief in the statement being true is only 1 in a thousand so that $Pr(G) = 0.001$ and therefore $Pr(\sim G) = 0.999$. Now given that God is creator there will be a reasonable probability of getting the evidence since there might be evidence of design in our universe. Being conservative, suppose $Pr(E_1 \mid G) = 0.01$ and $Pr(E_1 \mid \sim G) = 10^{-6}$ (which some would argue is even too large). Then substituting into our equation we find that $Pr(G \mid E_1) = 0.91$. If we use $Pr(E_1 \mid \sim G) = 10^{-7}$, then $Pr(G \mid E_1) = 0.99$. Using different numbers such as $Pr(G) = 10^{-8}$, Glass[152] came up with a probability near 0.99. The reader might like to put his or her own numbers in and see what comes out. If we reduce both $Pr(\sim G)$ and $Pr(E_1 \mid \sim G)$, we will tend to keep getting a high value for the posterior probability and a strong case for the existence of God.

Usually it is difficult to give actual values to some of the probabilities, especially prior probabilities such as the probability that God exists, and Licona[153] comments: "At this time, many are doubtful that Bayes' theorem can be employed effectively with most historical hypotheses." Logically, when it comes to the fine tuning of the universe for us to exist we cannot use a prior probability; it must be a conditional probability since it is the probability conditional on our existence and involves our ability to observe the universe. However, since everything is based on our existence and observation we can take it as an underlying assumption in all our calculations. Given these problems we can at least sometimes say that a probability is very small or one probability seems to be greater than another as one proposition is more likely than another to be true without postulating particular values. For example, we may wish to show that $Pr(G \mid E_1) > 1/2$, that is, it is more likely that

---

[152] Glass (2012: 170).
[153] Licona (2010: 116–117).

God exists given the evidence than God does not exist given the evidence. This is equivalent to $b < a$ in the above equation, or

$$Pr(E_1 \mid \sim G) < Pr(E_1 \mid G) \cdot Pr(G)/P(\sim G).$$

For the atheist $Pr(\sim G) \approx 1$, so that we then have the approximate inequality $Pr(E_1 \mid \sim G) < Pr(E_1 \mid G) \cdot Pr(G)$. There is good evidence to believe that this inequality holds as the left-hand side is very small. Sometimes one can work backwards with a given a lower value (bound) on the posterior probability and obtain a range of possible prior probabilities that lead to this lower value being exceeded. We may also be interested to see if $Pr(G \mid E_1) > Pr(\sim G \mid E_1)$. Replacing G by $\sim G$ in Bayes' rule above and taking the ratio of the two posterior probabilities we get

$$\frac{Pr(G \mid E_1)}{Pr(\sim G \mid E_1)} = \frac{Pr(E_1 \mid G)}{Pr(E_1 \mid \sim G)} \frac{Pr(G)}{Pr(\sim G)},$$

the so-called posterior odds ratio, which will be easier for the reader to compute. The first ratio on the right-hand side, called the prior odds ratio, will tend to be very large and will tend to dominate the effect of the second ratio thereby giving a high value to the left-hand side, the posterior odds ratio. We see that we go from our prior odds ratio (the second ratio on the right) to the posterior odds ratio on the left by multiplying by

$$\frac{Pr(E_1 \mid G)}{Pr(E_1 \mid \sim G)},$$

the so-called Bayes' factor, which we use below.

An important question is how does one combine different pieces of evidence. For example, how do we add the evidence for the existence of morality (E_2, say) to E_1? For the two events combined, we have the posterior odds ratio given by

$$\frac{Pr(G \mid E_1 \& E_2)}{Pr(\sim G \mid E_1 \& E_2)} = \frac{Pr(G)}{Pr(\sim G)} \frac{Pr(E_1 \& E_2) \mid G)}{Pr(\mid E_1 \& E_2 \mid \sim G)}$$

$$= \frac{Pr(G)}{P(\sim G)} \cdot \frac{Pr(E_1 \mid G)}{Pr(E_1 \mid \sim G)} \cdot \frac{Pr(E_2 \mid G \& E_1)}{Pr(E_2 \mid \sim G \& E_1)}$$

using the laws of probability. It can be argued that the existence of morality is more likely given the God hypothesis is true together with E_1, than it is if there is no God together with E_1, i.e.

$$Pr(E_2 \mid G \& E_1) > Pr(E_2 \mid \sim G \& E_1).$$

We visit these concepts again in Section 6.4.4 when we discuss the resurrection of Christ. The point of the above discussion is that if people want to use probabilistic arguments, then the probabilities should be assessed correctly using the above type of arguments.

2.10 CONCLUSION

The main focus of this chapter has been on what is usually referred to as "natural theology", which essentially involves philosophically reflecting on God's existence without referring to any kind of divine revelation or scripture. We have looked briefly at some of the powerful arguments for God's existence that perhaps explain the growing proportion of Christian philosophers. The atheist philosopher Quentin Smith[154] writes that "today perhaps one-quarter or one-third of philosophy professors are theists, with most being orthodox Christians."[155] However, a person can be a theist without embracing Judaism, Islam, or Christianity. Modern science has also had a big shift with the developments of chaos theory (discussed in the next chapter), relativity theory, and quantum mechanics, all showing us that our world is infinitely more complicated than we ever imagined. Scientists seem to spend more and more time looking at a smaller and smaller part of the world in an attempt to try and understand what is going on. As the magnifying glass gets more and more powerful, we begin to see more and more detail. The idea that pure chance is *soley* responsible becomes more remote.

Although we have not actually proved the existence of God in this chapter (though some of the philosophical arguments mentioned can be defended more fully),[156] we have considered evidence that strongly supports the idea that God exists. We have not considered what sort of God exists; at the very least such a God would be designer and creator of the universe, a first cause that is absolutely good.

Given the evidence, why don't more scientists believe in God? One reason, previously mentioned, is that we live in an age of specialization so that many scientists may not be up to date in other fields where evidence for theism has recently surfaced, or be aware of traditional philosophical arguments that are still valid today. Furthermore, materialism has almost become the default worldview today in some areas of science, and those who depart from it can face considerable hostility from the scientific community. However, many scientists in the past thirty or so years have begun to agree that the universe is not just one big cosmic accident. Licona[157] describes a number of survey results that show that atheists are very much in the minority throughout the world with the majority of countries having no more than about 8% in that group. In the next chapter we consider further evidence of a creator when we look at the human mind with its ability to think, to observe itself, to respond to emotion, and to be creative.

---

[154] Smith (2001).
[155] http://www.philoonline.org/library/smith_4_2.htm.
[156] See for example the articles by William Lane Craig at http://www.reasonablefaith.org/the-new-atheism-and-five-arguments-for-god, and http://www.reasonablefaith.org/dawkins-delusion, both accessed August, 2105.
[157] Licona (2010: 159–160).

CHAPTER 3

IS THERE A SPIRITUAL DIMENSION?

3.1 WHO ARE WE?

If you were told that you think like a computer, would you be flattered or insulted? In the past, computers were regarded as morons and you only got from a computer what you put in. We started with vacuum tubes, then followed transistors, integrated circuits, and now microprocessors. With the developments of artificial intelligence and bigger and faster computers we are realizing that computers have a much greater potential. For example, voice recognition has been one very important application. Unfortunately developments in artificial intelligence have been much slower than that predicted in the early 1960s as a major problem has been the storage and access of large quantities of information. However this is changing with such things as parallel processing and superconductors, quantum computation and nanotechnology; electronic implants in the brain will change the future of computing, and help people with artificial limbs by targeting particular brain cells. There are those who predict that machines will become more intelligent than humans early this

Can we Believe It? Evidence For Christianity. By George A. F. Seber
Copyright © 2016

century.[1] This raises two questions. First, what do we mean by intelligence as there are various kinds of intelligence, and second, where does human consciousness enter into this? The latter problem is discussed later in Section 3.3.2.

3.1.1 Nothing But ...

Given the developments in computing, it has been argued that the brain is just a powerful computer with a vast storage capacity. If we could build a computer with a similar capacity we could then reproduce human behavior. We see the beginning of this in robots. This kind of thinking leads to the view that our body is just a machine controlled by a computer called the brain. That is one description of a human being. I wonder how you would describe yourself? How about "a bipedal, featherless, tool-making, naked, ape-like, carnivorous, gene machine equipped with a language-programmed computer?" I feel like that first thing in the morning! In one sense it is a true description, but how complete is it? Such a definition reflects the tendency for scientists to describe humans as nothing but .... For example, the physicist and the physical chemist can describe a human as nothing but a collection of molecules and substances which can be reduced to a few dollars worth of basic elements and substances. Scientists can also measure the rates of many of the exchanges of energy that take place in the human body whether they be mechanical, chemical, electromagnetic, or electrochemical. At this basic level a human can be described very much as a complex machine, and the description seems to be much the same as for the other animals.

The molecular biologist could describe you as nothing but a product of your genes. The structure of DNA and operation of the genetic code seem to be the same for all organisms, whether they be bacteria, whales, or humans! However, humans are determined by specific DNA molecules (genes) derived from parents, and the purpose of mankind is to perpetuate these genes: the so-called "selfish gene" doctrine. You only exist to pass on your genes. If you are past child-bearing age then you are of no use any more!

Moving away from the microscopic, the biochemist can describe you as nothing but an interesting, and possibly good-looking(!), collection of certain organic molecules and biological systems that are common to all living organisms. Your personality is determined by the biochemicals (e.g., neurotransmitters) produced in your brain. Your glands and the hormones produced by them determine the type of person you are, that is, whether you are aggressive, sensitive, even-tempered, and so on.

Moving away further still from the microscopic, the biologist can regard you as nothing but another animal, and your behavior can be studied using the same criteria that you use for animals. Many of the traditional criteria

---

[1] See, for example, Kurtzweil (2002).

for distinguishing between humans and other animals have become blurred. For example it was said that animals cannot learn, plan or conceptualize; they cannot count, and have no artistic sense. However, evidence now seems to be available for rudimentary forms of all these activities. For example, at Auckland University in New Zealand Dr. Alex Taylor has been carrying out research on New Caledonian crows who can use tools in succession. For example a crow dubbed "007", featuring on a BBC video, had to to pull up a string to get a short stick tool and then use this tool to pull three stones from three separate boxes. These stones were then dropped onto a platform causing it to tilt, which released a long stick tool that was used to roll a piece of meat out of a hole. How long did 007 take to do all this—just under three minutes! These crows are famous for making wooden hook tools and using their beaks to carve wooden sticks into hooks to access food.[2] No other animal species makes this type of tool, and their intelligence rivals that of primates.[3] This topic is discussed in greater detail by Taylor (2009: chapter 9).

For another example, there is a species of a talking parrot that can answer questions about things like shape, color, and number.[4] I wonder what your cat, dog, bird, or goldfish would say if it had human vocal chords? In addition to vertebrates, one group of invertebrates (without backbones) that has received considerable attention are the coleoid cephalopods, namely octopuses, squids, and cuttlefish. They are are notoriously clever large-brained animals that are well-known for their camouflaging abilities and flexible hunting strategies. Psychologist Jennifer Mather[5] argues that they show many behavioral indicators of consciousness including complex learning and spatial as well as apparent play. Zoologist Donald Griffin in a series of books[6] gives further examples of communicative and problem-solving behavior by animals, especially under natural conditions.

The sociologist and anthropologist can describe us as nothing but a product of our society and the learning patterns imposed upon us by our family. We are fashioned by the customs and taboos of our social group. If I was brought up in a different culture I would be a different person. There is a breed of scientist called the sociobiologist who says forget sociology and anthropology: its just all biology. Society adds nothing new to a person that isn't already there. You don't need anthropology or sociology as subjects as they all reduce to just biology. Finally, the cosmologist may see us as an accident on planet earth or else the byproduct of some other terrestrial life form, or even belonging to one of several possible universes.

Dr Viktor Frankl, the famous Jewish psychiatrist who had a terrible time in concentration camps writes that

---

[2] I don't feel so bad if someone calls me a bird-brain!
[3] See www.psych.auckland.ac.nz/nc-crows, accessed August, 2015.
[4] See, for example, Alex the talking parrot on an internet video; cf. Pepperberg (1999).
[5] Mather (2008).
[6] Griffin (1976, 1984, 1992).

there is a danger inherent in the teaching of man's "nothingness", the theory that man is nothing but the result of biological, psychological and sociological conditions, or the product of heredity and environment. Such a view of man makes a neurotic believe ... he is the pawn and victim of outer influences or inner circumstances. This neurotic fatalism is fostered and strengthened by a psychotherapy which denies that a man is free.[7]

Are we more than the above? We now consider this question.

3.1.2 Looking for Meaning

We saw in the previous section that every branch of science ranging from looking at the very small (in microbiology) to the very large (in cosmology) has its own way of reducing a human being to nothing but ....This is not surprising as when we look closely at reality we find that it is in fact multileveled. Suppose I wrote a mathematical equation in ink on a piece of paper, such as

$$f(x) = \frac{1}{\sqrt{2\pi}} e^{-\frac{1}{2}x^2}$$

and looked at the letter "e" through a microscope. It would simply look like ink marks on paper. As I moved further away I would see the letter "e" appear, then even further away I would see several symbols, then finally an equation. To a child who has not seen mathematics before it would all look like scribble with no meaning. Would you believe me if I said to you that there is only ink there? If you didn't believe me I could invite you to take some of it away and analyze it. However, for a mathematician, at each observational position he or she would need a different framework or level of reality to describe what is seen, going from black marks to a symbol representing some mathematical idea, to a group of symbols, and finally to an equation used in Statistics. As we move further away, the level of reality becomes more abstract and more encompassing. Lower level language and concepts cannot be used to describe upper levels of reality. The same is true in science: a molecular description is inadequate for describing biological or social behavior. In the latter case we need a new "higher" language.

In Section 2.6.1 I gave the well-known example of a watch. The movement of molecules in my watch follow certain scientific laws and if I accidentally smashed my watch, the laws would be unchanged as they follow a "lower" level of reality. They still describe the way the molecules behave in my watch. However, the watch no longer fulfills its higher purpose or more complex level of reality of telling the time as the boundary conditions are changed. The function of the watch is not determined by the laws of molecular physics but by the language of time keeping and the boundary conditions.

---

[7] Frankl (1992: 132).

In both examples above we cannot determine the function of an object by just looking at the material that the object is made of. For example, my mathematical equation could be written in chalk on a blackboard or appear on a computer screen. It is the organization of the material, namely the boundary conditions and the associated level of reality, that describe the function of the object. We need a whole new level of description and a new language to describe function.

3.1.3 Are We More Than Molecules?

Materialism assumes that matter is all there is. This is usually referred to as *reductive materialism* as everything, including the so-called mind (or "soul") reduces to just matter or brain states. However, there is also *non-reductive materialism* in which the mind is regarded as an emergent property of the brain but is not reducible to just material (this is discussed further below). This variation of materialism tends to shift the problem elsewhere as we then ask how do things "emerge", as we begin as molecules and emerge to something else. When it comes to a scientific description of humans, we see that all the various descriptions that we have considered are helpful, and perhaps even true at that particular level of reality. We can study humans at the molecular, biochemical, animal, and social levels, with each level providing its own (limited) description.

When all is said and done, however, I feel that there is more to me than just molecules and that I am more than just an animal. Therefore reducing myself to just molecules and physics and chemistry, and thus explaining these feelings as simply emergent properties of my brain I don't find very helpful. If I am just an animal thrown up by chance, then, as I argued in Chapter 1, we have to ask where does the language of abstract mathematics come from? Mathematics is the underlying language of much of theoretical science. For example, although we don't understand the ultimate nature of the universe we seem to be able to get a workable description using mathematics. Why are we able to do this? If we are just a biological product of an evolutionary system, then where does something that is outside the system, namely abstract mathematics, come from? How can something which is concrete (us) produce something that is abstract, and we investigate this question further below. Some would argue that such a development arises as some kind of emergent property so that we are just molecules to start with. Well, if I am a mathematician as well as an animal, I believe there is more to me than meets the eye and I conclude from my own experience that reality is more than all the scientific descriptions. As a musician I like John Polkinghorne's comment when he talks about the effect of music:

> The scientific account is that music is neural response to the impact of sound waves on the eardrum. Of course that is true, up to a point, and even worth knowing, but it scarcely begins to do justice to the rich character of music in which a temporal succession of sonic wave-

packets evokes in us enjoyment of a timeless encounter with the realm of beauty.[8]

3.2 ARE WE SPIRITUAL?

When we look within, do we see a level of reality that we might call "spiritual?" An answer to this question will depend on what we mean by spiritual. There are a number of definitions of spirituality and they don't always refer to God in the standard sense. For example, at a counseling workshop I attended in 2008, Dr David Benner gave the following working definition, "Spirituality is a way of living in relation to something other (or larger) than ourselves that responds to our deepest longings for meaning, purpose and connection."[9] Spirituality is usually distinguished from religion as the former goes much deeper, so that spirituality can be religious or, as some would say, even non-religious depending on whether or not "other" is replaced by some "higher power." It is a lifestyle rather than a set of beliefs; it is being rather than doing. However, it could be argued that spirituality may not proceed very far without religion as there needs to be a communal and historical context for the spirituality. Separating the two is not easy[10] as they are distinct but overlap, and trying to measure the difference will depend on one's definition of spirituality.

Recently there has been a rise in aggressive atheism where its proponents such as Dawkins, Pinker, and others suggest that people who believe in the supernatural are deluded and are inferior intellectually. However, there has been a resurgence of traditional religions throughout the world including Christianity (e.g., in Asia, Africa, and South America), Islam, and Hinduism. For example, pastor Timothy Keller[11] refers to two fairly recent studies that say that in 1900 Christians were 9% of the African population and Muslims about 36%, whereas in the 1960s Christians outnumbered Muslims and more recently were 44% of the population. In China there has been a huge growth in Christianity at all social levels including the Communist party, and if the current growth rate continues 30% of the population will become Christians in about thirty years. The world is proportionately "becoming more religious"[12] even though the West is becoming more secular (see also Section 9.5.5). The secularization hypothesis, however, has been rejected by a number of contemporary scholars on the following grounds:[13]

(1.) Statistics show a continued strong interest in religion and low rates of atheism, even in technologically and scientifically sophisticated societies like the U.S.

---

[8]Polkinghorne (2012: 1).
[9]Reproduced with kind permission of Dr Benner.
[10]cf. Nelson (2009: 9–11).
[11]Keller (2008: 41)
[12]Norris and Inglehart (2004: 16).
[13]Nelson (2009: 14).

(2.) Trends in contemporary religion are not really that different from the past.

(3.) Trends away from religion are really part of a more general trend away from social involvement. It is more about social disengagement than a move away from religion and spirituality.

(4.) Secularization in apparently less religious areas such as Europe is an anomaly due to unusual socio-cultural factors. If the government and religion are closely linked, religious non-adherence may be a form of social protest and a swing towards a more personalized religion.

(5.) A decline in participation or membership rates does not necessarily imply that people are becoming less religious or spiritual.

D'Souza[14] comments that Christianity is the "fastest-growing religion" with Islam second, and argues that Christianity is the "only religion with a global reach." It has now become a universal religion with, apart from a few exceptions, Buddhism and Islam still remaining largely regional (though this is changing with Islam). He asks the question, "why humans would evolve in such a way that they come to believe in things that don't exist." He notes that secular oriented countries are in fact showing a decline in birth rates in contrast to the rapid growth of religious populations. It is therefore atheism "that requires a Darwinian explanation," not religion. As far as intellectual inferiority is concerned, there are many great intellects who are religious.

3.2.1 Neurotheology

Neurological science seems to have located a part of our brain that is spiritually oriented, and the study of this is sometimes referred to as neurotheology. In particular, some circuits in the temporal lobes and other parts of the brain are involved in religious experiences. Writer Claudia Wallis[15] summarized a number of body-mind links such as the limbic system being linked both to spiritual experiences and to relaxation. She listed studies showing the health benefits of spirituality and quoted from Benson[16] who believed from his research that "humans are also wired for God", and that "our genetic blueprint has made believing in an infinite Absolute part of our nature." Since then a number of books have been written on the subject, largely from a materialist point of view, that argue that spiritual experiences are not about connection with God but are simply created by our brains. This raises the question of whether God is a delusion created by brain chemistry, or is brain chemistry a necessary channel for people to reach God? Which is the cause and which is the effect?

---

[14]D'Souza (2007: 7, 13, 17).
[15]Wallis (1996).
[16]Benson (1996).

The fact that two things are correlated does not mean that one actually causes the other. One of the problems in statistical work is the tendency to equate correlation with causation, a fundamental statistical mistake. (So often A and B can increase together because they are both affected by a third factor C. For example the annual consumption of chocolate can probably be correlated with almost anything, for example, the annual number of crimes, as they both increase with a population increase; a third variable, time, is linked to both!) One exception to the materialist trend is by neuroscientist Mario Beauregard and journalist Denyse O'Leary.[17] Beauregard believed that religious, spiritual, and mystical experiences existed and the authors studied the spiritual experiences of Carmelite nuns, coming to the conclusion that it is more likely that these mystics are directly experiencing a reality outside of themselves.[18]

Philosopher Alvin Plantinga provides another explanation:

> To show that there are natural processes that produce religious belief does nothing, so far, to discredit it; perhaps God designed us in such a way that it is by virtue of those processes that we come to have knowledge of Him. Suppose it could be demonstrated that a certain kind of complex neural stimulation could produce theistic belief. Clearly, it is possible both that there is an explanation in terms of natural processes of religious belief (perhaps a brain physiological account of what happens when someone holds religious beliefs), and that these beliefs have a perfectly respectable epistemic status.[19]

Some might argue that there is perhaps a "God-space" in our brain because of our moral inclinations such as, for example, our desire for justice.[20] It might then mean that God is a figment of our imagination and has been thrown up by evolution. In fact, however, the God-space tells us nothing about the existence or nonexistence of God. Recent research has indicated that there isn't a particular "God spot" in the brain, but involves other parts of the brain as well such as those involved with self-consciousness and emotion.[21] The idea of being hard-wired for God is described in the Bible[22] where we read that God has revealed himself in the things that have been made and that we can know God.[23] But people choose not to do so,[24] and those people's views are based on moral and spiritual factors in their nature rather than the outcome of purely rational argument. I believe that the denial of God's existence, or degraded views of God's nature, can lead to conditions that are morally, spiritually, and socially harmful.

---

[17] Beauregard and O'Leary (2007); see the review by Mohrhoff (2007).
[18] See also Snowdon (2001).
[19] Plantinga (2000: 145).
[20] See Section 2.7.1.
[21] Beauregard and Paquette (2006).
[22] Romans 1:18–23.
[23] This topic of knowing God is discussed in Chapter 11.
[24] See also Acts 14:16–17, 17:26–28.

God Gene?

Geneticist Gene Hamer, a self-confessed materialist,[25] describes some interesting statistical research involving twin and other studies to see if there is a gene for spirituality that he calls the "God gene." In Zimmer's review of this book[26] he notes that Hamer's research was not published in a peer reviewed journal and not replicated. He shows that Hamer's measure of spirituality that Hamer calls "self-transcendence" (and which can be challenged on a number of counts) has a genetic link, namely the gene VMAT2. Zimmer notes that Hamer

> did not, for example, try to rule out the possibility that natural selection has not favored self-transcendence, but some other function of VMAT2. (Among other things, the gene protects the brain from neurotoxins.) Nor does Hamer rule out the possibility that the God gene offers no evolutionary benefit at all.

However, Hamer makes it clear that: "Genes explain only about half of the variation that is seen even for this one scale. And the single gene I've identified is responsible for less than that —a small percentage of variance at best."[27] Clearly other genes may possibly be involved, and there will be other measures of spirituality that one can construct. Hamer also says this work may help explain individual differences in spirituality, but not why humans are spiritual. He comments that:

> there is nothing intrinsically theistic or atheistic about postulating a specific genetic and biochemical mechanism for spirituality. If God does exist, he would need a way for us to recognize his presence.... Spiritual experiences, like all experiences, must at some level be interpreted by our biologically constructed brains.... What we do with our spiritual genes, however, is very much up to us.[28]

Even if religion arose through evolution there is the question of its adaptive benefit, and evolution tells us nothing about whether it is true or false, good or evil.[29] Some use such genetic research to describe us once again as nothing but ..., and our minds are nothing but an aspect of our brain rather than being something over and above our brain function.

Wired for God?

If we are wired for God, then the default position would seem to be a belief in a supreme being with atheism being the odd one out. Viktor Frankl wrote, "a religious sense is existent and present in each and every person, albeit buried, not to say repressed, in the unconscious."[30] An evolutionist might argue, however, that a belief in God arose through evolution as it has a

---

[25] Hamer (2004: 94).
[26] Zimmer (2004).
[27] Hamer(2004: 15).
[28] Hamer (2004: 211).
[29] For further comments see Section 10.1.3.
[30] Frankl (2000: 151).

beneficial effect (through optimism?) and enables one to endure hardship.[31] They might further argue that the next step in the evolutionary process is to move on to independence from God and ultimately to atheism. Like the gunman and the alien mentioned in Chapter 1, we cannot say that such an explanation cannot be true as some atheistic evolutionists can provide a hypothetical explanation for anything. But can it be tested? The question is, "How likely is the explanation to be true?"

All our experiences, especially vivid or repetitive ones, occupy a particular place or several places in our brain,[32] and similar spiritual experiences will tend to occur and link up together (neurons will "fire" together), with any collection of like experiences forming a so-called "ego state." If that part of the brain is stimulated by a hallucinogenic drug (e.g., psilocybin that mimics monoamine serotonin) or by epileptic activity, for example, it has been found that there is some type of "religious" response in some cases. Drugs, however, can produce all sorts of other responses as well through the release of neurotransmitters such as dopamine, serotonin, adrenaline, and noradrenaline. Nicotine, for example, mimics acetylcholine, an orchestrator of neurotransmitters in our brains, and stimulates the production of dopamine. Such drug reactions or illness responses do not negate the reality of such experiences that arise without the use of drugs or without the effects of illness. Showing that there is a natural explanation of an effect does not rule out its reality outside its physical effect. A brain scan showing an effect does not necessarily rule out an external being operating.

When my wife gives me a hug and I consequently get a shot of dopamine, in addition to the physical effect of pleasure I know that my wife loves me! It can also be said that we cannot always believe what our brain tells us. When I had life threatening open heart surgery because of a nasty super bug (staph. aureus) infection (MRSA) in 2010 I was lying in the intensive care room after surgery looking at the random pattern of black dots on the ceiling tiles. They turned into ants and began moving about! I had read that you can have hallucinations after heart surgery so my mind told me not to worry about what my brain threw up, and I watched with interest (no ants fell off the ceiling!)

3.2.2 Our Changing Brain

While we are considering the human brain I want to digress for a moment to discuss further this wonderful wrinkled object I have between my ears (and I

---

[31]There is evidence that religious involvement is associated with lower mortality, but whether a gene is responsible is another matter; see Section 9.3.
[32]Brain damage reports suggest destroyed memories can be recovered, indicating there is backup storage and more than one place for memory.

don't mean my face!) Psychiatrist Norman Doidge[33] has written a fascinating book about the new science of neuroplasticity in which our thoughts can change the structure and function of our brains, even into old age (there's hope for me yet!). It is true that certain specific mental functions like vision, hearing, and language are generally associated with certain regions of the brain. However, we find that the brain has a capacity to be reorganized and develop new ways to perform lost functions; one part can take over the function of another damaged part. For instance, Doidge mentions the story of a woman born with her left brain-hemisphere missing who was able to live a fairly normal life, as her right hemisphere took over many left-hemisphere functions.

Professor Richard Faull, a neuroscientist at Auckland University, New Zealand, provided the first evidence that the diseased human brain can repair itself by the generation of new brain cells. He said in 2007:

> I'm forever amazed at the complexity and beauty of the human brain. It's the last frontier and presents so many challenges. Despite all our research we still don't even know what constitutes an original thought. But we are starting to see through the haze, to begin to understand some of the marvels of the human brain.

Roger Lewin, science writer,[34] tells how John Lorber, a British neurologist, had examined many patients with hydrocephalus (water on the brain) and has found that some people with very little brain matter can still function very well. Lorber refers to a young student with an IQ of 126 who completed a first class honors degree in mathematics and was socially completely normal but had virtually no brain. Instead of a normal 4.5 centimeter thickness of brain tissue between the ventricles and the cortical surface there was a thin layer of a millimeter or so with his cranium being filled with cerebrospinal fluid.

Linguist Mark Baker[35] argues that the language problems caused by brain damage all have to do with words and grammar and that the brain damage appears not to affect the ability to think creatively, which suggests that the "mind" does not need the brain. New knowledge about the brain has been particularly helpful with, for example, stroke victims using so-called *constraint induced movement*[36] and in tackling some of the diseases of the brain like Alzheimer's and Huntingdon's diseases.

We also have electronic devices and probes that enable us to monitor individual or groups of neurons. For example, the so-called BrainGate research team have been developing a system that turns thought into action, which can help people with missing or disabled limbs. A person imagines they are

---

[33] Doidge (2007); see also more recent comments by Doidge in http://www.wsj.com/articles/our-amazingly-plastic-brains-1423262095?mod=trending_now_1, accessed August, 2015.
[34] Lewin (1980).
[35] Baker (2011).
[36] For details see the internet.

moving a paralyzed limb and the ensuing brain signals are picked up by an implanted device. The signals are then turned into a command to operate an external device such as a powered wheelchair, an artificial limb, or, in the near future, a functional electrical stimulation device that can move paralyzed limbs directly. I have mentioned these aspects of the human brain as I believe they show once again evidence of design—design for flexibility. Some have suggested that neurons are not very well designed but this begs the question of what they are designed for!

3.3 MIND AND BRAIN

The previous section raises an important question of whether the mind is simply a product of the brain and nothing more.[37] As we saw there, the brain can be reorganized, and this suggests that something is driving the reorganization. Why should the brain reorganize itself if materialism is all there is? Also there is the question of how can a group of brain cells that produce proteins generate thoughts when connected like an electrical network. Exactly what are thoughts? We know there are causal connections between neurons firing and hormones secreting, but why should there be a logical connection. For example, the propositions "it is raining outside" and "it is wet outside" are logically but not causally related in that one does not cause the other. How do we match up causally interrelated states with logically connected interrelated states and thus reduce the mental to the physical?

There is the added problem that we can have thoughts about non-existent objects (e.g., the tooth-fairy, unicorns) and future devices with imagined functions (e.g., an anti-gravity device), as well as events with which we have no casual connection. Philosopher Alan Thomas[38] comments that many philosophers believe that the mind possesses features that the brain does not possess. For example, as just mentioned, we are able to imagine something that is non-existent, whereas brain states simply exist, and in Chapter 2 I referred to the existence of abstract thinking. Related to our capacity to "represent", we are rational thinkers where a sequence of thoughts can be tied together by a rational chain of connections, and there is a sensitivity to rationality in our network of neurons (though I can get things wrong sometimes!).

3.3.1 Qualia

We also have subjective sensory experiences of color (e.g., red), taste (e.g., pepper) and smell (e.g., chlorine) technically called *qualia* (singular quale) by the philosophers that we can't seem to pass on to other people. Yet physical objects consist of particles that are intrinsically colorless, tasteless, and

---

[37] I found Custance (1980: chapter 3) and Feser (2006) very helpful in writing this section.
[38] Thomas (2002).

odorless so that such sensations, in some sense, exist only in our minds. But according to materialism, our minds are a function of our brains, and the latter also consist of the same colorless, tasteless, and odorless particles. No matter how you physically analyze the brain you will only find interacting fields and particles and not the qualia. As noted by philosopher Robin Collins,[39] if you asked a person what they experienced when a certain set of neurons are activated, you could obtain a connection between the qualia that are experienced and the pattern of neuronal firings. But this is not the same thing as being able to describe the qualia in purely physical terms. Even if we had complete maps of your brain and mine it raises the question of whether our experiences of a particular quale are the same. Such problems will remain, even with further advances in cognitive science and neurology. Robbins explains:

> The reason is that these sciences can only explain the physical abilities and functions of systems in the brain; the problem that consciousness and qualia pose, however, is the problem of why there is an inner experience at all in systems with certain physical abilities, functions, and physical structure.[40]

David Chalmers[41] calls this problem the *hard problem* and that it is beyond the explanatory scope of cognitive science and neurology. All of this poses a real challenge to the materialist. How can the behavior of neurons give rise to subjectively felt mental states? In the end consciousness cannot be demonstrated or related to anything observable, it can only be experienced.

We don't really know what the world is like as we only have our perception of it. Although sensations of heat and cold can be related to the average movement of particles, it is our subjective personal awareness of such sensations that vary from person to person, which is difficult to explain on a purely physical model. As Collins indicated above, why in fact should we even have a personal viewpoint? Furthermore, given we have all this variable activity in our brains with some neurons firing other ones, how is it that we are able to think in accordance with the strict laws of logic? There is an explanatory gap between the physical level of what happens and our conscious experience of it. For example we can describe red light by its wavelength and talk about the so-called red shift in cosmology, yet our experience of redness varies from person to person.

Linking Laws

Given the failure of materialism to explain our observational data, Collins indicates one particular consequence. There must be what he calls *linking laws* linking brain states to qualia to explain how activities in certain parts of the brain (the non-subjective states) are related to certain types of qualia (the subjective states). Such laws will specify for each brain state whether or not it gives rise to qualia and, if so, what type and intensity of qualia it

---

[39] Collins (2011).
[40] Collins (2011: 3).
[41] Chalmers (1997).

gives rise to. Also qualia need to be experienced by an "experiencer" so that non-reductive materialists, for example, must postulate a law or metaphysical principle that specifies which material systems constitute experiencers. Laws of nature, and in particular statistical relationships, exhibit two types of variable—dependent and independent variables.[42] For example, Ohm's Law (which can be expressed in a number of ways) states that for certain conductors of electricity, the current I through the conductor in amps is directly proportional to the voltage V in volts across the conductor, that is $I = cV$, where c is a constant for the conductor. If we choose two different voltages, we get two different currents, but we still have the same constant c so that $I_1 = cV_1$ and $I_2 = cV_2$. In fact $c = 1/R$, where R is the resistance in ohms of the conductor. Here I can be regarded as the dependent variable and V the independent variable which determines I.[43]

Relating this to qualia, Collins says that we have two dependent variables, the type of qualia and its intensity, and any independent variables will depend on the material aspects of the brain, with the more we find independent variables are involved the more complex the law. There will also need to be a law that determines when a brain state give rise to consciousness (we discuss consciousness further below). Collins refers to how the physical laws in the past became more and more complicated until some new entities were introduced, namely atoms (invisible at the time) and some fundamental variables, that allowed for the elimination of a large number of dependent variables with the result that the laws could be derived from a few basic laws. He then argues that we have a similar situation with qualia. To develop the linking laws referred to above would require an almost infinite number of independent variables, particularly as we don't fully understand how the brain works and what functions different parts of the brain engage in.

Collins suggests that the non-reductive materialist is in a similar situation as the former scientist before atoms came on the scene. He says that

> no simple relationship between the variables recognized by the physical sciences—such as energy, temperature, mass, and the like—seems to capture what differentiates those material systems that are conscious from those that are not.

What about emergent properties? He goes on to say:

> The introduction of emergent properties or structures, however, simply pushes the problem back to the laws specifying when those emergent properties or structures arise.

However, by postulating the existence of a new entity called the "soul", which has both subjective and non-subjective properties, provides simplification rather than further complications. He goes on to use an analogy from

---

[42]These terms are a bit misleading and are going out of fashion in Statistics.
[43]To verify Ohm's law one could plot V against I for different I and see if the graph is a straight line through the origin.

string theory to suggests that various modes of vibration could be linked mathematically to the types of qualia. I am not able to do full justice to the arguments that Collins presents and the reader is referred to Collins' (2011) online preprint of his article for details.[44]

3.3.2 Consciousness

Having already referred to consciousness above we shall explore this topic further. We note that we are conscious subjects and we can be aware or not aware of certain states we are in. If someone looked at our brain states, would they be able to explain the difference between those we are aware of and those we are not? More importantly, what is it like to be me, that is a conscious person who can stand apart from myself, see myself as an object standing alone in this vast universe, and think about the fact that I am thinking. If I shut my eyes I can see myself sitting here. This other self that transcends space and time has been called the "transcendental self." This transcendental experience is even greater when I worship God in the spirit, which for me is definitely a non-machine experience, though a skeptic would no doubt give an alternative explanation of the experience.

Self-Consciousness

The someone called "I" and the transcendental self mentioned above relate to my self-consciousness, and there is the question of where did it come from. One of the problems with self-consciousness is that it can only be experienced and not demonstrated, so can it ever really be explained? It seems clear that many animals as well as humans have consciousness, whereby the animals perceive and hence respond to particular features of their environments, thus making them conscious or aware of those features. This idea has even been extended by some to single-celled organisms.[45] However, a human has self-consciousness as well that may be absent in animals (though we cannot be sure).[46] By self-consciousness we mean paying close attention to what is going on inside of us, that is, intense self-awareness and, as mentioned above, thinking about the fact that we are thinking. (Some authors use the term "consciousness" instead of self-consciousness when referring to humans, which is confusing.)

One of the problems of experiments with animals is that we tend to impose a human interpretation on what we see; for example, Herbert Jennings[47] described amoebae as exhibiting attention, desire, frustration, established habits, and even some intelligence. Jay Best[48] also found similar responses in

---

[44]For further discussions about the existence of the soul see the essays in Baker and Goetz (2011).
[45]Margulis (2001); or plants, Nagel (1997).
[46]For a very detailed discussion of the topic see Allen and Trestman (2014).
[47]Jennings (1917).
[48]Best (1963).

experiments with planarian worms, thus demonstrating apparent mindedness in lower life forms. More recently we have found that bacteria can communicate through chemical signals and possibly touch, and one aim seems to be to find out the current size of the bacterial population (called "quorum sensing").[49] There is a huge range of views on consciousness ranging from the view that only humans are conscious to almost all animals including simple invertebrates are conscious and able to experience the world.

Another problem with animal experiments is that animals often detect things humans are unaware of (e.g., sounds out of human range) so we are not sure exactly what an animal is responding to. Geneticist Robert Berry[50] makes the following observation:

> Indeed it is difficult to identify absolute differences between the minds of animals and man: animals are capable of some degree of aesthetic appreciation and abstract thought; they can have "nervous breakdowns"; they may "play" elaborate games; and show considerable community and family care.

Some animals have self-recognition in a mirror (e.g., chimps, bonobos, orangutans, dolphins, and elephants) and some do not (e.g., dogs, monkeys, and other primates).[51] Some animals such as dogs for example seem to experience shame and learn to change. In what sense then do we differ from other animals? I like the questions posed by geneticist Robert Berry and psychologist Malcolm Jeeves

> Are we apes on the way up? Angels on the way down?... Are we embodied souls or mere DNA reproducing machines driven by deterministic physico-chemical reactions?[52]

Such questions about humanness are discussed further in Section 10.3. I would argue that it is particularly in the realm of the spirit where there is a difference between humans and other animals; according to the Bible we can be subject to "spiritual death." Theologian Nicola Hoggard-Creegan[53] makes the following insightful comment:

> If humans alone can repent, then the lack of repentance may lead to spiritual death in which animals do not participate. Nevertheless, when chimpanzees seek brutal conflict with other chimps, and perhaps even in predation, there is a yawning gap between the way things are and the vision of the peaceable kingdom, and in this sense all animals must participate in the spiritual death that humans experience so acutely.

The New Testament sums it up well[54] when it speaks about the whole of creation groaning together until the redemption of those who are led by the Spirit of God when there will be freedom from bondage to decay.

---

[49] See "bacteria communicate" and "quorum sensing" on the internet.
[50] Berry (1988: 66).
[51] Gallup et al. (2002).
[52] Berry and Jeeves (2008: 3).
[53] Hoggard-Creegan (2012: 37).
[54] Romans 8:21–23.

Emergent Properties

With regard to the mind and brain, neuroscientist Antonio Damasio says that,

> it is probably safe to say that by 2050 sufficient knowledge of biological phenomena will have wiped out the traditional dualistic separations of body/brain, body/mind and brain/mind.[55]

Will this happen given what we know about quantum mechanics? We know that there is a strong correlation between certain mental processes and parts of the brain that are shown to be activated but, as I said above, correlation does not imply causality. Those that argue that mind emerged as a natural process from molecules must assume from continuity that matter was both physical and mental at the same time right down to fundamental particles,[56] but there is the problem of where such pre-emergent mental properties came from in the first place. You can't get something from nothing. There is also the problem of believing that what the brain produces is rational or true if the mind simply emerged without the input of a superior intelligence. Would you trust the output from a computer programmed by random forces or by non-rational laws? Cosmologist Paul Davies puts it this way:

> Mindless, blundering atoms have conspired to make, not just life, not just mind, but *understanding*. The evolving cosmos has spawned beings who are not merely to watch the show, but to unravel the plot.[57]

In discussing this problem, James Moreland, philosopher and theologian[58] pointed out that it is hard for atheists to deny the possibility of the emergence of a super-mind from our collective minds that we can call God, as a lot of people believe that they have had religious experiences with God. It may not be the God of Christianity but it still poses a problem for the atheist.

Soul?

Another problem for the materialist presented by Moreland is the so-called "binding" problem that goes like this. When we observe our surroundings and what we are doing, different parts of the brain become activated and no single part of the brain is activated by all the sensory experiences. If I was just a brain I would be a crowd of different activated parts each with its own sense of awareness. However I am an integrated whole that binds all the experiences together because my sense of self and internal unity is different from my brain. We also note that our minds (souls) are not located in any one place in our bodies so that if I lose part of my body I don't lose my soul. We can say that God occupies space everywhere as my soul occupies all of my body. Since we are told in the Bible that we are made in the image of God, we can perhaps expect some parallels. John Polkinhorne, physicist and theologian[59] discusses personhood and the soul and believes that:

---

[55] Damasio (1999).
[56] For example, Chalmers (2014).
[57] Davies (2006: 5, author's italics).
[58] Strobel (2004: 314–338).
[59] Polkinhorne (2002: 105–107).

> Whatever the human soul may be, it is surely what expresses and carries the continuity of living personhood.... If that carrier of continuity is not a separate spiritual component, what else could it be? It is certainly not merely material.... It is this information-bearing pattern that is the soul.

He prefers

> a thorough going psychosomatic picture of human nature, in which the preservation of the soul depends only on divine faithfulness.

3.4 DUALISM OR MONISM

From the above discussion we see that there are two broad possibilities in explaining consciousness: dualism or monism, or some kind of combination. Cartesian dualism, introduced by Descartes,[60] describes self-consciousness as something created external to matter (either from the beginning or at a certain stage of organic development). Here the mind and the body are not identical and the mind is non-physical. Monism assumes that consciousness and self-consciousness arose through evolution out of nonliving matter because matter contains the potential for it. As noted above, this implies that by continuity fundamental particles must have the potential for self-consciousness as well; albeit in an elementary form. Otherwise there is a point of discontinuity, as in the case of non-life becoming life.[61] It is interesting that Sir Charles Sherrington, the father of the modern understanding of brain function, finally supported a kind of dualism. Five days before his death he said, "For me now the only reality is the human soul."[62] One of his famous students, the neurosurgeon Wilder Penfield, carried out an extensive study of hundreds of patients afflicted with epilepsy and came to a dualist position with the words:

> In the end I conclude that there is no good evidence, in spite of new methods, such as the employment of stimulating electrodes, the study of conscious patients, and the analysis of epileptic attacks, that the brain alone can carry out the work that the mind does. I conclude that it is easier to rationalize man's being on the basis of two elements than on the basis of one.[63]

Another famous pupil of Sherrington was the neurophysiologist Sir John Eccles who described experimental evidence of the existence of "interactionism," that is of mind/brain interaction. Both he and the prominent philosopher Sir Karl Popper concluded that dualism is supported,[64] though they disagreed over the origin of the mind and its destiny after death. Popper avoided any transcendental notions (i.e., the mind just evolved out of brain activity but

---

[60] Technically referred to as "substance dualism."
[61] This is referred to as "panpsychism" where matter is not just physical stuff but contains something else. This is not true materialism.
[62] Popper and Eccles (1977: 558).
[63] Penfield (1975: 113)
[64] See Popper and Eccles (1977).

then gained a measure of independence), while Eccles believed in God the creator of the mind and in the continuation of the soul after death. They saw the mind governing and using the brain for its own conscious purposes, but with limitations depending on the capacity and efficiency of the brain as a machine—each affecting the other. The Jewish psychiatrist Viktor Frankl[65] described the brain as conditioning the mind, but not causing it.

Six authors[66] from different backgrounds have written a large book that examines findings from nearly a century of psychical research. They present empirical studies related to all kinds of phenomena including psychosomatic medicine, placebo effects, near-death and out of the body experiences, mystical experiences, physical effects induced by hypnosis, trances, lucid dreaming, and the existence of creative genius, to argue for a "strongly dualistic theory of mind and brain." Their book depicts the mind as being independent of the brain but causally interacting with it and surviving death. It argues that properties of minds cannot be fully explained by those of brains. Norman Geisler, philosopher and theologian,[67] raises a number of questions about monism and a materialistic approach to the problem. For example, how can I know I am nothing more than my brain unless I am more than it? I cannot put my brain under a microscope and analyze it unless I (my mind) am standing outside the microscope. Matter has a space-time limitation but the mind is not so limited as it can roam the universe without leaving the room. Also, it can be argued hat a materialist should have no discrete thoughts, but simply experience a stream of particles.

One objection to mind-body dualism is the problem of how an immaterial mind can influence the brain without violating the Principle of Energy Conservation (PEC). How can a non-physical mind provide the energy to get a human body to carry out an action? This question is answered in detail by Robin Collins[68] who first considers the fact that PEC does not apply to all known physical interactions. He points out that, according to many modern scientists, energy is not conserved "in general relativity, in quantum theory, or in the universe taken as a whole." For example, the non-conservation of energy in general relativity is due to gravitation. The discussion is very technical so I won't go there!

One view that supports monism but endeavors to give a different perspective without giving a clear explanation is proposed by psychologists David Myers and Malcolm Jeeves.[69] They describe the mind-brain system as not being reducible to its physical parts, and give the following analogy from the behavior of the social insects—the ants, the bees, and the termites. An ant colony has a collective intelligence when it comes to growing, moving,

---

[65] Frankl (1969: 254).
[66] Kelly et al. (2007).
[67] Geisler (1999: 351).
[68] Collins (2008).
[69] Myers and Jeeves (2003: 24).

and building, but this intelligence is not reducible to individual ants as a solitary ant has little going for it mentally! This idea of a collective consciousness has already been mentioned above and brings an aura of mystery to the mind/brain problem. We now look for further evidence of a mind.

3.4.1 Near Death Experiences

One area of research that imposes a challenge for the materialist is the so-called "near-death" experience or NDE. Elements of such an experience include:[70] awareness of being dead, positive emotions, out of body experience, moving through a tunnel, communication with light, observation of colors, observation of a celestial landscape, life review, indescribable music, experience of overwhelming love, and presence of a border. Also people meet a variety of greeters in an NDE such as light beings, angels, deceased loved ones, God's presence, religious figures, and even animals.[71] However some of the NDEs (less than 10%) have been been distressing.[72] The literature on NDEs is very extensive and there are very many examples available.[73] According to a Gallup poll in 1982, about eight million people in the U.S. have had an NDE, while a U.S. News Report in March 1997 found that 15 million people had the experience. Also surveys from the U.S., Australia, and Germany estimated that between 4 and 15% of the population have an NDE.

In spite of all the evidence about NDEs, past evidence has tended to be anecdotal, and opinions are divided on these experiences. Some maintain that these experiences are a product of our brain playing tricks on us and can have a natural explanation such as rapid eye movement (REM) activity and lucid dreaming, lack of oxygen, cortical disinhibition, hallucinations, depersonalization, birth memory, endorphins, and denial or fear of death. Dr Eben Alexander III, a neurosurgeon who had an NDE was interviewed about his experience.[74] He said that he was absolutely certain that it is not all brain chemistry and that "consciousness is the thing that exists." During his NDE he said that he experienced wisdom, guidance, and unconditional love, and believes that such things can be experienced without an NDE through meditational techniques and centering prayer.

It has been argued that because the term "near death" is somewhat vague and ill-defined a more direct approach has been to look at cases of cardiac

---

[70] See van Lommel et al. (2001). An example is given there of an out-of-body experience.
[71] See, for example Atwater's summary statistics at
http://www.near-death.com/experiences/evidence06.html, accessed August, 2015.
[72] See, for example,
http://www.dancingpastthedark.com/dndes-by-the-numbers/ by Nan Bush, accessed August, 2015.
[73] See the collection at http://www.near-death.com/experiences/evidence08.html, accessed August, 2015.
[74] A video of the interview is at http://www.btci.org/consciousness/video_archive.html; see 2012 no. 3. Accessed August, 2015.

arrest when people have actually died and then been brought back to life.[75] A very large international program called AWARE that looks at cases of cardiac arrest had been ongoing in a number of countries at the time of writing this. It has been found that about 20% of such cases had an NDE. If the NDE was due to purely physiological factors we would expect most people to have an NDE, which is not the case. Small children, who have no preconceived ideas as to what to expect, have also had similar experiences. The following are facts that need to be explained.[76]

Some Facts About NDEs

We now consider five facts about NDEs. First, people who are brain dead can have an NDE. One well-known recorded incidence is the story of Pam Reynolds who underwent a procedure called hypothermic cardiac arrest in which surgery was carried out after her vital signs were stopped, but she was able to report later what happened and what was said.[77] During her "stoppage," her brain was found "dead" by all three clinical tests—her electroencephalogram was silent, her brain-stem response was absent, and no blood flowed through her brain. This raises the question of how could somebody experience clear consciousness outside of one's body at the moment that the brain no longer functioned during a period of clinical death and with a flat EEG? In fact the level of consciousness and alertness during an NDE is usually greater than that experienced during everyday life.[78]

Second, people have known things they could not have possibly known when unconscious. For example Raymond Moody, medical doctor and philosopher,[79] who introduced the term "near death experience", documents cases where people see verified events while out of their bodies. He records the case of a woman blind from birth who regained her sight during her NDE, and she was able to accurately describe the instruments and techniques used during the resuscitation of her body. After revival, she was able to tell her doctor who came in and out, what they said, what they wore, what they did, all of which was verified.[80]

Third, life reviews experienced in NDEs include real events that took place in the person's life, even if the events were forgotten. When they encounter beings they knew from their earthly life, they are virtually always deceased, usually deceased relatives.

---

[75] For simplicity I will continue to use the commonly used term NDE.
[76] See also Long and Perry (2010).
[77] Sabom (1998); see also http://www.near-death.com/experiences/evidence01.html, accessed August, 2015.
[78] Long and Perry (2010: 201).
[79] Moody (2001); see also http://www.near-death.com/experiences/evidence02.html, accessed August, 2015.
[80] For further examples relating to blind people see Ring and Cooper (1999).

Fourth, NEDs are remarkably consistent around the world irrespective of whether the person is from a Western or non-Western country, or whether religious or not.

Fifth, people who have NDEs are convinced of the reality of their experience and often their lives are radically changed. As an example, an atheist who was pronounced dead and in a morgue for three days came to life after a NDE that led him to becoming a pastor.[81]

Neurologist Kevin Nelson, who suggested that NDEs experiences can have a natural explanation, however writes:

> Do these cold, hard clinical facts suck the divine nectar from our spiritual lives? My answer is an [emphatic] NO! We are poised on the threshold of a new era that holds tremendous promise for a new level of spiritual exploration.[82]

Peter Fenwick, a leading British authority on NDEs believes that such natural explanations fall far short of the facts. In a documentary entitled "Into the unknown: Strange But True" he described the state of the brain during an NDE as follows:

> The brain isn't functioning. It's not there. It's destroyed. It's abnormal. But, yet, it can produce these very clear experiences ... an unconscious state is when the brain ceases to function. ... The memory systems are particularly sensitive to unconsciousness. So, you won't remember anything. But, yet, after one of these experiences [a NDE], you come out with clear, lucid memories ... This is a real puzzle for science. I have not yet seen any good scientific explanation which can explain that fact.[83]

Shared NDEs

What is more incredible is the phenomenon of shared death experiences where bystanders around a dying person can have out-of-the-body experiences, for example, seeing relatives come to meet the dying person or seeing the spirit of the person stand up after death. Such experiences are documented by Raymond Moody and Malcolm Perry[84] and suggest that the experiences have nothing to do with physical processes as the bystander is not having a near death experience.

3.4.2 Mind Over Matter and OCD

As I mentioned in Chapter 2, we see that materialism explains nothing if our brains are just a collection of molecules. As a trained counsellor, my experience is that Cognitive-Behavioral Therapy (CBT) can be very effective in

---

[81] See http://www.near-death.com/experiences/evidence10.html, accessed August, 2015.
[82] Nelson (2010: 258–259).
[83] http://www.near-death.com/experiences/skeptic04.html, accessed August, 2015.
[84] Moody and Perry (2011); see also Moody's other books and his interview at http://www.youtube.com/watch?v=DvNDrZv8HwE, accessed August, 2015.

reprogramming the brain, where a change of beliefs can bring about a change of mental state and a change of emotions and behavior. For example, CBT has been used effectively in dealing with "brain" problems such as depression, phobias, and obsessive compulsive disorder.[85] Psychiatrist Jeffrey Schwartz has authored or coauthored a number of books that describe how Positive Emission Tomography (PET) scans and functional magnetic resonance imagery (fMRI) have shown how a change of mindset with thoughts and behaviors can cause physical changes in the brain. In studying OCD, Schwartz and Begley explored

> emerging evidence that matter alone does not suffice to generate mind, but that, to the contrary, there is exists 'a mental force' that is not reducible.[86]

They found that certain regions of the brains of people with OCD had abnormal patterns of brain activity with repeated bouncing around the "the worry circuit", namely the orbital frontal cortex, anterior cingulated gyrus, caudate, and thalamus (got that!). This repetitive biochemical firing triggers an overwhelming sense that something is wrong, followed by compulsive attempts to make it right. The intrusive thoughts and ritualistic behaviors are typically described by patients as observing activities of a third party, leading them to having "brain lock." However, the authors showed that after intense mental effort, willpower, and psychotherapy, patients had relief from their symptoms with corresponding permanent changes in their brains. Something had changed them and it certainly wasn't the brain unlocking itself. The authors called it "the power of mental force" and argued that this showed that we are more than just our brains. They also described how changes could be introduced not only with human stroke victims and those with OCD but also with monkeys and other primates.[87] Another possible example of mind over matter is the so-called placebo effect where a belief in a certain medication, even if it is just an inert substance, can bring about an improvement in health. However, other explanations of this may be available.

The above evidences suggest that the mind is not the brain, though damage to a part of the brain can inhibit the mind as the mind uses the brain as a vehicle of communication. People with brain damage or even a brain tumor can experience a complete change in personality, which reverts back to what it was before when the tumor is removed.[88] Conversely, anything that happens to us can upset our mind, which can then affect our brain and lead to physiological expression. Sometimes it is a question of which is the cause and which is the effect. The main difficulty with dualism is the question of how the mind and brain are connected.

---

[85] Seber (2013).
[86] Schwartz and Begley (2002: 52).
[87] Other books by Schwartz on the subject are Schwartz(1997, 2011).
[88] For an example see Burns and Swerdlow (2003).

Arising from the above evidence it can be argued that we have some sort of existence after death due to our mind. This raises the question of immortality, even if the NDE experience is only temporary. If death destroys the brain it may still be possible to have access to the world through another body or even have access to other worlds. There is no argument to say that we cannot be conscious without some kind of body. An after life is particularly appealing from a justice perspective since justice may not be achieved in this life but it will be in the next if there is a continued existence. Something we might expect from a totally just God. This question is discussed further in Chapter 8 on suffering and evil.

There is one idea that I find fascinating. Every seven years every cell in my body is replaced by a new cell,[89] yet in spite of this change I experience a continuity of personality, both conscious and unconscious. There is a blueprint or set of information defining who I am that continues. When I talk to my elderly friends we all agree that we feel just as young as ever inside, but things like memory recall are not so good (and the outside doesn't look too good either!). This continuity suggests the existence of something outside my brain but which unfortunately can be heavily affected by my brain, as we find with brain-damaged people and degeneration from small strokes with some people as they age.

Abstract Ideas

I conclude this section by referring to an argument for the mind not being just material from the philosopher Edward Feser[90] that follows from the existence of abstract ideas discussed in Section 2.3. For example, the number five and triangularity are abstract ideas with material representations. The form of a triangle (a universal) that exists in our intellects or minds is of the same form as actual material triangles, otherwise we wouldn't be really thinking about triangles at all but about something else. However, if our intellect was just material and simply part of the brain, triangularity would exist in material form and the corresponding part of the brain would essentially become a triangle, which being material contradicts our abstract nature of triangularity. Also the idea of triangularity leads to theorems like the theorem of Pythagoras about right-angled triangles, which do not exist (exactly) in the material world and are even independent of our minds and what we believe. The same idea applies not only to mathematical propositions in general but also to other abstract concepts like so-called "universals" (e.g., humanness, justice, beauty).

---

[89] Except those that die because the telomeres on the ends of chromosome are now too short; it is called aging!
[90] Feser (2008: 120–142).

3.5 BIBLICAL CONCEPTS OF BODY, MIND, AND SPIRIT

Although the Bible is an ancient book, and we examine it more fully in Chapter 5, it is interesting to see what it has to say about who we are, especially as some people have misconceptions about what the Bible actually teaches. Although I have used the term "mind" above, the New Testament uses the terms "spirit" or "soul" and refers to a duality of body and spirit.[91] There are also many references about being in-dwelt by the Holy Spirit of God.[92] The Bible makes it clear that when the spirit or soul leaves the body, the body is dead,[93] but if the spirit returns, the body comes back to life.[94] Jesus told his disciples not to fear those who kill the body but can't kill the soul.[95] When he was on the cross he told one of the thieves being crucified with him that the thief would be with Jesus immediately after his death.[96]

Biblical Models

For the interested reader I want to enlarge on the biblical models of who we are. For example, we are material but also immaterial (e.g., soul/spirit);[97] we are body, soul, and spirit;[98] and we have the more holistic Hebraic way of thinking of a person as an integrated unity, where both soul and spirit together refer to the person, the latter idea being more in keeping with today's knowledge. We have to be careful how we interpret biblical words like the Hebrew *nephesh* and the Greek *pseuche* as they can have many different meanings.[99] For example, the word translated "a living soul" in Genesis 2:7 is used in earlier verses with regard to beasts, birds, and every creeping thing, that is every living thing, so that it does not appear to be a unique characteristic of humans. The above three models all focus on different aspects of our existence and indicate that we are more than just material, yet we are a unity that is brought out in the emphasis on the resurrection of the body in the Bible, rather than continuing as a disembodied spirit.

Another biblical facet of who we are is that humans are described as being made in the image of God (Imago Dei) and we are to have dominion over the earth.[100] Although various possible meanings of this image concept have been discussed down through history, it does suggest that humans somehow have the stamp of the divine upon them, thus establishing their rationality, their

---

[91] Matthew 26:4; Romans 8:10; 2 Corinthians 7:1.
[92] For example 1 Corinthians 6:20.
[93] James 2:26.
[94] Luke 8:55.
[95] Matthew 10:28.
[96] Luke 23:43.
[97] For example Genesis 2:7 and Ecclesiastes 12:7. These two references use the Hebrew words *nephesh* and *ruasch*, respectively.
[98] For example, Hebrews 4:12 and 1 Thessalonians 5:23. These texts use the Greek words *pseuche* (soul) and *pneuma* (spirit), respectively.
[99] See the words on the internet.
[100] Genesis 1:26–27.

worth, their guardianship of our planet, their community, and their spiritual connection. It is not clear how we actually received the image, but in the words of pastor Chris Wright "The image of God is not so much something we possess, as *what we are. To be human is to be the image of God.*"[101] We have a God-likeness, that is a spiritual nature, and because God is just and righteous, we have an awareness of right and wrong and yearn for justice and fairness. When someone does us wrong we feel violated, because a moral law has been broken. We also have the capacity to communicate with God.[102]

There is also a part of us that has the potential to be eternal, when we are resurrected with a new body. There is a "lower level" part of us which dies, but there is another part that can have a continued new existence in some other state not necessarily bound by time. A well-known example is that of a computer program that can be stored for example on a hard disk drive, CD, or USB stick. In the old days (and that is not very long ago!) it was stored on floppy discs, punched cards, and punched paper tape. The medium may change but the program continues. The program is more than just molecules of a certain type, and it is even more than just the boundary conditions that I talked about in Chapter 2 as these vary with the medium. The program has its own existence, which is ongoing. There is a programmed part of me that is my essential being, and that has an independent existence.

Another aspect of having the image of God would mean that we are different from other animals and we can, in a limited way, understand God's creation. This enables us to make up laws about the consistency that we see so that we can do science and advance in our knowledge of the world. We can also link up things that have never been linked before. Einstein predicted how to split the atom and convert matter into energy simply through mathematical equations that he worked out in his own mind. I believe that such scientific inspiration has it source from God, though sometimes we abuse this privilege. Every discovery can be put to a good or bad use, for example, atomic energy (radio therapy and the use of radioactive tracers in medicine versus atomic weapons). The Genesis account also mentions that we are to have dominion over the all the earth and every living thing. These two facts together mean we can exercise some control over the environment and we have some responsibility for it. On an individual level we have some control over our bodies. Our bodies can heal themselves and we have developed medicines and techniques to accelerate the process. We have the potential to live a long time.

3.6 EXTERNAL EVIDENCE OF A SPIRITUAL DIMENSION

So far we have looked mainly at some "internal" evidence of a spiritual dimension. We now endeavor to look at some other evidences. In doing so we face

---

[101] Wright (2004: 119; his italics).
[102] As described in Genesis chapter 3.

the problem discussed above that you cannot prove something about a given level of reality from studies based on lower levels. You cannot use the laws governing the movement of molecules to prove that the watch has the function of time-keeping. I therefore cannot use science and material experiments to prove or disprove the existence of a spiritual world. However, "higher" levels of reality affect "lower" levels. For example, society can affect what I do as an animal, which is lower down. And what I do as an animal affects my body chemistry, which is again "lower down" the scale. We get indirect evidence of the spiritual realm by observing what happens in the physical realm. Some evidence has been given in Section 2.8.

3.6.1 Spiritual Forces

How would one answer the question, "Is there evil in the world?"[103] A philosopher would probably ask for a definition of evil, and this is not easy. To see why, suppose I asked you for a definition of a curved line. You would probably respond with "a line that is not straight" so that we now have to define a straight line, an idea referred to in Section 1.2.2. There are two ways of approaching this. One way is for me to give you the mathematical formula for a straight line (which represents an abstract version of a "perfect" straight line) and tell you to go away and plot it on a computer to obtain a concrete representation of it. The second way is to point to objects with "straight" edges and say that a straight line is like those edges and is an abstract representation of the family of edges. A curved line is then a line with the absence of straightness. Therefore returning to the question of what is evil, we can say that evil is the absence of good (not the opposite as we would then have explain the word "opposite!"). Evil then is not something but the absence of something, and we can point to actions that most people regard as evil.

Goodness

So what is goodness? A materialist would have a problem with this as he or she would have no absolutes and no absolute version of "good." They might argue for moral relativism that everybody's beliefs are equally worthy of respect (except the belief that there is absolute morality!). They are not morally neutral if they complain about being treated unfairly or complain about evil in the world; good is meaningless. The only neutral position is silence, with all language about being wrong being given up as, for example, injustice and fairness since they imply moral judgement. In fact relativism is not moral as you cannot, for example, demand tolerance. Labeling someone as intolerant means making a judgement, and requiring tolerance refers to a moral rule. The theist however can say that God is absolute goodness. What does this mean in practice? For the Christian it means loving God with all

---

[103]This question will be discussed further in in Chapter 8.

our being and our neighbors as ourselves. As mentioned in Chapter 2, there seems to be a universal morality that can help us to make good decisions.

Do all decisions have just good or bad consequences or are there neutral ones as well? Everything we do is for a purpose (even to do nothing) and that purpose is either "good" or "bad" for us and others, depending on the consequences of our action. I could not think of anything that I did that I could label as being completely neutral (e.g., getting up in the morning is good for me!) However, determining whether something is good or bad will have cultural overtones, and sometimes we have to choose between two decisions that both have bad consequences. We see then that the actual practice of morality can lead to all sorts of philosophical questions. However, I won't go down that path because I am interested in the question of a spiritual dimension—hence the next question.

Demonic Influences

Coming back to the question of evil, the reader might ask: "Does Satan exist and are there other spirits (or demons) who are evil?" as the New Testament has over eighty references to demons. If these entities are spiritual and not flesh and blood, then we cannot prove their existence, only look for any evidence. I believe that there is plenty of evidence of so-called "demon possession" where we encounter strange (supernatural?) phenomena, although it tends to be very rare (except in some societies). How this is interpreted is where the debate starts. Some would say that such phenomena are the result of mental illness, and this could be true in many cases, especially in the past with schizophrenia, Tourettes syndrome and tics, and various forms of psychosis, but it leaves other cases unanswered. With some people, possession may occur along with mental illness and may cause the mental illness. Manifestations of demon possession can occur in three general ways.

Mental changes: These can be as follows: changes in personality (or multiple personalities) and behaviour (e.g., more hostile, violent, or abusive for no reason); cursing a lot; the destruction of religious objects; severe nightmares or night terrors; knowing things about the past or future that they are not expected to know; unusual changes in diet (eating what was originally detested); and changes in hygiene.

Physical changes: Examples are non-blinking, hair or eye color changes (e.g., almost black pupils), appearing catatonic, being completely rigid, having contorted features, speaking an unknown language, speaking with the voice of the opposite sex, having writing or symbols on their body in the form of welts or scratches (especially in places they cannot reach), levitation, unusual movement of the person, unusual strength, or slow deterioration and sickness.

Environmental changes Examples are objects moving themselves, objects disappearing and appearing elsewhere (I think this happens with my glasses or my car keys!), and strange noises.

There are many different symptoms of demon possession,[104] and clearly some of the above can be due to mental or even physical illness such as a brain tumor. However, there are some aspects that are out of the ordinary and could be labelled supernatural. We find that traditional medicine fails to help the genuinely possessed, whereas exorcism through prayer can be successful. There are many examples of demon possession in the New Testament.[105]

Angels

We have talked about the bad guys, but what about the good guys! Do angels exist? This is a difficult question as the Bible recounts incidents when angels have appeared in human form. For this reason it is difficult to know whether angels still do this or not. There are current stories of encounters where there has been something strange and even miraculous about the encounter.[106] Angels are mentioned in most religions but they generally are regarded as invisible, which makes their existence difficult to prove!

Faith Healing

Finally there is the topic of faith healing. Some have argued that faith healing is evidence of the supernatural. However we have to be careful here as such healing occurs not only in various religions, but also with non-religious people. The problem is confusion between psychological healing and supernatural healing and also the role of faith, that is having a belief that a person would be healed. Looking first at the question of faith, it appears that faith isn't always necessary in the case of supernatural healing; a person can be healed without having faith. Looking at the thirty-five miracles of Jesus, in only ten is the faith of the recipient mentioned and generally not explicitly demanded. The three people he raised from the dead certainly weren't able to exercise faith! He performed some miracles when there was unbelief on the part of the disciples.[107] It is clear that the performance of supernatural miracles did not depend on belief or the exercise of faith, though a hostile environment might prevent some miracles.[108]

It is however a different story with psychological healing as what we believe can affect our health, whether it is belief in ourselves or belief in another person. We know from the placebo affect and other studies that a positive attitude and self-belief will promote self-healing, as I have found in my counseling practice. It would appear that much of the healing that occurs today would be psychological healing, especially in the case of functional diseases where a

---

[104] For one documented case see http://www.newoxfordreview.org/article.jsp?print=1&did=0308-gallagher, accessed August, 2105.
[105] Mark 5:1–20 and 19:17–29 are good examples of what happens.
[106] See, for example, Morgan (2011).
[107] Luke 8:25.
[108] Matthew 13:58.

person is not able to do something physical (e.g., paralysis). Sometimes such a healing does not last.

On the negative side, in some cultures a witch doctor may put a curse on someone and the recipient will fall ill and even die because he or she believes in the curse. But what about healing by other people when often the recipient has little or no faith but is still healed? I hear about stories of healing by local pastors in some difficult countries all the time, though I cannot vouch for their validity. Of particular interest is instantaneous healing through prayer (and sometimes through fasting as well) such as the recovery of sight and a short leg lengthening. Sometimes prayer has to persist for some time as healing takes place slowly, perhaps depending on the source of the illness, but the healing is still spectacular. Such events suggest supernatural intervention, indicating a spiritual dimension.

I personally believe that in the same way that natural laws operate in the physical world there are spiritual laws that operate in the spiritual domain. Faith seems to have a part to play. The skeptic may perhaps argue that if God is doing the healing, why doesn't everyone get healed? This is an interesting question and will be considered in Chapter 8. Miracles are discussed in Chapter 7.

3.7 CONCLUSION

This chapter asked the initial question of whether or not there is a spiritual dimension. We have seen some evidence that we are more than just molecules and that there is an added spiritual dimension to our lives that may be an empty space waiting to be filled. Summing up, I believe that materialism does not provide answers, as there exists another level of reality, which we might call a spiritual dimension to life, that adds meaning to the flesh. For me there is a message in us and in the world about us to be read. What I see about me and within me cannot be reduced to nothing but .... I believe that we are more than just machines or animals; we are spiritual beings as well. This gives special dignity to humanity.

For an adequate model of reality I therefore believe we need both the "lower" storey physical side of the world and the "upper" storey unseen spiritual realm. One without the other will give an incomplete picture. It is like tearing a picture in half and giving a person one half! Problems like suffering, for example, can be best understood within this two-level model of reality considered in Chapter 8. However, we have only considered part of the story as many questions may have risen in the reader's mind about such things as free will, the mind-brain connection, and the impact of science on such topics. We deal with these in the next chapter when we consider how new scientific developments, like quantum theory, provide a a new way of looking at reality and whether or not we have free will.

CHAPTER 4

DO WE HAVE FREE WILL?

4.1 WHAT IS FREE WILL?

The reader may respond to the title of this section by saying "Of course I know what free will is. I exercise it every day." However if you believe in materialism as considered in the previous chapter, namely that material is all there is, then free will does not exist but is simply the product of predetermined activity in our heads. I realize that this an understatement of the problem given that this is an extensive topic and has exercised and still exercises philosophers and psychologists today. What do we mean by free will? Collins dictionary defines it as " the apparent human ability to make choices." The Stanford Encyclopedia of Philosophy describes it as "a philosophical term of art for a particular sort of capacity of rational agents to choose a course of action among various alternatives."[1] The long debate over centuries is about what sort of free will we are talking about. For example, free will tends to be related to the concept of moral responsibility and being responsible for one's actions. We can also distinguish between freedom of action and freedom of

---

[1] http://plato.stanford.edu/entries/freewill/#3.3, accessed August, 2015.

will, as a decision to carry out a particular action may be thwarted by external constraints outside our control. There may be other factors in addition to physical constraints such as psychological, biological, or even theological constraints on our freedom. I shall therefore approach this topic from several points of view.[2] The first question I shall address is the role of chance in our lives. I discuss it first from a statistical point of view, and then move into its role in modern physics and quantum mechanics. This will show up some of the short-comings of materialism as well as leading into the question of the role of determinism and finally to a discussion on free will itself.

4.1.1 Does Randomness Exist?

The question of free will arises if we take a totally materialist view of who we are. Do we have free will or not, and if we do how did it arise if our origins were purely mechanistic? If my thinking is predetermined, then what I think about determinism is predetermined— end of story. In grappling with this question of freedom we have to first consider whether or not there is such a thing as "randomness" and how do we define it. Dictionaries defines random as having no definite aim or purpose; not sent or guided in a particular direction; without method or conscious choice; and haphazard. A major problem is how do we determine whether something is random or not, and this leads us into a more technical mathematical definition of a random variable defined in the subject of Statistics. Such a definition is outside the scope of this book, though most people have an intuitive idea of randomness when it comes to such things as tossing a coin, rolling a dice, choosing a card from a well-shuffled pack, or choosing a name out of a hat. What is confusing is that there are sequences of numbers that can be generated using nonrandom methods that statistically have the properties of random numbers when various statistical tests for randomness are carried out.[3] (For those in the know, π and $\log 2$, for example, can only be expressed as numbers with an infinite number of decimal places with sequences that behave as though they were randomly generated.)

Randomness can be regarded as a model for unpredictability, and that avoids the question of whether randomness actually exists. Determinists would say randomness doesn't exist. When we toss a coin, if we knew everything about the coin, how it was tossed, and under what experimental conditions, we could perhaps work out whether it would land heads or tails. However we don't have all this information and all that we can do is try and carry out our toss in such a way that either heads or tails is "equally likely" to turn up. If the coin was "perfectly" made we could then assume that the chances of a head or tail were "50-50", which reflects our assessment

---

[2]For further comments about free will and its practical outworking see Section 8.5.
[3]These are called pseudorandom numbers.

of the unpredictability of the outcome. As we shall see next we can have a deterministic process that is unpredictable.

Chaos Theory

To make things even more complicated (time for another coffee break?) we have the mathematical topic called "chaos theory" that refers to a property of dynamical systems.[4] The topic seriously began to develop with Edward Lorenz developing a weather model on his computer in 1960 followed by a publication on fluid dynamics in 1963. As the subject is quite mathematical I don't want to bore the reader with too many details, so I shall endeavor to give just a brief overview.[5] Before giving an example to explain the mathematics that the reader might wish to skip, I shall summarize the essential idea of chaos theory. We start with a particular deterministic mathematical process that follows a predetermined path that depends on the values of certain constants (parameters) associated with the process. When the parameters take on certain values, the process becomes chaotic and wanders all over the place. Because there is a limit to the number of decimal points that we can work with even with a computer it turns out that we don't know where the process will end up.

For the reader who wants more mathematical detail I now give a simple example based on the so-called *logistic map* that arises in population ecology, namely[6]

$$x_{t+1} = kx_t(1 - x_t),$$

where we shall assume k can be any number from 0 to 4. If we choose x_0 in the interval between 0 and 1 we get a sequence of numbers in this interval, namely x_0, x_1, x_2, …. For example, given $k = 1$ and $x_0 = 0.5$ we compute the next value in the sequence, namely $x_1 = kx_0(1 - x_0) = 0.5 \times 0.5 = 0.25$. Substituting x_1 into the right-hand side of our equation we then calculate $x_2 = kx_1(1 - x_1) = 0.25 \times 0.75 = 0.1875$, and so on; we put one number in and get the next number out. The set of all the possible values of x_t for any t is called the *phase space*, and a plot of x_t versus x_{t+1} is called a two-dimensional *phase diagram* that can be useful to describe the behavior of the sequence x_t.

The question is what happens to the number x_t as t increases. Well, the answer depends on the values of k and the starting value x_0. However, before pursuing this further we introduce the idea of an *attractor*; a rather misleading term as there is no actual physical attraction as say in magnetism. Attractors come in three species. The simplest or *fixed point* attractor is a single number that x_t gets closer and closer to as t gets bigger and bigger.[7] A simple example

---

[4] See, for example, Stewart (1997), Smith (2007), and for a deeper look at the subject see http://plato.stanford.edu/entries/chaos/, assessed August, 2015.
[5] I will not try and give a technical definition of "chaos" as there isn't agreement on this.
[6] Technically the equation is known as a *difference equation*.
[7] Referred to as a *limit point* in mathematics.

is to consider the sequence 1, $\frac{1}{2}$, $\frac{1}{3}$, $\frac{1}{4}$, ... i.e. $x_t = \frac{1}{t}$ ($t = 1, 2, \ldots$) that gets closer and closer to 0 as t gets bigger and bigger: 0 acts like a fixed point attractor. A second kind of attractor is a set of points arising from periodic behavior where, as t increases, the numbers eventually go round and around a fixed cycle with a given period. An example of this is the set of numbers 1.2, 3.2, 5.6, 2.1, 1.2, 3.2, 5.6, 2.1, ... where four numbers in succession get repeated again and again as t increases, giving us a period of 4 and an attractor set of four numbers. The third kind of attractor is called the *strange attractor*, and which is a part (subset) of the phase space, where x_t wanders around in an apparently haphazard or chaotic manner. Once x_t is in this subset it does not escape from it.

We now investigate what happens to x_t as t increases for our logistic map as k ranges from 0 to 4. The following hold for either all or almost all values of x_0 between 0 and 1:[8]

$k = 0$ **or 1:** $x_t = 0$ for all t.

$0 < k < 1$: x_t approaches zero, i.e., has a fixed point attractor of 0.

$1 < k < 2$: x_t rapidly approaches the value of $(k-1)/k$, a fixed point attractor.

$2 < k \leq 3$: x_t eventually approaches the value of $(k-1)/k$, but initially fluctuates around that value for some time.

$3 < k < 3.44949$ (approximately)[9]: x_t will approach a permanent oscillation of period 2 between two values that depend on the value of k.

$3.44949 < k < 3.54409$ (approximately): x_t approaches a permanent oscillation of period 4.

$k > 3.54409$: As k increases, x_t approaches oscillations of periods 8, 16, 32 etc. with the periods doubling,[10] i.e., the repeating sequences get longer and longer as the period approaches infinity.

k approximately 3.56995: At this point we get "chaos" and we have a strange attractor where the values x_t appear to behave like random numbers.

$k > 3.56995$: Here x_t is generally chaotic except for certain isolated ranges of k that show non-chaotic behavior, for example oscillations of periods 3, 6, 12 etc. In fact all oscillation periods occur for some values of k.

There are several important features of chaotic behavior.

(1.) Nonlinear processes can lead to chaotic behavior. A linear process takes the form $x_{t+1} = kx_t + c$, where, if we ignore the constant c, x_{t+1} is proportional to x_t with k being the constant of proportionality. Anything

---

[8] http://en.wikipedia.org/wiki/Logistic_map, accessed August. 2015.
[10] Called a period-doubling cascade.

which is not linear is said to be nonlinear. The process we considered above is $x_{t+1} = kx_t^2 - kx_t$, which is quadratic because of x_t^2, and is therefore not linear.

(2.) The process is deterministic, that is every value is determined by the previous values with no random elements involved. Logically then we must be able to determine what happens in the long run and therefore be able make an accurate prediction of what will eventually happen. Right? Unfortunately this is not necessarily so, as we see next.

(3.) A chaotic sequence is extremely sensitive to even the tiniest change in the starting value of x_0 (e.g., even to a change of say 0.000000001 or smaller). It can send the sequence on a completely different path. A good example of this sensitivity is the case of a bead threaded at exactly the highest point of a smooth circular wire. The slightest shift can send the bead dropping in one of two entirely different directions. We note that if measurements are involved in beginning a chaotic sequence, then we know that there is a limit to how accurately we can measure something so that we can't be sure where the process is going as a slight change can send the process off on a different path. As we shall see next, there is a further complication in that the slightest input of randomness can completely change the path of the process.

(4.) Numbers must be calculated exactly. This puts a limit on how big t can be as eventually x_t requires an accuracy that is beyond even the most powerful computer. Eventually the computer has to round off a number to so-many decimal places, for example 0.2374581 may get rounded to 0.237458. This can change the process completely because of (3) above, and imposes a limit to the number of terms of a chaotic sequence that can be generated exactly on a computer. We find that the path of the process, although theoretically completely determined, is eventually unpredictable.

(5.) One characteristic of a chaotic process is that the paths ("trajectories") taken by x_t when two starting values of x_0 are very close together diverge substantially from each other. In fact they cannot cross, as at any point of intersection there is a choice of two paths; this is not possible with a deterministic process (that has no choices).

(6.) The above logistic map is said to be a discrete process as t takes values 1, 2, ... However there are also so-called continuous processes where t can take any number in an interval, and $x_{t+1} - x_t$ in the discrete case is replaced by a derivative dx_t/dt, which in some situations can be interpreted as a velocity. We won't take this idea any further.

(7.) The above logistic process is one dimensional. However the idea can be extended to two or more dimensions, for example x_t and y_t, and the number pair (x_t, y_t) can then be plotted on a graph. In the case of two dimensions there are some chaotic systems that lead to amazing patterns. Of particular note is the sequence defined by $z_{t+1} = z_t^2 + c$ where, without going into mathematical details, $z_0 = 0$ and c is a so-called complex number. In this case it transpires that c and z_t are each represented by a number pair. The set of all complex numbers c (that is number pairs giving us points on a graph) such that the sequence is bounded and does not head off the page to infinity, is known as the Mandelbrot set. The overall picture that we get is known as a *fractal* and it repeats itself in exactly the same pattern (so-called "self-similarity") at smaller and smaller scales down to the infinitely small. Numerous pictures are given on the internet.[11]

The reader may wonder why I have wandered off into this esoteric subject where a strictly deterministic process leads in practice to uncertainty. In the first place it transpires that the above fractal kind of behavior occurs throughout nature where, for example, we get approximately self-repeating patterns at smaller and smaller scales for at least several levels of scale so that these patterns resemble fractals. These physical as opposed to mathematical fractals are therefore only approximately self-similar at certain scales of magnitude. Examples of chaos and fractals have arisen, for example, in relation to galaxy formation, weather, snowflakes, clouds, botany (plants, leaves, and ferns), eroding coastlines, biology, ecology, cardiac medicine, epidemiology, electronics, economics, fluid dynamics, physics, engineering, sociology, urban development, organizational behavior, and psychology. It should be noted that in reality, a small change does not necessarily mean a large change elsewhere; it depends on the type of processes involved.

A famous example, related to the chaotic nature of weather is referred to as the "butterfly effect." It comes from Edward Lorenz's idea that a butterfly flapping its wings in Brazil may make a tiny change to the atmosphere and the path the weather now follows, and some time later sets off a tornado in Texas.[12] (Now I know why it rains!).

Another well-known example involves three or more stars interacting gravitationally. For instance the planet Pluto has a chaotic orbit, with a "divergence time" of about ten to twenty million years. In English this means that astronomers cannot be certain whether Pluto will be on this side of the sun (relative to Earth's position) or the other side ten or so million years from now![13] Although we can provide a mathematical model for such processes, it must be remembered that reality is more than just the equations that endeavor to represent it. Mathematical chaos is only a model for physical systems that

---

[11] e.g., http://www.fractalsciencekit.com/gallery/gallery.htm, accessed August, 2014.
[12] Lorenz (1972).
[13] We also have the earth, moon, and sun, the three atoms in a water molecule, and the three quarks in a proton as three body systems.

appear to be chaotic. It stems out of deterministic classical mechanics which we know is only an approximation for reality. Another reason for looking at chaos theory is its interaction with quantum theory, referred to as quantum chaos, which is discussed briefly below.

4.1.2 Heisenberg's Uncertainty Principle

Our discussion about randomness would not be complete without looking at its role in physics, mentioning in particular Heisenberg's principle of uncertainty that arises in quantum mechanics, discussed in more detail in the next section. Heisenberg concluded that, in effect, one half of all information required for the prediction of the future from the past is unavailable, not just in practice but also in principle. The Uncertainty Principle states that it is impossible to know both the exact position and the exact velocity of an object at the same time. To measure the position and velocity of any subatomic particle, you shine a light on it and then detect the reflection. Since light consists of packets of energy called photons, these photons hit the particle and cause it to move significantly so that although the position has been measured accurately, the velocity of the particle will have been altered. By learning the position, you negate any information you previously had on the velocity so that the observer ends up affecting the observed. On a larger (macroscopic) scale, the effect of photons on an object will of course be insignificant, but not zero.

Technically, since momentum is mass times velocity, the Principle further states that product of the uncertainties in position and momentum is always greater than or equal to a certain number. This number is one half of the reduced Planck's constant $\hbar$, which is defined as the re-scaling $h/(2\pi)$ of Planck's constant h.[14] The uncertainty extends to all subatomic phenomena. For example, if we measure the exact time when we measure a particle's energy, we will not be able to measure the exact value of that energy and therefore predict its exact subsequent value.

In dealing with these phenomena, Heisenberg had to develop a new statistical kind of mechanics where, instead of calculating the precise positions and speeds of atomic particles, we calculate the relative probabilities of particular kinds of atomic events such as electron impact, photon emission, and the like. In order to calculate these probabilities, physicists now talk in terms of "probability waves" traveling from place to place. This does not mean that electrons or photons are waves, but simply means that the probabilities of the atomic events can be calculated using a wave model. In a sense we can now dispense with classical particles and consider any physical system in terms of waveforms alone. From a practical point of view we only need the probabili-

---

[14]$h = 6.626068e^{-34} m^2 kg/s$, which shows we are working at such a microscopic level.

ties as these can be directly compared to experiment. We now discuss further aspects of quantum mechanics.

4.1.3 Enter Quantum Mechanics

The introduction of quantum mechanics has completely changed the way we look at matter and the mind-brain problem. Classical physics on which materialism is based has now been superseded for atomic events. It was restrictive as it was reductionistic (everything reduces to neurochemical reactions and numbers, and matter to simple microscopic parts), deterministic (so our brain activity follows a predetermined course of events), and local (everything is reducible to a collection of simple local parts where we know where everything is). Psychologist William James[15] realized the mismatch of the view of science in the 1890's with what he observed at the time about the mind, namely that each conscious thought, although containing parts that can be later examined, was essentially a complex unified whole that cannot be reduced to a set of parts without destroying its basic essence. He believed that the classical concept of matter and classical physics left no room for human thought and was therefore flawed, even though he was at odds with nineteenth century psychologists. At the time there was no room for human consciousness and this had led Descartes in the 17th century to introduce so-called Cartesian dualism, which treated mind and matter as two completely separate entities that did not connect in any way. We now find that in contrast to classical mechanics, quantum mechanics is irreducible, probabilistic, and nonlocal, all of which will be enlarged on briefly below. Mathematical physics has undergone a major transformation with quantum mechanical ideas still evolving.[16] Physicist Henry Stapp commented that in Descartes's view of nature

> the essence of man, namely his consciousness, is torn from his body and forced to reside, impotently, outside the world described by physicists ... This failure of classical mechanics at the foundational level removes all justification for retention in philosophy of Descartess dualistic conception of man.[17]

The connection between mind and matter is far more complex, as we shall consider later.

Particles or Waves?

Light has always been an enigma for the physicist as it was found to behave like a string of particles (packets of energy called photons) or as an electromagnetic wave, depending on the experiment performed. In the famous double-slit experiment,[18] light from a very tiny monochromatic source is allowed to pass through a first screen containing two very narrow slits close

---

[15] James (1950, 2001).
[16] For a helpful, historical description see Malin (2001: Part 1), and for aspects of quantum mechanics see Stapp (2009).
[17] Stapp (2009: 172).
[18] Malin (2001: 44–46).

together and then fall on to a second screen. What is seen on the second screen is very different from what one might expect from each slit if it was opened separately. Each photon of light is recorded on the receiving screen device as a dot, thus indicating light's particle nature. The recorded arrival position of any photon varies from photon to photon and the photons form a cloud of dots on the receiving screen if just one slit is used.

In the case of the two slits, as more photons arrive we see an interference pattern consisting of spaced lines of dots on the screen indicating the wave nature of an individual photon and the interference of photon waves. What happens is that when two waves are combined they add to each other when in phase, but tend to cancel out when out of phase giving rows of light and dark that make up the interference pattern. This dual or complementary nature of light also applies to other subatomic particles like electrons, and it has now been understood through quantum mechanics that the state of every subatomic particle such as an electron, for example, can be described by a wave of probabilities.

Wave Function

Mathematically we have something that is referred as a wave-function due to Schrödinger. This is a mathematical equation that can be used to calculate the probability that the "particle" will turn up in a given location or state of motion if measured in some way. A particle can turn up in one place with a given probability and in a different place with another probability; we don't know where the particle's exact location will finally be. When left alone, a subatomic particle could be said to not exist bodily in space or time but simply has the potential for existence as a kind of wave. However, once we interrogate the particle by measuring its location, the particle is forced to show up so that in a sense an observation brings a particle into concrete existence. This idea is referred to as "the collapse of quantum states" as the probability distribution has "collapsed" to a single location rather than having a range of possible locations. An implication of this means that we can no longer think of an observer as being independent of the observation or experiment being carried out, as the observer is part of the system and effects the outcome. Dean Overman[19] argues that the observer cannot just be material as he or she would then consist of atomic particles satisfying the same probability laws. There must be a mind that knows the outcome of the experiment.

We have in fact three systems: the experiment, the observing or measuring device, and the observer who also acts in a sense like a measuring device, although the brain is infinitely more complex and consisting of atoms obeying quantum mechanics. Although this very small observer effect applies to so-called "quantum experiments" with subatomic particles, the quantum rules apparently apply to larger systems containing billions and billions of electrons

---

[19] Overman (2009: chapter 8).

for example with the result that an averaging process takes place so that our overall perception is actually described in terms of classical physics. Observed physical objects then appear to occupy definite locations even though quantum effects are essentially global as waves can spread out indefinitely. Henry Stapp stated that:

> There is, in fact, in the quantum universe no natural place for matter. This conclusion, curiously, is the exact reverse of the circumstance that in the classical physical universe there was no natural place for mind.[20]

It would appear that most physicists believe that, at least on a small scale, nature is not totally determined or totally predictable.

Quantum theory has some concepts that are counterintuitive and mind bending! We shall look briefly at just two to highlight the strange world of quantum mechanics, namely non-locality and entanglement. Our discussion is somewhat oversimplified, but it is hoped that reader can at least get a rough idea of these concepts. We have already considered the so-called *non-locality* nature of particles, with each particle behaving like a probability wave spreading out globally into space and time. We need to envisage the particle as not following a certain path but rather a family of possible paths, with each path having its own probability of taking place. Once it reaches the recording device and is observed, we know which path it has taken. With the double-slit experiment mentioned above, a particle effectively passes through both slits at the same time giving two families of paths that interact like two waves.

Entanglement

The notion of entanglement and arises when there are two or more particles. We know that each quantum particle such as a photon or electron has its own quantum state, but sometimes two or more particles can act on one another and become what is called an entangled system. When this happens, the group of particles can only be described by one big quantum state for the group as a whole, rather than as a bunch of separate quantum states put together; the particles are then said to be *entangled*.[21] If nothing else is acting on the particles, certain things have to stay the same both before and after the interaction when the particles become separated such as the properties of momentum, spin, polarization, and so forth. Heisenberg's uncertainty principle tells us that we can never know the exact state such as the momentum of a particle before we measure it or even the total momentum of the pair. What we do know is that with two particles the conservation law requires the total momentum of the two to remain unchanged when the particles interact. We therefore have to consider the pair as a single quantum system.

Quantum particles also have spin (angular momentum) so if one of the pair is measured and found to have a clockwise spin then the other particle, when measured at any subsequent time, will have an appropriately correlated

---

[20] Stapp (2009: 195).
[21] Technically the wave function of the group is a linear sum of the individual wave functions.

value, in this case a counterclockwise spin. As a further example, a photon spins horizontally and vertically (different polarizations) at the same time, but when you actually measure the photon, it fixes on a single state. With entanglement, when you measure one half of the entangled pair, the other half assumes the exact opposite state. Hence if one photon is vertically polarized, its entangled partner will be horizontally polarized. The particles are always connected and can behave as one. It thus appears that one particle of an entangled pair "knows" what measurement has been performed on the other, and with what outcome. This happens even when there is no known means for such information to be communicated between the particles.

When a laser beam is fired through a certain type of crystal it can cause individual photons to be split into pairs of entangled photons. Another example is the decay of a particle called a neutral pion into two entangled photons. We don't know their individual spins as either type of spin is equally likely, but if we measure one we know the other other will be opposite no matter how far away it is. What is striking about this is that the "correlation" is observed even though the entangled pair may be separated by an arbitrarily large distance consisting of many light years for example. Experiments have verified that this phenomenon happens even when the measurements are carried out more quickly than light can travel between the measurement sites. It appears to be instantaneous so how does one of the pair influence the other? No influence traveling slower than the speed of light between the measurement sites can pass between the entangled particles in time to create the "correlation" effect. Here one object affects another that is separated from it without any intermediate agency causing the interaction. This strange phenomenon is called "non-locality" or action at a distance and perhaps suggests that timelessness is somehow involved.

The theory behind non-locality is the violation of mathematical inequalities due to physicist John Bell in 1964, later clarified by physicist Bernard d'Espagnat in 1979,[22] and then added to by other authors; nature is non-local. A two-photon experiment in 1982 verified the inequality, thus demonstrating non-locality. One different example of this is the so-called Aharonov-Bohm effect in which an electrically charged particle is affected by an electromagnetic field even though the particle is in a region where there is no such field. Technically, the underlying mechanism is the coupling of the electromagnetic potential with the complex phase of a charged particle's wave function (got that?). Crudely it seems to be a matter of probabilistic fields of potential extending out into space and time and interacting before measurements are taken, after which the particles "materialize" and the quantum fields collapse. For example, a photon of light from outer space comes to the earth in the form of an ever-widening probability wave which reaches both me and my friend,

---

[22] d'Espagnat (1979).

who is distant from me. However, I see it first because I am looking up so that it becomes visible to me, but my friend then does not see it.

What is even more startling is that quantum physicists in Israel carried out an experiment in which they successfully entangled two photons that didn't even exist at the same time.[23] They created one photon, measured its polarization, destroyed it, and created another photon that turned out to always have the opposite polarization to the first. This showed that the two photons were entangled even though they never coexisted. Well that's enough mind bending for the moment. What is the relevance of all this? Clearly the quantum world is a strange world and we have to ask where it came from. Its existence has created a number of philosophical problems. For example, it has been used as an argument for multiverses, a topic discussed in Section 2.5.6. When an electron hits a target we don't know in advance whether it will bounce off to the left or to the right; we only have the unknown probabilities of p and $1-p$ for the two directions before impact. On impact, nature has to decide—left or right. However, if we observe the electron we will find out its actual direction and which of the alternatives is our actual world.

A multiverse view of this experiment is that prior to impact there are two identical copies of the universe, but on impact both copies of the electron and its observer split off to form two separate worlds, each observer thinking his or her world is real. Since we can apply this thinking to all particles, such subatomic activity will lead to an enormous number of possible universes, including multiple versions of us! (I could therefore be beside myself.) However, physicist Nicholis Gisin[24] argues that non-locality together with the existence of free will is incompatible with the multiverse view of quantum physics. Multiverses are discussed in Section 2.5.6

Brain Quantum Effects

Quantum effects will also occur in the human brain with the "firing" of neurons so that a certain amount of neuronal activity and the opening of ion channels are stochastic, that is involve probabilities. What standard neurobiology tells us is that tiny vesicles in the nerve endings contain chemicals called neurotransmitters (e.g., serotonin, adrenaline), and in response to an electrical impulse from a neuron (brain cell) some of the vesicles release their contents that cross the synaptic gap and transmit the impulse to the next neuron via calcium ion channels. What then determines the final state when probabilities are concerned? In 1986 Eccles proposed that the probability of a neurotransmitter release depended on quantum mechanical processes that can be influenced by the intervention of the mind. Although we still don't know how the mind influences the brain, quantum mechanics allows for the possibility of it happening.

---

[23] Megidish et al. (2013).
[24] Gisin (2013).

For a detailed (and complex) examination of this idea the interested reader is referred to Henry Stapp.[25] His picture of nature is that atoms are not "actual things" but the physical state of an atom or collection of atoms represents what he calls just a set of "objective tendencies" with certain probabilities of actually occurring; the process does not control the occurrence of actual things. He describes what happens in terms of two processes of which the first is a continuous, orderly, deterministic process controlled by fixed mathematical laws that are direct generalizations of the laws of classical physics and which control the probabilities but not the occurrence of the actual things. The second consists of a a sequence of what he calls "unruly quantum jumps" that although collectively following statistical rules are not controlled by any known law of physics.[26] Consciousness can therefore operate through this second process of indeterminism. He argues that the possibility of a particular brain event must be formed by unconscious brain activity before the event is actually selected, and once selected other possibilities are eliminated. Such an approach also allows for the role of directed attention and willful effort.[27] It highlights the interaction between the mind and brain.

Quantum Chaos

The subject of *quantum chaos* endeavors to link up the two topics of quantum theory and chaos theory discussed above. It is a complicated subject and has become a focussed area of research since the latter part of the twentieth century, but it is still in its infancy and there are problems in trying to define it. What it does is to look at the chaos that can exist in large (macroscopic) systems as discussed above but through the eyes of the laws of quantum mechanics that affect atomic (microscopic) systems. In general quantum mechanics behaves very much like classical mechanics with macroscopic rather than microscopic systems, as with quantum systems small changes tend to get averaged out. In fact Bohr's correspondence principle claims that classical mechanics is a limiting feature of quantum mechanics when objects become much larger than the size of atoms; any random effect tends to be swamped as systems get bigger. However, in trying to marry quantum mechanics and chaos theory together we run into a number of difficulties. We saw in Section 4.1.2 that because of the Uncertainty Principle we cannot know exactly both the velocity and position of a particle at any time. This means that we cannot talk about sensitivity to initial conditions as in chaos theory, nor follow the trajectory of such a process because of a random component.[28] Also John Polkinghorne, physicist and theologian, says,

---

[25] Stapp (2009).
[26] Stapp (2009: 41).
[27] The concept of willpower and its likening to a muscle that gets tired was investigated by Baumeister and Tierney (2011).
[28] For further comments see http://facultyweb.berry.edu/ttimberlake/qchaos/qchaos.html, accessed August, 2015.

chaos theory and quantum theory are incompatible with each other, since the former has a scale set by Planck's constant while the latter's fractal character means that it is scale-free[29]

This is a technical comment but it indicates that there are problems. Ilya Prigogine, author of a theory of dissipative structures in thermodynamics, considers that the universe is neither totally deterministic nor totally stochastic, contrary to materialism. Because of the unpredictability of chaos systems and the inherent uncertainty of quantum theory, Polkinghorne believes that it leaves room for God (and us) to act.

4.1.4 Relativity theory

Not only are there strange phenomena in quantum mechanics, there are also some strange phenomena relating to Einstein's theories of Relativity. In particular his Special Relativity is based on two postulates: (1) All motion is relative and (2) the speed of light is constant irrespective of the speed of the light source and the speed of the observer measuring it, which also leads to the postulate that nothing can move faster than light (called "locality"). The second principle seems contrary to commonsense as we might expect the light from a source moving towards us to have a speed consisting of the speed of light plus the speed of the source. From the two postulates it can be shown that if the length of an object is measured when the object is moving relative to me, the length is less that its measurement when the object is at rest relative to me. The faster an object moves relative to me the shorter it appears to get. Of course the object has to be going very fast, say close to the speed of light, for the difference in length to be noticeable.

If the object is a spaceship, we also find that time slows down inside the space ship relative to me. This throws a big question mark on what it means to have two events being simultaneous. The enigma of time is seen clearly in Einstein's famous twin paradox where one of the twins who is an astronaut flies near the speed of light into deep space while the other twin stays at home. When the traveling twin returns home, he discovers he is younger than his brother. (I need to get moving).

4.1.5 Short Comings of Materialism

What has the above discussion got to do with previous arguments about materialism? It first shows that materialism is in difficulty on three counts. First, because of quantum theory, is matter really real or does it become real only when observed by a perceiving mind? If the physical world exists independently of humans then perhaps it exists only in the mind of God, but its form is veiled from our understanding. Second, the world does not

---

[29]Polkinghorne (2012: 10).

appear to be deterministic as probability is now involved at a fundamental level. Third, there is no definite instant when all matter is simultaneously real. Because time can depend on the observer, the division of time into the past, present, and future raises the question of whether time is just the creation of our minds as temporal beings. It opens up all kinds of philosophical ideas with regard to God, eternity, timelessness, and time itself. For example, entanglement (cf. Section 4.1.3) suggests the existence of something that is timeless (a timeless dimension?), and is independent of space. At the speed of light time ceases to exist. It can then be argued that perhaps our minds can interact with this timeless dimension, and the transfer of thoughts can occur instantaneously and independently of time and distance.[30]

4.1.6 Uncertainty and Determinism

Continuing with the above theme, the relationship of quantum uncertainty with the concept of determinism is a matter of contention among physicists. Mathematician Ian Stewart[31] conjectured that perhaps quantum chaos might be responsible for the observed randomness in quantum mechanics so that everything might be deterministic. However, Polkinghorne questions whether we are right to stipulate deterministic laws with unpredictable behavior following from them. "Which is the approximation and which is the reality?" he asks.[32] He argues that deterministic laws may be no more than an approximation to reality[33] since we believe we experience free will and that it is not an illusion. However, because we are part of the universe, when it comes to predictions our efforts to predict something may interfere with what the universe is doing (as we saw in quantum mechanics above), and we get into a loop: we need to predict our efforts as they are part of the system, then we need to predict our efforts at predicting our efforts, and so on! A prediction can only be made by someone outside the system such as God, though it seems that God is playing a deeper game than a game of chance!

We are in deep (and controversial) waters here and I want to stay in the shallows! I guess that the point I am trying to make is that probability exists at a very fundamental level, bearing in mind that the Heisenberg Principle is perhaps more about unpredictability than randomness. We can still have causality but not predictability. One of the problems is that randomness is not an easy concept to pin down as philosophers sometimes misuse or don't define the word "chance." We saw above with the butterfly effect that a tiny input at some stage into a deterministic process can lead to unpredictability, so it would seem then that modern science is a mixture of deterministic and statistical ideas. It should be clear that I haven't actually proved the exis-

---

[30] Perhaps prayer works this way?
[31] Stewart (1997: chapter 15).
[32] Polkinghorne (1991: 41).
[33] Polkinghorne (2005: 35–36).

tence of randomness, only unpredictability. In the following sections below we address the question of how free we are to make decisions as the atomic particles in our brain are also subject to the uncertainty principle. (That's why I can't make up my mind!)

4.2 DETERMINISM, INDETERMINISM, AND SELF-DETERMINISM

After a long preamble we now consider the question asked in the title of this chapter about free will, and begin with the contentious topic of determinism versus free-will. How choices are made can be described under three categories: determinism, indeterminism (related to chance), and self-determinism (related to free will and the technical philosophical term libertarianism). Here determinism means that every action is determined by all previous events (and the laws of nature), and it can be labelled as naturalistic or theistic. Naturalistic determinism implies that it is a waste of time philosophizing and trying to change another person's view as he or she is not free to change. Also with determinism applied to the brain, thinking does not have a rational basis since everything is determined by non-rational forces so that determinism itself is irrational. Such determinism destroys human responsibility so that criminals are not responsible for their crimes, and it leads to fatalism or "whatever will be will be."

There are those that argue that (causal) determinism is somehow compatible with free will and moral responsibility, referred to as *compatibilism*, but their philosophical arguments are difficult to follow and some of the examples they use to prove points are somewhat esoteric.[34] They do point out that we believe ourselves to be free and we wish to be free from constraints such as those due to coercion, compulsion, and oppression, which is different from causal determinism. In this case free will means unencumbered choices.

Theistic Determinism

This type of determinism arises from the belief that God is the ultimate cause of all human actions (as in Calvinism). However, contrary to this, it has been argued that God is not bound by time although He entered time (became temporal) when He created the universe so it is possible for God to "see ahead" and know for certain what free people will do with their freedom so that their actions are only determined in that sense. God's sovereignty of the universe still holds along with free will. The apparent conflict between God's sovereignty and human free will arises because we are creatures of time but God is in some sense not. I return to this theological issue in Section 4.3 when I consider various views on the question of free will and God's sovereignty.

---

[34] See, for example, the four views of Fischer, Kane, Pereboom, and Vargas (2007), and also Mckenna and Coates (2015) in http://plato.stanford.edu/entries/compatibilism/, accessed August, 2015.

Indeterminism

Indeterminism is a rather nebulous concept relating to unpredictability. We have seen above from quantum mechanics that probability underlies the way we see our physical world, though some could argue that it does not rule out determinism as such because of the difference between unpredictability and randomness. In studies of the brain, neuroscience uses probability and stochastic processes (called stochastic neurodynamics) to model neural processes such as the release of neurotransmitters and the opening and closing of ion channels. This randomness or "noise" in our brain could have a positive function such as preventing deadlock in decision making. However, instability could lead to mental problems like cognitive decline or schizophrenia, while over-stability could lead to such problems as obsessive-compulsive disorder. As I have previously mentioned, our brains are every adaptable.[35]

If indeterminism is the main factor in our brains then we might believe that that there are some events, particularly some human actions or decisions, which have no cause. They could be caused by random quantum jumps in atoms and therefore just happen. According to this idea, some free choices could be uncaused events so that humans would not be morally responsible for their ensuing actions. This runs contrary to the principle of causality that states that all events have a cause, even if self-caused. Undetermined events in our brain or body would be contrary to having free choices as we would have no control.

Self-Determinism

What about self-determinism, usually referred to as libertarianism? As already mentioned, if we are not constrained in any way we have the strong belief that we have freedom of choice. There is also a popular belief in luck which, however, is fed by the gambling media. People with a gambling addiction may believe that a certain gambling machine is a "lucky" one and give machines personalities. A strong case for libertarianism is given by philosopher Robert Kane.[36] He comments that sometimes we are unsure which decision to make, but we are willing to stand by our eventual choice and take responsibility for the consequences. He argues that although there are times when we may believe we have no choice as to what decision we make, we are still ultimately responsible because of free choices made sometime in the past. Kane notes that "if any libertarian theory of free will is to succeed there must be some genuine indeterminism in nature to make room for it."[37] However, quantum indeterminacies tend to be small and, combined with their large numbers, tend to be "damped" out. They would therefore have negligible effects on the larger activity of the brain and body and, as noted above, indeterminism on its own does not lead to free decision making. Kane suggests that the theory of chaos, which we have seen as being regarded as deterministic, can be brought into the picture alongside quantum uncertainty and says:

---

[35] Doidge (2007).
[36] Fischer et al. (2007).
[37] Kane (2005: 133).

> There is growing evidence that chaos may play a role in the information processing of the brain, providing some of the flexibility that the nervous system needs to adapt creatively—rather than in predictable or rigid ways—to an ever-changing environment.[38]

Even if there is some indeterminism taking place in our brains caused by chaotic thinking combined with quantum effects to become magnified, Kane goes on to argue that we can still overrule it. Some upset childhood parts of the brain or so-called ego states can sometimes interfere with a person's adult daily living and lead to different parts of the brain having different or contrary agendas. Here psychotherapy/counseling can help such a person to freely make a decision to integrate those parts; something (inner child therapy) that I have used myself with counseling clients. Another issue is whether freewill is uncaused. My answer is no. God is an uncaused being and it can be argued that the power of freedom that we have is caused by God, but the exercise of that freedom is caused by the person.[39]

4.2.1 Libet's Experiments

Benjamin Libet carried out experiments in the late 1970s and early 80s that have stirred up considerable controversy about the nature of free will and whether it is actually "free." In his initial experiments, subjects were to choose when to flex their wrists while he monitored their brain activities. What he found was that before a person pressed the button there was some brain event involving an unconscious build up of electrical activity, a so-called "readiness potential" or RP that occurred about 550 milliseconds before the wrist moved in the supplementary motor area of the brain.[40] Also the subject had an awareness of his decision to move his finger about 200 milliseconds prior to finger movement. This means that there was a gap of about 350 milliseconds between the brain activity and the actual awareness. In further experiments Libet discovered that subjects still had the ability to veto their decision and not push the button even after the initial brain event, and they were aware of their decision to press the button. A subject's belief that the action occurred as a result of the subject's will was therefore in retrospect.

Although Libet himself considered the results were compatible with the existence of free will, others have used such experiments to say there is no such thing as free will and that the brain event caused the voluntary action. There is certainly no agreement about the meaning of the experimental results. For example, some believe that unconscious processes may play a bigger role

---

[38] Kane (2005: 134).

[39] For a philosophical overview of two-stage models, which shows the diversity of view points about free-will see
http://www.informationphilosopher.com/freedom/two-stage_models.html, seen August, 2015.

[40] This electrical phenomenon was discovered by Kornhuber and Deecke in 1964.

than previously thought in our behavioral responses, and that our conscious self is somehow alerted to what the rest of the brain and body are going to initiate.

Psychologists Judy Trevena and Jeff Miller carried out slightly different experiments where they compared the electrophysiological signs before a decision to move with the signs present before a decision *not* to move. They state:

> There was no evidence of stronger electrophysiological signs before a decision to move than before a decision not to move, so these signs clearly are not specific to movement preparation. We conclude that Libets results do not provide evidence that voluntary movements are initiated unconsciously.[41]

Their study suggests that the readiness potential (RP) signal does not represent a decision to move, but that it's just a sign that the brain is paying attention. In contrast, neuroscientist Patrick Haggard,[42] who referred to literature that distinguishes two different circuits in the brain that lead to action, namely a "stimulus-response" circuit and a "voluntary" circuit, suggests that researchers in applying external stimuli may not be testing the proposed voluntary circuit. Others have suggested that our behavior may also be the result of the previous training of unconscious habits. It could be that the power of intention and action can be independent of awareness.

A recent study[43] referred to research indicating that a gradual increase in neural activity preceding spontaneous movements appears to be a very general phenomenon, common to both vertebrates and invertebrates. Their experiments produced evidence that the readiness potential build-up seen before voluntary self-initiated movements does not necessarily determine the action, and they considered a two-stage model that closely predicted the exponential shape of the RP curve. One of the problems in assessing neurological experiments is that we cannot assume that only physical stimuli are affecting the mind.[44]

We might ask where did the initiating electrical brain activity come from? Does it have to lead to any particular action? Libet knew that there were very likely other times when the RP rose, but which did not lead to a flick of the wrist, so his experiment could not detect them. We note that only very simple behaviors have been studied such as flicking a wrist, and the free will decisions used have had a short time frame (seconds) rather than "thoughtful" decisions made over a much longer period of time. It can be argued that the above experimental results can be reconciled with a dualistic view. The mind works and thinks through the brain rather than independently of it, and because neural signals take time to travel at a finite velocity we can expect

---

[41] Trevena and Miller (2010).
[42] Haggard (2008).
[43] Schurger, Sitt, and Dehaene (2012).
[44] Goetz (2011).

there to be a time lag between the mind's decisions of them and the awareness of them. It takes time for a decision to surface in our consciousness, which is what Libet's experiments show, and such an approach supports the idea that decisions are conscious rather than unconscious, but delayed. There is therefore no reason to believe that that we are not free to make decisions; our mind just needs to be conscious of all the facts before making a decision.

We may compare what happens with our minds with the fact that we only see events in the past tense as light takes time to reach us; there is a delay. It is interesting that Libet wrote:

> As a neuroscientist investigating these issues for more than thirty years, I can say that these subjective phenomena are not predictable by knowledge of neuronal function. This is in contrast to my earlier views as a young scientist, when I believed in the validity of deterministic materialism.[45]

Because of quantum brain effects, the early activity could have some random components when various alternatives arise before a decision is made. We all have "random" thoughts at times (for example, so called "intrusive thoughts" discussed below), and dreams seem to contain some random elements. We know that when we sleep the brain mops up some of the "rubbish" both physical and psychological that accumulates during the day. Clearly further research will throw up interesting results that will no doubt create even more questions and debate.

Determinism and Logic

There are some arguments that show that there are problems with determinism. For example, if everything I do is completely determined, then I am not responsible for any of my thoughts or actions; there is no moral "I" only some sort of spatial "I". Therefore any conclusion that I come to must also be determined so that if I believe that I am determined then that belief must also have been determined. Therefore why should I value that belief more than any other belief? An argument against determinism due to physicist Donald Mackay[46] goes something like this. (You might like to skip next paragraph!)

In a few minutes time I have to make a decision between accepting or rejecting a consulting contract. Suppose that a detached observer who has complete knowledge of me and my brain states secretly predicts my decision to accept the contract without telling me what it is and writes my decision on a piece of paper. After I have made the decision he shows me the paper confirming that he was right. Now from the standpoint of the observer the outcome was inevitable. However if he told me the prediction in advance, my brain state would now change and he would have to take into account this added new information. Unfortunately no completely detailed description of the present or immediately future state of my brain could be equally accurate whether I believed the prediction or not. If it was accurate before I believed

---

[45] Libet (2004: 5).
[46] Mackay (1974: 78–79).

it, then when I believed it my brain-state would change so that the description would be out of date, so it would be a mistake to believe it as it is not based on all the information. Now my observer may be very cunning and make adjustments to his information so that he allowed for the effect on my brain state of my believing his prediction so that the changes to my brain state would make it correct if and only if I believed it. However, if I didn't believe it I still wouldn't be wrong as my brain would not be in the correct state for making the prediction. What it means is that I would be making up my mind in advance of the point at which it described me as doing so. However not all descriptions of my future are indeterminate like this. Mackay goes on to say,

> A description of a brain-state which is vague enough, or sufficiently far into the future, or sufficiently unrelated to the parts of your brain that would be affected by your believing it, may have as good (or almost as good) a claim to your assent as to that of a detached observer.

My personal view is that life is a mixture of determinism, indeterminism, and free will, namely limited freedom, though it depends to some extent on how you define free will, as mentioned above. There are times when I feel I have little or no freedom, but seem to be caught up in the circumstances of life where decisions are forced upon me and alternatives are limited or nonexistent. My experience as a counsellor has brought home to me how much our genetic background and our upbringing (nature and nurture) affect our current behavior. If someone knew our background very well, he or she could probably predict with high probability how we would act in a given situation; my wife certainly could do so with me! However I don't believe that the probability is one, which implies certainty. When faced with various expectations, which may be randomly generated, we have a choice.

For all of us the future is unpredictable for some of the reasons given above, so that we can be somewhat constrained by the unknown future. Events can happen to us that completely change the course of our lives, so we try to minimize this by actually limiting what we do sometimes. (As a statistical person I often say to my wife, "We need to minimize the risk.") In Chapter 2 we discussed the moral faculty that we have within us that brings a universal sense of right and wrong. This sense would be pointless if we couldn't choose!

An important aspect of free will that will be further discussed in Chapter 8 is that in order to exercise our freedom, which includes freedom to believe in God or not, our world must be a neutral place with regard to belief in God. There is a sense in which God is hidden and we have to make an effort by faith to find God.[47] There are clues given in the world that God has made but not conclusive proof, which allows us to exercise freedom of choice concerning God as I believe that God that given us this freedom. We must seek God to find God, though sometimes God finds us first!

---

[47] Hebrews 11:6.

Intrusive Thoughts

As an aside I want to to say something about so-called "intrusive thoughts" mentioned earlier in relation to free will. Most people, and not just those with Obsessive Compulsive Disorder (OCD), have intrusive and irrational thoughts at times. In fact "about 90% of the general population report intrusive thoughts in the absence of OCD and that the form and content of normal intrusive thoughts and obsessional thoughts is indistinguishable."[48] Examples of such thoughts are: inappropriate aggressive thoughts (e.g., about violence against the elderly, someone close, children, or animals; rude or abusive behavior); sexual thoughts (e.g., about sexual violence, sexual activity with all sorts of of people including children, or painful sexual practices); and blasphemous religious thoughts (e.g, about obscene images, blasphemous words or acts, or performing a ritual incorrectly). What counts, however, is the significance attached to the thoughts; usually they are dismissed as meaningless. In some ways such thoughts are a bit like random thoughts, though they may be linked to other aspects of our lives (e.g., a religious person may be more likely to have blasphemous thoughts). People with OCD, however, may have compulsive thoughts that lead to compulsive actions (e.g., checking and rechecking that the front door is locked, or repeated hand washing). They may feel locked into a deterministic process. However, they can learn to override such thoughts through Cognitive-Behavioral Therapy and take back control of their lives. This seems to me to be another good example of the mind being more than just the brain and of using free will to change the brain, as mentioned in Section 3.4.2 where OCD is discussed.

4.3 GOD'S SOVEREIGNTY AND HUMAN ACCOUNTABILITY IN THE BIBLE

A question that often comes up is that free will conflicts with the biblical idea that God knows everything, including the future. With regard to determinism and free will, the Bible holds in tension three ideas. The first is God's sovereignty, even with regard to our salvation (predestination).[49] The second, which runs right through the Bible, is that we are free to choose whether to follow God or not, and we are accountable for our own actions.[50] There are even some verses having both sovereignty and accountability together in the same verse.[51] The third, which is related to sovereignty, is that God has foreknowledge and is able to see into the future and control the future. We

---

[48] Marks (2003: 276).
[49] Genesis 45:7–8; 50:20; Proverbs 16:9; Matthew 10:29; John 6:44, 65; Acts 4:27–28, 13:48; Romans 8:29, 9:14–18; Ephesians 1:4–5, 11; Philippians 1:29; 2 Thessalonians 2:13–14: 2; Timothy 1:9; 1 Peter 1:2.
[50] Deuteronomy 30:19; Joshua 24:15; Luke 13:34; John 3:16–17. For a study of these ideas see Carson (1981).
[51] For example, Luke 22:22; Acts 2:23; 1 Corinthians 10:13; Philippians 2:12–13.

see this particularly in the book of Isaiah[52] where God declares the end from the beginning and brings things to pass that He had previously purposed.

In the New Testament we again have references to God working out His purposes.[53] From time to time God made predictions such as the Israelites going into captivity for four hundred years,[54] the eminent death of Moses,[55] the future destruction of some nations and cities (e.g., Tyre),[56] some of the actions of people before they were born,[57] and the details of Jesus' death and resurrection mentioned in Section 6.4.3.

Some might argue that because we are somehow in bondage to things that tend to drag us down[58] we are not free to respond to God. However, God calls on people to repent,[59] which means to change direction, and believe,[60] so we have the responsibility to respond resting with us. We note that salvation is not earned as such, but is a gift of God's grace,[61] and working for God is a partnership rather a dictatorship.[62] We see the interplay between God's overshadowing and freedom of expression in Christ's attitude to his own sufferings. On the one hand he was betrayed by one of his disciples Judas, was brutalized by soldiers, suffered the cowardice of Pilate, and experienced terrible pain and suffering, all of which would have been contrary to God's will and be the result of human sin and failure. All of this was foreseen[63] and foretold by Jesus, but Jesus faced it for the benefit of mankind even though he did not want to go through with it.[64] However, his concern was to carry out the will of God. On the other hand, Jesus told Pilate that Pilate had no power over him unless it had come from above.[65]

4.3.1 Four Answers

I now want to consider four possible answers to the apparent contradiction between God's sovereignty and human accountability.[66] The first is referred to as the open-theism view, where although God knows where history is going it doesn't necessarily mean that everything leading up to the prescribed end result is foreknown by God, though some of the future may be settled in

---

[52] For example, Isaiah 46:9–11, 48:3–5, 14:27.
[53] Ephesians 1:11; Romans 8:28.
[54] Genesis 15:13–15.
[55] Deuteronomy 31:16–17.
[56] Ezekial:7–21.
[57] 1 Kings 13:1–2; Isaiah 45:1.
[58] Romans 3:23, 5:12; Acts 2:39.
[59] Luke 13:3; Acts 2:38.
[60] John 3:16, 3:36; Acts 16:31.
[61] Ephesians 2:8–9.
[62] Philippians 1:12.
[63] Acts 2:23, 3:18; 1 Corinthians 15:3.
[64] Luke 22:42.
[65] John 19:11.
[66] Beilby and Eddy (2001).

advance. It is argued that God would not need to know every individual decision to make a prediction about a group of people. However, because God knows us so well individually,[67] God would have a very good idea as to how we would act under certain circumstances. This means that God would need to have perfect knowledge of the past and the present, and know about all the future possibilities. As I mentioned above, life seems to be a mixture of free will and determinism in that we live as though the future is partly settled and partly open. Open-theism supports this view of God.

Open-Theism

Greg Boyd, theologian,[68] gives the following arguments in support of open-theism:

God confronts the unexpected. God expresses His surprise three times at Israel's behavior in Jeremiah.[69]

God experiences regret. God regretted making man and making Saul king.[70]

God expresses frustration. God found no one to intercede on behalf of the land.[71] Also the Holy Sprit can be grieved[72] and be resisted.[73]

God speaks in conditional terms. God makes a large number conditional statements about the future (Jeremiah 18:1–10) and uses words like "may" and "if... then" when trying to convince Moses to be His representative in Egypt. There are too many examples to list so I just give a few instances. God gave Moses three successive signs to convince the elders in case the first two didn't work (Exodus 4:1–9). God instructed Ezekiel to symbolically enact Israel's exile as a warning so that they "perhaps" will understand (Ezekiel 12:3), but they didn't. Jeremiah was commanded by God to preach to the Judeans so that "may be" they will listen and then God may change His mind about bringing a disaster (Jeremiah 26:3). God did not lead the children of Israel via a shorter route near the Philistines in case the Israelites faced the possibility of war and as a result return to Egypt (Exodus 13:17). Finally, Jesus asked God if it were possible for him to avoid all that the crucifixion would entail, even though the events were predicted (Matthew 26:39).

God tests people "to know" their character. Abraham passed the test concerning his son (Genesis 22:12), and Hezekiah was allowed to do his

---

[67] Psalm 44:21, 139:1–6.
[68] Boyd (2001).
[69] Jeremiah 7:31, 19:5, 32:35.
[70] Genesis 6:6; 1 Samuel 15:10, 35.
[71] Ezekiel 22:30–30.
[72] Isaiah 63:10; Ephesians 4:30.
[73] Acts 7:51.

own thing so God could observe him (2 Chronicles 32:31). God also tested the Israelites.[74]

God changes His mind. There are numerous occasions when God is portrayed as changing His mind and showing flexibility, because He is gracious and merciful.[75]

It can be argued that the above passages are metaphorical and look at God through human eyes, i.e., they are anthropomorphic and not to be taken literally. In fact there are some passages that suggest that God does not change His mind,[76] and Boyd endeavors to interpret these passages as still fitting with the open-theism theme. It should be noted that because God acts in a certain way on occasions it doesn't mean that He always acts that way.

Divine Foreknowledge

A second contrasting view is the so-called Augustinian-Calvinist view, which is considered by philosopher and theologian Paul Helm[77] and which supports the concept of divine foreknowledge. The argument roughly says that if God is perfect He must know everything about the future, and if He doesn't He cannot be perfect. Helm argues that salvation and the resulting freedom is by God's grace alone. How his deterministic approach fits with human freedom is not clear. He distinguishes between what God causes and what God permits.

Simple Foreknowledge

The third view I consider, discussed by philosopher David Hunt,[78] is called the simple-foreknowledge view where God simply knows the future. His discussion is mainly philosophical and technical. He raises the question of how much control does God exercise even though He has the ability to totally control. We have already considered verses that support God's foreknowledge at the beginning of this discussion. In fact it is this ability that distinguishes God from false gods.[79] We see Joseph having access to God's foresight in the story of his life.[80] The Psalmist says that God knows completely what he is about to say.[81] Hunt mentions that one way God could have foreknowledge is that if He exists outside of time He can see what decisions we will freely make. This would imply that there is no before or after with God so that there is no

---

[74] Deuteronomy 8:2, 13:1–3; Judges 3:4.
[75] Exodus 32:14 and Psalm 106:23; Deuteronomy 9:13–29; Jeremiah 18:4–11, 26:2–3, 13, 16–19; 1 Samuel 21:27–31; 1 Kings 21:21–29; 2 Kings 20:1–6; 1 Chronicles 21:15; 2 Chronicles 12:5–8; Ezekial 4:9–15: Joel 2:12–13; Jonah 4:2; Amos 7:1–6.
[76] 1 Samuel 15:29; Ezekiel 24:14; Zechariah 8:14; Numbers 23:19.
[77] Helm (2001).
[78] Hunt (2001).
[79] Isaiah 41:22–23.
[80] Genesis 40 and 41, and 50:20.
[81] Psalm 139:4.

temporal instant when God knows everything, and we cannot use the words "at once" or "simultaneously" with God. In the end Hunt does not elaborate on how God might know the future and leaves this as an open question.

The idea of a timeless God probably originated with the 6th century philosopher Boethius, and was used by Aquinas as one of his solutions to the dilemma. Although it is also supported by a number of contemporary philosophers, there are some difficulties with this approach. For example, we have the problem of a timeless God who became involved with time at the instant He created the universe, and He continues to sustain the universe moment by moment or at least coexist with the universe. Further, according to Christianity, God entered into time through Jesus Christ, the incarnation of God. It is also argued that the idea of timelessness may seem to be incompatible with other religious characteristics of God such as God's personhood, though philosopher and theologian William lane Craig[82] argues that timelessness and personhood are not incompatible. Michael Robinson, philosopher and theologian,[83] presents a case for the timelessness of God that goes into considerable philosophical detail. He argues that God does not know the future before it happens, but simply knows the future as it happens in an eternal present.[84] Craig[85] on the other hand argues that God was timeless outside of creation, but became temporal with creation. So the arguments continue! The current view among philosophers seems to be that God is everlasting, but exists within time.[86]

Molinism

A fourth approach to the dilemma is discussed by Craig,[87] and is usually referred to as Molinism. It uses the notion of a "counterfactual" in which a statement is true, conditional on another statement being true, for example, a counterfactual of freedom is: "If I go into my computer I will freely choose to check my e-mail." Molina, a Jesuit theologian in the post-Reformation period, argued that God has three logically (not time) ordered moments of knowledge prior to creating the universe as follows:

(1.) God has "natural knowledge" in which He knows every possible state of affairs that can occur and all possible decisions that are open to free creatures, given a particular state of affairs. He knows, from an eternal standpoint, all things that are possible and are logically necessary. This includes all worlds He might create. Such knowledge is a natural attribute of God's existence. It is essential to God and does not depend

---

[82] Craig (1998c).
[83] Robinson (2001).
[84] His case is summed up on the last page of his book.
[85] Craig (2000).
[86] For a discussion of the various viewpoints see http://www.iep.utm.edu/god-time/, accessed August, 2015.
[87] Craig (2001b).

on the free decisions of His will. He knows what any free creature *could* do in any set of circumstances.

(2.) God has "middle knowledge" where He knows the counterfactuals of freedom and therefore knows what any free creature *would* freely do in any particular set of circumstances. This is not because the circumstances causally determine the creatures choice, but rather this is how the creature would freely choose, given the circumstances. God knows that if He brings about a certain state of affairs, certain consequences would definitely follow. For example, if an agent A was placed in circumstances C, then A would freely perform action X. However, even God in his omnipotence cannot bring about the alternative scenario that if A is placed in C then A will not perform action X. Middle knowledge is limited and does not include all possibilities. Middle knowledge might explain how a "random process" like evolution for example went down the right path to produce us.

(3.) God has "free knowledge" that consists of conditional (contingent) truths that depend on God's creative will. After God's free act of creating a particular world, He knows all things that are going to happen so He knows what every free creature *will* do.

Considering now our dilemma, if God has middle knowledge, then He knows what an agent A would freely do in a particular situation. For example, if agent A is placed in circumstance C, then A would freely choose option X over option Y. Therefore, if God wanted to accomplish X, all God would need to do was use His middle knowledge to bring into being the world in which A was placed in C, and A would freely choose X. God retains an element of providence without nullifying A's choice, and God's purpose (the actualization of X) is fulfilled. The Molinist therefore believes that God, using his foreknowledge and middle knowledge, surveyed all possible worlds and then actualized a particular one. God's middle knowledge of counterfactuals then plays an integral part in this "choosing" of a particular world and the middle knowledge must be logically prior to God's creative decree.

Molinists have argued that their position is supported biblically.[88] Craig[89] discusses 1 Samuel 23:6–12 and Jeremiah 38:17–18, which spell out very clearly the options God presents. He argues that although the test of a true prophet is the fulfillment of a prophecy,[90] some prophecies from true prophets can be regarded as counterfactuals. They are conditional statements and don't necessarily take place because people might respond to a warning and repent.[91] Craig also argues that many of Christ's statements seem to indicate

---

[88] 1 Samuel 23:6–12; Jeremiah 38:17–18; Ezekiel 3:6–7; Matthew 11:23, 12:7; Luke 22:67–68; 1 Corinthians 2:8.
[89] Craig (2001b).
[90] Deuteronomy 18:22.
[91] Isaiah 38:1–5; Amos 71–6; Jonah 3:1–10.

middle knowledge and cites a number of passages.[92] He notes that the most these texts indicate is that God has counterfactual knowledge. In order for this knowledge to be middle knowledge, it must be logically prior to God's free knowledge, something the Biblical texts I have mentioned do not seem to affirm or deny. We see that a combination of middle knowledge and God choosing a particular world leads to divine foreknowledge. Craig illustrates how this approach makes good sense of divine providence by referring to the preordained crucifixion of Jesus referred to in Acts 2:23 and Acts 4:27-28, whereby God is sovereign over the affairs of many individuals, especially Herod and Pontius Pilate. Craig concludes:

> Philosophically, omniscience by definition entails knowledge of all truth and, since counterfactuals of creaturely freedom are true logically prior to God's creative decree, they must therefore be known by God at that logical moment.[93]

A natural question to ask is how does God choose a world based on middle knowledge. Clearly there will be conflicting desired goals and some people will invariably make a personally detrimental choice, so the world created will be a compromise and exhibit suffering and evil. Does God then actualize evil or simply permit it? This question is taken up further in Chapter 8. Although the whole issue of God's sovereignty and human responsibility is complex and there are various versions of the above approaches, the reason for going down this path is to indicate that the problem is not unsolvable. Answers are available, even if hotly debated by philosophers!

4.4 CONCLUSION

This chapter asked the initial question of whether or not we have free will. By moving away from classical mechanics and materialism we can argue a case for the existence of free will. The developments in chaos theory and quantum theory open the door to the existence of unpredictability and probability in our world and in our brains. We saw in the previous chapter that materialism does not provide answers, and this chapter reinforces the ideas expressed there.

---

[92] Matthew 17:27, John 21:6; John 15:22-24, 18:36; Luke 4:24–44; Matthew 26:24.
[93] Craig (2001b: 143).

CHAPTER 5

IS THE BIBLE RELIABLE?

5.1 GENERAL PRINCIPLES

Given the above question, you will no doubt ask what is meant by "reliable." For Christians this raises all kinds of issues such as "infallibility" (e.g., infallible as originally written), "inerrancy", and "inspired by God", and there is the question of which translation should be used. Some would argue that the Bible is the word of God or contains the word of God, and this raises further questions about what this means. For me, reliability of the Bible means many things, including at the very least being historically accurate, being dependable in providing a faithful copy of the original documents, and giving accurate information about how God has interacted with the world. It also means being trustworthy in the sense that if we use the Bible as a guide book and follow Biblical principles it will enable us to live a life pleasing to God, and it also faithfully describes in the New Testament how we can obtain "salvation." As I do not wish to get caught up in the debate about words I shall let the evidence for reliability speak for itself.[1] There is a huge literature on the Bible so that at best I shall only be able to give just a broad overview with a focus on topics such as historicity and archaeology, and accuracy on how the texts were copied.

In reading the Bible, a number of general principles need to be taken into account. The Bible is a library of books, rather than just a single book, and it consists of 66 books covering a period of over 1500 years with a 400 year silence between the two testaments, apart from the collection of books called the Apocrypha. Although compiled by 33 authors, there is a central theme running throughout giving an amazing unity. It contains various kinds or genre of literature such as narratives, poetry, history, etc., in fact at least twenty, and it can be argued that it points to a divine mind inspiring its writers.

John Polkinghorne[2] explains why the Bible is so important to him in his religious life. For him it is not a divinely dictated textbook, but more like a laboratory notebook that gives practical ways in which God has disclosed Himself through the history of Israel and the life of Jesus Christ. It also shows a steadily growing understanding about God that reflects the culturally-determined viewpoint of the author. For example, in the early history there are described some disturbing incidents that I consider below where worshipping one God meant the extermination of unbelievers. Later we see God's mercy and love extended to all peoples. There are timeless truths, but also time-bound cultural aspects (like women should wear hats in church).

If the Bible is inspired by God as it claims to be, then we can expect that it has multiple layers of depth. For example, in one sense it is a historical compilation, but with a difference; it has been called "salvation history." For example, the Old Testament describes God's dealings with a chosen nation, the Jews, leading up to the arrival of Jesus Christ. Some critics have said that myths and legends gave rise to the Biblical stories and the Jewish religion in the Old Testament. However the reverse can be said to be true; actual miraculous events gave rise to the stories and the Jewish religion. The same is true in the New Testament; miraculous events in the life of Jesus gave rise to the Christian church. In particular, the bodily resurrection of Jesus from the dead was what led to Christianity arising from Judaism. New Testament scholar Tom Wright asked the question:

> Why did Christianity even begin, let alone continue, as a messianic movement, when its Messiah so obviously not only did not do what a Messiah was supposed to do but suffered a fate which ought to have showed conclusively that he could not possibly have been Israel's anointed?[3]

---

[1] For a discussion on the inerrancy of the Bible see Geisler (1999: 74–80), and for an appraisal of Biblical inspiration see Geisler and Nix (2012).
[2] Polkinghorne (2012).
[3] Wright (1998).

His answer was that Jesus, following his execution on a charge of being a would-be Messiah, had been raised from the dead. The apostle Peter sums it up when he says that they didn't make up clever myths when talking about the coming of Jesus, but they were eye-witnesses to the majesty of Jesus.[4]

Another feature of the Bible is that contrary to modern historical books, the space given to a particular historical period in the Bible does not depend on the number of years in the period but rather to the spiritual and moral significance of the period and the lessons the reader can learn from it.[5] For example, the last 14 chapters of the book of Genesis are given to the life of Joseph, while Mark's gospel focussed on just a three-year period with half his gospel devoted to the events leading up to and culminating in Jesus' last week. Another feature of the times was that it wasn't deemed necessary to tell the Biblical story in strictly chronological order (e.g., Joshua 1 is after the death of Joshua while Joshua 2 is before his death) or even quote people verbatim as long as the essence of what was said was preserved. There weren't even symbols for quotation marks in ancient Greek and Hebrew. It is therefore inappropriate for biblical critics to expect New Testament citations of the Old Testament to always be exact quotations.

One important aspect of the Old Testament is that it was written by writers who believed that God is sovereign and everything is due to God.[6] For example, we read that God tempted Abraham,[7] which really means that God tested Abraham by allowing him to be tempted by Satan.[8] The Lord's prayer asks God not to lead us into temptation,[9] whereas God does not tempt anyone.[10] We are told that God clogged or damaged the chariot wheels of the Egyptian chariots when they tried to follow the Jews across the parted waters of the Red Sea.[11] Although it says that God hardened Pharaoh' s heart,[12] we also read that Pharaoh hardened his own heart[13] or his heart was hardened,[14] so we see the interchange between free will and God's sovereignty, with God acting in accord with Pharaoh's free will.

God is also at work in nature[15] and gives life to everything.[16] In the New Testament this idea of sustaining all things shifts to Christ the Word of God.[17] We note in passing that this immanence of God rules out the need

---

[4] 2 Peter 1:16.
[5] This was typical of Jewish history at the time.
[6] Isaiah: 45:5-7.
[7] Genesis 21:1.
[8] Matthew 4:1-10; James 4:7; 1 Peter 5:8-9.
[9] Matthew 6:13.
[10] James 1:13
[11] Exodus 14:25.
[12] Exodus 7:3, 9:12.
[13] Exodus 8:15, 32, 9:34.
[14] Exodus 9:7, 34-35.
[15] See Job 9:5-6, 9:7, 26:8, 26:9, 38:22.
[16] Nehemiah 9:6; Act 17:25.
[17] John 1:3, 14: Colossians 1:17.

for any God-of-the-gaps arguments. Denis Alexander comments that "if all things hold together in Christ, then it is unclear why current human ignorance should have any particular theological relevance."[18]

In the Old Testament there are a number of passages where evil is attributed to God because of His sovereignty when in fact God allowed evil to happen to achieve His purposes. For example, we read that God sent an evil spirit to Saul in 1 Samuel 18:10, which means that God allowed an evil spirit to achieve His purposes, as Saul had already rejected God and God had rejected him from being king. Ultimately the action led to David becoming king. In 2 Samuel 24:1 we read that God was angry with Israel and moved David to number Israel, whereas in 1 Chronicles 21:1–4 it was Satan who moved David to number Israel with the consequence of many dying from pestilence. It appears that the reason for numbering was perhaps to gratify David's pride to find out the number of warriors he could muster for future conquest, or more likely to set up a taxation system to provide for the monarchy. The latter would have been regarded as tyrannical and David took full responsibility for his sin in doing this.[19]

In the story of Job, God allowed Satan to hurt Job in various ways and thus provide us with a story of immense value about suffering that has helped people down through the ages. We sometimes read that God repented of the evil he was going to do when a nation turned from evil. We see this in Jeremiah 18:5–11 where the word "evil" might be better translated as "calamity." The execution of God's justice is sometimes called "evil" by those undergoing it! God may discipline us for our own good, and we are not too happy about it.[20] It can be argued that using the word "repent" with regard to God, like the other terms above are anthropomorphisms, and it is the writer's way of trying to understand God's actions in human terms. It should be noted that the Old Testament shows a deepening understanding of the nature of God, as for example in the books of Amos and Isaiah. Revelation in the Old Testament was progressive.

We now look specifically at the two Testaments with regard to their reliability. This is a big subject, and because of its size I will be able to give only an overview. I note that with the passing of years the evidence for the reliability of the Bible has grown.

5.2 OLD TESTAMENT

We begin with the Old Testament and first take a look at early manuscript evidence of reliability. This leads on to looking at internal evidence with regard to authorship and we focus particularly on the first five books, the

---

[18] Alexander (2012 : 238).
[19] 1 Chronicles 21:8.
[20] Hebrews 12:11.

Pentateuch. Several other books that involved some controversy are then considered. How we got the so-called "canon" or list of books that we now have is described. Some archaeological evidence supporting the Old Testament is presented, along with a short diversion into the so-called "Bible codes."

5.2.1 Earlier Manuscripts

We do not possess the original Hebrew manuscripts of the Old Testament but must rely on hand-written copies of copies, especially as papyrus and skin deteriorate. Since copying by hand can introduce copying or intentional errors, we would want to know how much confidence we have in the accuracy of the copies. Added to to this problem, different copies have led to different "textual traditions" over the years. However, the Hebrew Bible was meticulously standardized and copied by a group of Jews called the Masoretes in the eight and ninth centuries giving us the Masoretic Text (MT). They used the best possible documents and, because of their extreme reverence for scripture, strived to preserve document purity using all kinds of very detailed checks (e.g., the middle letter of a book had to be the same otherwise a new copy was made). In fact the new copies were believed to be so accurate that older copies that were damaged or incorrect in any way were destroyed. This is not a promising start with the lack of earlier manuscripts and the huge time span between the original documents (1400-2300 BC) and the earliest existing copy of the MT (about 900 AD).

The discovery of the Dead Sea Scrolls[21] in caves by the Dead Sea in 1947 revolutionized our ideas about the Old Testament as we now have copies of Old Testament books that are a 1,000 years earlier. The scrolls date from the third century BC to the first century AD and are dated using such methods as carbon 14 dating, paleography (ancient writing forms), orthography (spelling), and archaeology. They provide an excellent opportunity to check up on how much copying the text has led to changes, as well as providing some insight into Jewish life and customs; Jesus is not specifically mentioned. Also, some of the fragments have been identified as the earliest known pieces of the New Testament. The scrolls were hidden by a Jewish Community called the Essenes at Qumran when their community was destroyed by the Romans during the Roman-Jewish War of AD 66–70. From the 40,000 inscribed fragments recovered, over 500 books have been identified with a 100 of these being copies of the 39 books of the Old Testament.[22] Much of this material pre-dates the community such as a text of Exodus dated about 250 BC.

Such was the accuracy of the scribes that the introduction of the Isaiah scroll for example led to only 13 minor changes in the Revised Standard Version of the Bible in 1952. By comparing the standard Hebrew Old Testament

---

[21] See Charlesworth (1992) for details.
[22] Only Esther is not represented among them.

used today with the Dead Sea Scrolls, theologian Gleason Archer found that the two copies of Isaiah:

> ...proved to be word for word identical...in more than 95% of the text. The 5% variation consisted chiefly of obvious slips of the pen and variations in spelling."[23]

If the text changed so little in its second thousand years, then we can have confidence that it changed very little in its first thousand years. Also there is no gap in the scroll between chapter 39 and 40. This and other arguments[24] support the fact that Isaiah is a single book and not two books, which strengthens its case as a prophetic document predicting before and not after some events.

A further conformation as to faithfulness of the Masoretic tradition is obtained by looking at the Septuagint and the Samaritan versions of the Jewish Pentateuch. The Septuagint (or LXX) is the Greek translation of the Old Testament made between 250–150 BC, and the Samaritan Pentateuch is in Hebrew. The oldest existing manuscripts date back to the fourth century AD. There are some Dead Sea Scrolls that can be identified as belonging to the separate textual traditions that produced the Septuagint and the Samaritan Pentateuch, respectively. These Scrolls can then be used as independent sources to check on the fidelity of the textual tradition behind the Masoretic Text. Both the Septuagint and the Dead Sea Scrolls reveal an amazing consistency with the Masoretic Text, and although there are variations indicating different traditions, overall they are essentially the same. All the major historical facts and almost all the minor details are the same. It could be argued that God was divinely protecting the Old Testament through thousands of years of copying and translating. Because of the need to establish dates when prophecies were made and the fact that the New Testament often cites the Septuagint, the reliability of the latter is important.

5.2.2 Internal Evidence

Although we have established above that there is good evidence for the faithful reproduction of the original manuscripts, the next question to consider is whether the contents are historically reliable. Because the Old Testament covers a huge period of history it is not possible to go into a detailed examination in this chapter but simply take a broad brush approach.

Let us begin with the first five books or the Pentateuch (also called the Books of the Law or Torah), which according to internal claims[25] and Jewish and Christian tradition, were written by Moses, except possibly for the last

---

[23] Archer (1994).
[24] Geisler and Howe (1992a: 265–266).
[25] See Exodus 17:14, 24:4, 34:27; Leviticus 1:1, 6:8; and Deuteronomy 1:1, 4:44, 29:1, 31:9, 31:24–26 in the Pentateuch. Also Joshua 1:7–8, 8:31–34, 22:5; Judges 3:4; 1 Kings 2:3; 2 Kings 14:6; 2 Chronicles 34:14; Ezra 3:2, 6:18; Nehemiah 1:7, Daniel 9:11; and Malachi 4:4.

chapter of Deuteronomy concerning his death.[26] The unity of the Pentateuch suggests a single author, and Jesus referred to the Pentateuch attributing the source to Moses.[27] Luke, in Acts 3:22, comments on a passage in Deuteronomy 18:15 and credits Moses as being the author of that passage. Paul, in Romans 10:5, talks about the righteousness Moses describes in Leviticus 18:5. We also note that Moses was most unlikely to have been illiterate and would have certainly been up to the task, having been well educated in an Egyptian court. Because most of the many miracles in the Pentateuch are said to have been carried out through Moses or in his presence, the anti-supernatural critic has to either accuse Moses of lying or deny that Moses was the author, the latter being the usual path chosen. In this case it was proposed that the Pentateuch is a compilation of four different documents (whose physical existence has not been established) labelled J, E, D, and P, all composed long after Moses died and brought together by an editor. This is known as the JEDP theory or documentary hypothesis and is still popular among academics.

Support for the JEDP theory comes from a number of features of the Pentateuch such as the existence of duplicate and triplicate passages about the same event. For example:

- Two creation accounts in Genesis 1 and 2.

- Two interwoven flood stories (Genesis 6 and 7).

- Two stories about the naming of Isaac (Genesis 7:19, 21:3).

- Two flights of Hagar (Genesis 16:4–14, 21:9–21).

- Two stories about the renaming of Jacob as Israel (Genesis 32:28, 35:10).

- The twice naming of Bethel (Genesis 28:19, 35:15).

- The twice naming of Beersheba (Genesis 21:31, 26:33).

- Three versions of Abraham and Isaac denying their wives (Genesis 12:10–20, 20:1–18, 26:6–11).

- Two versions of the Ten Commandments (Exodus 20 and Deuteronomy 5).

- Two accounts of Moses striking the rock at Meribah (Exodus 17:7, Numbers 20:13).

In answer to this, narratives of the same event may simply provide different perspectives on the same event rather than being contradictory. Also similar acts may be repeated but under different circumstances, as for example in

---

[26] This was more than likely to have been written by Joshua.
[27] See, for example, Mark 7:10, 12:26; Luke 20:37; John 5:46.

the cases of Bethel and Beersheba, where the second narrative refers back to the first. A key argument against multiple documents is that the duplications don't always occur in the different JDEP documents but can occur in the same document. There is also the question of how did the so-called scribe select the different parts of the documents to combine.

Other comments supporting the documentary hypothesis relate to the use of parenthetical remarks,[28] but these do not necessarily mean the result of later additions or a different author, or if so, deny the "essential authorship" of Moses. (I sometimes have a sentence in parenthesis, though these days we tend to use footnotes.) There is also the question of the use of different names for God (e.g., Yahweh and Elohim) and different writing styles suggesting different documents. There are a number of problems with this. First, different names are sometimes used in the same sentence or paragraph, and different names can indicate the different roles of God. In the Qur'an, known to have one writer, Muhammad, two different words are used for the name of God (Allah and Rab) in different "chapters." Second, variations in writing style do not necessarily mean different authorship, as style can vary with subject matter and over time.[29] It can be argued that we shouldn't be surprised if Moses had a different style when he is writing history (Genesis), writing legal statutes (Exodus, Deuteronomy), and writing intricate details of the sacrificial system (Leviticus). For example, we have census lists, an organization manual for encampment and march, regulations for the priesthood and Levitical orders, laws of sacrifice and ritual, instructions about the conquest and division of the land, laws regulating inheritance, poetry, songs, and the various prophetic oracles of Balaam.

Late place names can be explained as later interpolations, where later copyists would update some place names for the sake of the reader.[30] Clearly the problem of authorship is a big subject leading to endless microscopic analyses and debates. What is important is that although the Jews were not learned in science or philosophy and prone to idolatry, they generally produced an advanced doctrine of God, which provides support for the supernatural nature of Moses' teaching.[31]

I now want to look specifically at some of the Old Testament books. It is not my intention to go through them all but rather focus on the ones where there have been some questions about authorship.

Genesis

Beginning with Genesis, which was before the time of Moses, it would be reasonable to assume that Moses, as a leader, compiled Genesis from family

---

[28] For example Deuteronomy 2:10–12 and 2:20–23.
[29] This has been my own personal experience of over 40 years of writing, having written on Statistics, Counseling, and the topics in this book.
[30] For example, Joshua 14:15.
[31] For further discussion see, for example, Geisler (1999: 586–588).

documents and oral history as there are several references to family genealogies and records throughout the book.[32] The question of how to interpret the first few chapters of Genesis is discussed in Sections 5.2.5 and 8.9.1.

Exodus

Exodus is all about Moses and has the marks of eyewitness testimony to events like the crossing of the Red Sea, the receiving of the commandments, and the desert wanderings. The author shows a detailed knowledge of the wilderness geography that Moses would have had after 40 years as a shepherd and 40 years as a national leader in the region. We note that Moses had an anger management problem when he killed an Egyptian, and then spent 40 years languishing in the land of Midian. In the story of the burning bush we see Moses reluctant to obey God in going to the Egyptians to obtain the release of Israel and in putting up all sorts of excuses. We also have some comments about his time management problems.[33] Given the importance of Moses in Jewish history, it is unlikely that later writers would include such unflattering details if they were trying to produce a legend about Moses. It sounds like an honest personal account. Events from every one of the first eleven chapters are referred to by Christ and New Testament writers.

Leviticus

Leviticus, which forms a connection between the narratives of Exodus and Numbers has been ascribed by some to the priestly or P version of the JEDP editors; for example chapters 1–16 are ascribed to Moses while chapters 17–26 (the so-called Holiness Code) are ascribed to a priest at a later date. However, we note that Moses is usually the recipient of the frequent phrases "God said" or "God commanded." Although Moses is frequently referred to in the New Testament, there are few actual references to verses in Leviticus.[34] According to Leviticus 26:46 and 27:34, God revealed some contents of Leviticus to Moses on Mt Sinai. Leviticus can be regarded as a kind of reference manual for Levitical priests. However, many people struggle with this book as it has some hard sayings such as the deaths of Aaron's sons in Leviticus chapter 10, all the strange kinds of offerings, some of the unusual ordinances especially relating to purity and sexuality, and the apparent acceptance of slavery and rape. Some of these questions are discussed later.

The reader needs to take into account the culture of the day and the reason why the book was written. The nation of Israel was surrounded by hostile nations, and time and time again Israel drifted away from the worship of God (Yahweh) through intermarriage and contact with these nations. God wanted to set apart Israel as a holy nation; judicially, ethically, culturally, and geographically. Through Moses, God introduced a number of strict regulations

---

[32] For example, Genesis 2:4; 5:1; 6:9, and so forth.
[33] Exodus 18:13–27.
[34] For example Jesus acknowledged the authorship of Moses in Matthew 8:4 when he referred to Leviticus 14:2–4.

applicable to just Israel to preserve the nation in many ways such as preserving the national health with regard to eating and hygiene, the preservation of land inheritance, and the good treatment of foreigners living in their boundaries. Penalties were tough as Israel had a past history of continually wandering off the straight and narrow.

Most important of all was the emphasis on holiness[35] through the complex Jewish sacrificial system involving five main offerings—burnt, grain, peace, sin, and guilt offerings. The idea of holiness appears eighty seven times in Leviticus, and sometimes it refers to ceremonial holiness (ritual requirements) and at other times to moral holiness (purity of life). The first fifteen chapters deal with sacrificial principles and procedures relating to the removal of sin (priestly holiness), and the last eleven chapters emphasize ethics and morality (practical holiness). Four times in Leviticus God commanded the people through Moses to be holy as God is holy. It is interesting to note that archaeological research has found that all of Israel's neighbors in the ancient Near East had some form of sacrificial worship to appease their gods, using similar sacrificial terminology. Sacrifice was therefore part of Israel's cultural environment as it was established long before the time of Moses.[36]

Israel's system was however very different as it was based on their covenant relationship with God. Without going into details, a big difference was the fact that their neighbors had no concept of "holiness", which was a huge part of Israel's sacrificial system: there was an absence of degrading elements. Keeping the ceremonial laws was not so much a matter of earning salvation, but were rather a way to belong to God's people. Actually God preferred Israel not to sin at all so that sacrifices would not needed.[37] In Isaiah 1:11–17 God says he is tired of burnt offerings and the rituals, and wanted the people to live righteously instead.

According to Gary North, economist and historian,[38] Leviticus is about boundaries, with most of Israel's economic boundaries based on geography (land laws), tribal membership (seed laws), and ritual requirements (laws of sacrifice). Also, the system of offerings had an economic impact on the Israelites, as adults who were circumcised males living inside the land's boundaries had to make several visits every year to a local center and in particular to Jerusalem three times where they would spend some time. The sacrifices and the traveling were very costly in time and money, particularly as most people either travelled by donkey or walked. During this time God promised to protect their borders from invaders and wild animals and bless their land

---

[35] Leviticus 19:2, 20:7.
[36] For example, the story of the offerings by Cain and Abel in Genesis 4:3–4; Noah's offering in Genesis 8:20; Abraham's offerings in Genesis 22:13; Jethro's offerings in Exodus 18:12; and offerings in Job 1:5, 42:8.
[37] Jeremiah 7:21–23; Psalm 51:16–19; Micah 6:6–8; Amos 5:21–24.
[38] North (1994).

with rain and good crops if they kept God's commandments.[39] As a result, the ceremonial laws led to the Israelites moving off the land into the cities and from there out to the world.

What about the New Testament? We find that offerings were made at the time of Christ.[40] After the death of Jesus, offerings were no longer needed,[41] although participation was not prohibited.[42] We note that not everything in Leviticus is about ritual as many of the requirements are valid today: for example in chapter 19 a number of things are prohibited such as theft and lying, defrauding a neighbor, rendering unjust judgement, partiality, slander, grudge bearing, making a prostitute of one's daughter, the mistreatment of resident aliens, and using false weights and measures. Finally, the theology and symbolism of the sacrificial system play an important part in understanding the theology of the New Testament.[43]

Numbers

The Book of Numbers essentially bridges the gap between the Israelites receiving the Law (Exodus and Leviticus) and preparing them to enter the Promised Land (Deuteronomy and Joshua). It is about both the severity and the goodness of God; severity in the eventual death of the rebellious and complaining generation who also refused to undertake the conquest of Canaan so that they never entered the Promised Land, and the goodness of God in protecting the new generation as they possessed the land. The people challenged the authority of Moses, rejected the spies reports about Canaan, rejected the priestly role of Aaron, and committed idolatry and immorality concerning Baal Peor. Consequently there were some traumatic deaths resulting from God's judgment in this book because of the rebelliousness and idolatry of the people, as explained by Paul in 1 Corinthians10:6–12 and by the author of Hebrews in 3:7–4:13, who describes them as warnings.[44] Even Moses was not exempt from God's judgment.[45]

Although it seems unlikely that Moses wrote Numbers 12:3 about his humility, and there is only one reference to Moses recording material, namely the itinerary of the Israelites in their desert trek from Egypt to Moab (his "log" book),[46] there is no reason to doubt that this book was mainly written

---

[39] Leviticus 26:3–6; Deuteronomy 28:1–14.
[40] For example, there were burnt and sin offerings made by Mary and Joseph (cf. Leviticus 12:1–8) and in Luke 5:12–14 Jesus commanded a healed leper to give an offering in accordance with the law.
[41] See the book of Hebrews.
[42] For example Paul's participation in Acts 21:18–26 as he did not want to create a stumbling block for Jews., cf. 1 Corinthians 9:20; Acts 16:1–3.
[43] See, for example, https://bible.org/seriespage/1-learning-love-leviticus by Bob Deffinbaugh, accessed December 2012.
[44] The incidents are described in Exodus 32:4–6 and Numbers 25:1–18, 21:5–6, and 16:41,49.
[45] Numbers 20:19 and Deuteronomy 34:1–8.
[46] Numbers 33:2.

by Moses with possibly some later priestly additions.

Deuteronomy

With regard to Deuteronomy, Moses is identified as the book's author in 1:1 and 31:9, and his name appears around forty times, as well as there being a continued use of the first person. There is widespread consensus that the book is structured after well-known ancient Near Eastern treaty forms, but with added exhortations and other additions that might belong to a farewell sermon of Moses. The New Testament contains more than 80 quotations from Deuteronomy, and Jesus often quoted from this book identifying the words as the words of Moses. For example, we have the verse about loving God totally[47] as well as the command from Leviticus to love your neighbor as yourself.[48] Jesus also quoted scripture four times during his temptation with three quotes from Deuteronomy.[49] Moses was the first to prophecy about the coming of the Messiah, a prophet like Moses himself,[50] and he also prophesied the scattering of Israel followed by its repentance and collective return to its own country.[51] He was the only person that Jesus compared himself with.[52]

Joshua

Although the Book of Joshua does not explicitly name its author, there is no reason to suppose it was written by anyone other than Joshua the son of Nun. We read that God spoke to Joshua as he did to Moses. The book describes the entering and conquering most of the Promised Land and redistributing it among the tribes. It should be noted that the Edomites, Moabites, Ammonites, Ishmaelites, Midianites, and Amalekites were close relatives of Israel, all named after their forefathers. The Amorites and the Canaanites were, however, not related to Israel, and were not even Semitic in origin, coming from Noah's son Ham. Although some Semites retained a knowledge of the true God, the Hamites had corrupted themselves with abominable religious practices, including infant sacrifice and ritual prostitution. Understanding this background history helps explain some of the activities and God's judgment on some of the nations mentioned in the book.

Judges

We are not sure who wrote the book of Judges, Samuel or Solomon, though tradition sides with Samuel. Choosing Samuel as the author explains the detail known, while Solomon as the author explains why it was written as a recollection of a distant event. One of the key lessons of Judges is that in spite

---

[47] Deuteronomy 6:5; Matthew 22:37; Mark 12:30; Luke 10:27.
[48] Leviticus 19:18; Matthew 22:39; Mark 12:31; Luke 10:27.
[49] Matthew 4:4, Luke 4:4, Deuteronomy 8:3; Matthew 4:7; Luke 4:12, and Deuteronomy 6:16; Matthew 4:10; Luke 4:8; Deuteronomy 6:13.
[50] Deuteronomy 18:15 and Acts 3:22–23, 7:37.
[51] Deuteronomy chapter 30.
[52] John 5:46–47.

of Israel continually being disobedient and idolatrous, God took them back by raising up fifteen consecutive judges when they repented. This happened again and again. The lesson from this is that God is faithful to His covenant with Abraham and shows patience and long suffering.[53]

Isaiah

Isaiah is said to have been written by more than one person, in particular by two (or three) people at different times because of a perceived difference in style between the first 39 chapters and the last 27, with the first part containing some personal details; the so-called "Deutero-Isaiah" theory. Some scholars have tried to use this argument to explain a fulfilled prophecy concerning Cyrus mentioned in Section 5.5. A strong argument against this theory is the finding of the book of Isaiah as a book without a break among the Dead Sea Scrolls and also its place in the Septuagint, translated 300 years earlier. The apostle John quoted from both chapters 53 and 6 (John 12:38, 40) and linked the two in verse 39 with the words "For Isaiah again said." Jesus quoted verses from both sections of the book and attributed both to Isaiah.[54] Some of the differences can be explained by the different subject matter; chapters 1 to 35 are preparatory chapters that warn of the Assyrian menace, with chapters 36 to 39 forming a transition to chapters 40 to 66. There are common phrases in both parts: for example the title "Holy one of Israel" is found twelve times in the first 39 chapters and fourteen times in the remaining chapters. Norman Geisler[55] notes a number of similar pairs of verses. [56] Whatever one's view of the theory it does not detract from the prophecies concerning Jesus.

Daniel

Daniel was a noble Jew exiled in Babylon and the book is about his activities and visions. Chapters 1–6 contain six tales of Jewish heroism set in the Babylonian court, and chapters 7–12 contain four apocalyptic visions. Chapters 1 and 8–12 are in Hebrew and 2–7 in Aramaic. The story of Daniel begins when Nebuchadnezzar, king of Babylon, besieged Jerusalem during the third year of Jehoiakim king of Judah in 605 BC. Because of battle details given later in the book (e.g., chapter 11) and other arguments about supposed inaccuracies or errors, a later date of the second century BC has been given for when it was written. However, there is pervasive evidence for the earlier date of the sixth century BC.[57] For example, the earlier date is supported by Josephus in his Antiquities and by the Jewish Talmud. Fragments of Daniel from possibly the second century were found among the Dead Sea scrolls,

---

[53] Psalm 86:15, 103:8–18.
[54] For example Mark 7:6–7 and Matthew 12:17; he quoted from the Septuagint, which differs slightly from the Hebrew version.
[55] Geisler (1999: 367).
[56] Isaiah 1:15b and 59:3a; 28:5 and 62:3; 35:6b and 41:18.
[57] Geisler (1999: 178–180) and in particular http://www.jeramyt.org/papers/daniel.html (accessed August, 2015) for a helpful summary.

and being a copy would indicate an earlier date for the original document.[58] Ezekial, who is regarded even by the critics as being contemporary with the sixth century BC, refers twice to an important person, on par with Noah and Job, called Daniel and once refers to Daniel's wisdom.[59] If this is the same Daniel, which is a reasonable assumption, then if he came from the sixth century BC there is no reason to reject his prophecies as coming from the same period as well.

Belshazzar was mentioned as the last king of Babylon[60] when secular sources originally stated that Nabonidus was the last king of Babylon. However, the discovery of the Nabonidus Chronicle[61] showed that Daniel was correct and that Nabonidus entrusted the kingship to his eldest son Belshazzar[62] who was there when Babylon fell in 539 BC. Since all knowledge of Belshazzar was lost by at least 450 BC until the discovery of the Chronicle, how did Daniel know about this unless he was there, if he is given a later date.[63] This earlier date would indicate that Daniel was writing prophecy, not history, thus supporting the supernatural nature of the predictions. Even a later suggested date of about 170 BC would still imply some predictions that were later fulfilled. Other criticisms of Daniel have been replied to from more recent findings already footnoted above. Finally we note that Jesus confirmed that Daniel was a prophet and used the example of a prediction made by Daniel that was yet future in Jesus' day.[64]

5.2.3 Old Testament Canon

My reason for including this topic is that critics of the Bible have suggested that the books of the Old Testament have been arbitrarily put together and the church cannot agree as to which books should be in the canon. Anything that is a bit different is left out! If God inspired the Old Testament, then it seems reasonable that he would make sure that the right books were selected. This is a big topic so I can only give a brief overview.

The inclusion of which books should be in the Old Testament was not a decision made lightly. Many books were immediately accepted as authentic while others took some time, with the belief that the Holy Spirit of God guided the scholars. According to the Jewish historian Josephus, the main criterion for selection seemed to be whether a book was written by a prophet

---

[58] http://www.biblearchaeology.org/post/2012/07/31/New-Light-on-the-Book-of-Daniel-from-the-Dead-Sea-Scrolls.aspx#Article, accessed August, 2015.
[59] Ezekial 14:14, 20; 28:3.
[60] Daniel 5:30.
[61] It apparently surfaced in the 19th century.
[62] cf. the Nabonidus cylinder from Ur.
[63] Unless of course there were other documents that have since been lost.
[64] Matthew 24:15.

whose prophecies turned out to be correct. Other criteria (for both the Old and New Testaments) are related to:[65]

(1.) Divine confirmation of a prophet through the performance of miraculous acts.

(2.) The prophet tells the truth about God and agrees with previous revelations.

(3.) The writing had transforming power.

(4.) The book was accepted by the people of God and in particular by those to whom the book was addressed.

The Old Testament books were organized as the Law, the Prophets, and the Writings,[66] and were generally accepted by the Jewish community in the Hebrew Bible (Tanakh). The books were dated from approximately the 13th - 3rd centuries before Christ, except for a group of books known as the "Apocrypha." The Jewish canon was compiled by the Talmudic sages at Yavneh around the end of the first century after Christ. The Old Testament apocryphal books, written during the period between the two testaments and referred to as "Deuterocanonical" (meaning "secondary canon"), are placed between the New and Old Testaments in some Protestant Bibles and interspersed with rest of the text in Roman Catholic Bibles. There are a number of other apocryphal texts that did not make it into the canons as they lacked authority and divine inspiration, though they shed some light on the Bible and its history and have some devotional content. There were some other omitted books called "pseudepigrapha" (meaning "falsely attributed") that were purported to have come from biblical characters. Apocryphal books were included in the Greek (Septuagint) and Latin (Vulgate) translations of the Old Testament. However Josephus, a famous Jewish historian, did not support the inclusion of the apocryphal books.

How did the New Testament writers treat the Old Testament canon? The apostles often used the Septuagint in their quotations as that translation was generally known to those in the church.[67] However, the apostles never quote from the apocryphal writings, and there is no reason to believe that they regarded them as Scripture. The New Testament as a whole quotes from 34 of the Old Testament Books, while Jesus quoted from 24 different Old Testament books[68] using a variety of textual sources.[69] There is strong evidence that the

---

[65] Geisler (1999: 81–85).
[66] Often the books were divided into just two collections, the Law and the Prophets, though sometimes the Psalms were highlighted separately.
[67] For a list of quotes see http://www.bible-researcher.com/quote01.html, accessed August, 2015.
[68] For a list of quotes see http://www.jewsforjesus.org/publications/newsletter/2008_09/05, accessed August, 2015.
[69] See http://www.bible.ca/b-canon-jesus-favored-old-testament-textual-manuscript.htm, accessed August, 2015.

apocryphal books are not prophetic and none of them claim to be written by a prophet which, as already noted, is a major criterion for canonicity. Not once is one of the books cited authoritatively by a prophetic book written after the apocryphal books; neither Philo, a well-known Jewish teacher, and the historian Josephus refer to them as scripture. Even the Jewish community that gave rise to these books acknowledged that their prophetic line ended after the fourth century BC before the apocryphal books were written. The New Testament regards the Jews as the custodians of scripture and their canon,[70] and the Jews rejected the Apocrypha.

5.2.4 Archaeological Evidence for the Old Testament

There is a huge amount of archaeological evidence supporting the historicity of the Old Testament (OT), and a large book and could be written about the thousands of finds supporting the biblical record of events. Several historical identities, for example Sargon (Isaiah 2:1), Belshazzar (Daniel Chapter 5), Nebo-Sarsekim (Jeremiah 39:3), and certain peoples (the Hittites and the Canaanites) were unknown outside the Biblical records for many years, and their existence was doubted until archaeological finds proved otherwise. Forty kings listed in the Bible between 2000 BC and 400 BC are judged by external sources to be in correct chronological order. Many a time the OT was thought to be wrong, especially in the 19th century when skepticism was rife, but new evidence would come to light verifying the events recorded and providing increasing corroboration. For example, the Moabite stone discovered in 1868 and now in the Louvre, records the defeat of Omri, King of Israel whose God was Yahweh, by Mesha king of Moab; the conduit built by Hezekiah[71] was rediscovered in 1880; and the shaft up which Joab climbed[72] was discovered and subsequently climbed by British officers in 1910.

There is increasing support for the existence of the patriarchs (Abraham, Isaac, and Jacob), the destruction of Sodom and Gomorrah (mentioned below), the crossing of the Red Sea in the Gulf of Aqaba (not the place originally thought),[73] the conquest of Jericho,[74] the existence of king David, the Jewish captivity, and Solomon's temple, to name a few major events, as well as many minor details.[75] There are so many documents from ancient cultures (e.g., ancient Egyptian, Hittite, Canaanite, Assyrian, and Babylonian cultures) which parallel accounts from the Old Testament that Rabbi Nelson Glueck, a prominent Jewish archaeologist, stated,

---

[70] Romans 3:2.
[71] 2 Kings 20:20.
[72] 2 Samuel 5:8.
[73] Remnants of chariots have been found on either side of the gulf in deep waters.
[74] See Wood (1990).
[75] For a number of references see "Old Testament archaeology" on the internet.

> No archaeological discovery has ever controverted a biblical reference. Scores of archaeological findings have been made which confirm in clear outline or exact detail historical statements made in the Bible.[76]

In fact Glueck used the Bible as a guide to find over 1000 ancient sites in Transjordan and 500 more in the Negev.

5.2.5 Early Chapters of Genesis

A question often raised is how much of the early chapters of Genesis, if any, is mythological rather than historical. Some have tried to to argue that the Genesis creation story arose out of earlier mythical versions such as the Babylonian and Sumerian creation stories found in the ancient Near East. However these latter accounts are completely fanciful, describing creation as the result of a conflict among finite gods. Normally a later account would be embellished, but instead the Biblical account is simple, monotheistic, and free from embellishment suggesting that the other accounts are more likely to be corruptions of the Genesis story. Support for this idea comes from the recent discoveries of creation accounts at Ebla. Sixteen thousand clay tablets from the third millennium BC were discovered at Ebla in modern Syria. They predate the Babylonian account by some 600 years and provide support for the Genesis account, mentioning Adam, Eve, Noah, and other Genesis names. In particular there is the belief in creation out of nothing by one God that fits with today's understanding of how the world began, and as astronomer Robert Jastow comments, "Now we see how the astronomical evidence leads to a biblical view of the origin of the world."[77]

The theory that monotheism evolved out of polytheism is not supported as the Ebla tablets indicate that monotheism predated polytheism. Even if we don't accept this, we ask how did monotheism originally arise out of a world immersed in polytheism at the time? We note that the book of Job, the oldest biblical book after Genesis, also reveals a monotheistic view of God (Job 1:1, 6). God is described as personal (Job 1:6, 21), moral (1:1; 8:3–6), sovereign (42:1–2), almighty (13:3), and Creator (4:17; 38:4–5). Philosopher John Mbiti studied three hundred traditional African religions and showed without exception that they revealed an explicit monotheism.[78] This is also the case with primitive religions around the world. Even with polytheistic religions there is usually a special (high) god indicating a latent monotheism.

Other questions that might be asked are: did Adam and Eve actually exist, was there actually a garden of Eden, was there a universal flood, did some people really live to a great age, and so forth. Whatever one's view about evolution there is no reason why God didn't create a "human" couple, or even

---

[76]Glueck (1959: 136); for much more detail see Kitchen (2003).
[77]Jastrow (2000: 14).
[78]Mbiti (1990).

a group of people, in His image, as well as there possibly being other human-like creatures around at the time. He may have even drawn the couple from this group to establish a special communication with them. Modern genetics tells us that there were initial male and female ancestors through looking at mitochondrial DNA in women and the Y-chromosome in men, but their initial existence is still a mystery and highly controversial, which I won't pursue here.

We note that in the beginning of Genesis, the word Adam means mankind, but then it comes later to refer to an individual. Adam and Eve are portrayed as real people and important events in their lives including the births of their children are mentioned in Genesis. Later Old Testament chronologies[79] describe Adam as the first man, and there is a sequence of descendants listed going from Abraham to Jesus (Matthew 1:1–17) through his mother Mary's line, and a sequence going back from Jesus to Adam through his father Joseph's line (Luke 3:23–38). However there are gaps in both lists as the main aim was to show an ancestral connection. We note that Jesus[80] refers to the Genesis story about male and female as though they were real people while Paul refers to Adam.[81]

What about the Garden of Eden, the temptation, and the "Fall?" We are now in controversial waters as there are many differences of opinion about the story. I believe the key message here is that man (and woman) wanted to be their own bosses apart from God; this brought about separation from God described as being thrown out of the Garden of Eden and being subject to (eventual) death. As a result the world was no longer the same for the couple, but was "fallen" in some sense. This is described, for example, as the couple now having to face some difficulties, presumably existing outside the garden, such as thorns and thistles. It can be argued that there were also anatomically modern humans outside of the garden. For example Michael Jarvis zoologist notes that

> The existence of such people may also be supported by biblical references, including the origin of Cain's wife, the people he feared would kill him when he fled from the Garden of Eden, the people he was building a city for, and the descriptions of 'sons of God', 'daughters of men' and the 'Nephilim'.[82]

The Garden of Eden is described as a real place in Genesis 2:10–14, somewhere in modern Iraq. If the garden existed we wouldn't expect to find it as it was sealed off after the expulsion of Adam and Eve[83] and a flood has intervened, which we now consider.

As with the creation accounts, there are other versions of a flood from the Babylonians and Sumerians that are much more mythical, but the basic

---

[79] Genesis 5:1 and the first eight chapters of 1 Chronicles.
[80] Matthew 19:4–6.
[81] Romans 5:12–14; 1 Corinthians 15: 45.
[82] Jarvis (2008: 18).
[83] Genesis 3:24.

story is the same, suggesting an older source such as the story of Noah. The common element is that a man is told to build a boat for himself and his family because the gods were going to destroy the world through a flood, and the man then offers a sacrifice after riding out the storm and leaving the boat. In the Sumerian King List the flood is used to divide its history into pre-flood and post-flood periods with the pre-flood kings having enormous life spans, but the post flood lifespans being much reduced. Genesis refers to a number of early patriarchs prior to the flood living to great ages. Philosopher Norman Geisler[84] argues that there are good reasons to believe that Genesis gives the original story. Only in Genesis are dates given relative to the chronology of Noah's life, and the story reads like a straightforward narrative.

Geisler notes that the cubical Babylonian ship would have been disastrous if it had been used, but the biblical ark being rectangular and low, long, and wide would have coped with the raging waters, with its dimensions or proportions being similar to those of modern ocean liners. It was made of gopher wood, a flexible material, which would provide an extremely stable boat in very turbulent waters. The ark was very big, being three stories high[85] and estimated to have a volume of 1.5 million cubic feet, but there is some question as to how it could accommodate the numbers of animals including the small and microscopic creatures and the quantity of food and water needed to be taken on board.

It is argued that the concept of "species" was not the same as "kind" in the Bible; eggs of species could have been used, and sea creatures stayed in the sea. However there is the question of food for the animals after the flood. Given the time on the ark and the need for fresh water it would appear that the ark floated on fresh water. Also the water receded (where to?) and land plants immediately sprang back into life, including olive trees. Soon after leaving the Ark, Noah planted a vineyard, which would have been impossible if the oceans were involved and everything had been covered with salty seawater for twelve months. Also, the problem of accommodating all flora and fauna in the ark does not arise if the flood was just local.

It is interesting that flood accounts have been found in other parts of the world including, for example, China, Hindu mythology, Greece, and Mexico, and there is also a deluge story in North America.[86] There has therefore been some debate as to whether the flood was local or global, as for example the silt deposits in the Mesopotamian Valley being described as from local flood(s). We note that the phrase "all the world" sometimes means "all the known world." The Hebrew words *adamah* and *erets* translated as "earth" or "world" are translated differently elsewhere in the Bible and rarely as "world."

---

[84] Geisler (1999: 49).
[85] Genesis 6:16.
[86] For a list of so-called flood myths see http://en.wikipedia.org/wiki/List_of_flood_myths, accessed August, 2015

Theories have been put forward as to how the flood might have occurred by some natural disaster as water is described as coming from beneath the earth, and there is the question of very distinct types of animals and plants on different continents, with some animals being unique to their part of the world, and how they got there.[87] A world-wide catastrophe would have changed the world landscape and there are signs of such dramatic changes.

I will revisit Genesis later in Chapter 8 on suffering and evil as the Bible regards what happened in Genesis story as the source of all our problems.

5.3 OLD TESTAMENT CODES

Before moving to the New Testament I wish to refer to a somewhat controversial topic, the possible existence of hidden codes in the Old Testament. Ancient Hebrew is an interesting language in that there are no gaps between the words, the vowels are only inferred, letters are read right to left, and like Greek the symbols are used for both letters and numbers. In contrast to English, there is little redundancy in the use of Hebrew, i.e., there are fewer "wasted" words. My attention was drawn to the subject of patterns in Genesis by a paper in a good statistics journal by Witztum, Rips, and Rosenberg (1994). Previous to that, Jewish scholars including mathematicians, had been searching for codes in the Old Testament.

Chuck Misler, engineer and biblical scholar,[88] has written more extensively about the topic giving some references to further work in this area, which I shall refer to briefly. In particular there is the so-called equidistant letter sequence (ELS) where, starting with a given Hebrew letter, the letters corresponding to given a interval from that letter form a word. For example, starting with particular letter, one might find that every 5th letter going in one direction or the opposite direction spells out a word. Witzum et al. used a computer search to find the names paired with their birth and death dates of 66 anticipated famous rabbis in the Genesis Hebrew text, and concluded with the following comment:

> We conclude that the proximity of the ELS's (Equidistant Letter Sequences) with related meanings in the Book of Genesis is not due to chance.

Misler mentioned that one mathematician calculated that the odds of the find occurring at random were one in 2.5 billion, and other mathematicians have confirmed the huge odds. Since Hebrew can lend itself to this kind of find, the method has been applied to other Jewish writings and translations into Hebrew without success. Misler also mentions clusters of codes relating to the Holocaust in Deuteronomy; words like Hitler, Auschwitz, Eichmann etc. appearing as ELS's. Care is needed in dealing with such findings as one can

---

[87] For one view see Jarvis (2008: chapter 7).
[88] Misler (1999).

always find patterns in a string of numbers, say, if the string is long enough. In a very large set of numbers, the sequence 1,2,3,4,5,6,7,8,9,0 will eventually turn up, even if the numbers are created at random. However the statistics suggest that the findings are not due to chance.

5.4 NEW TESTAMENT

In this section we continue with our theme about reliability and begin with the topic of the authorship, along with the dating of the gospels and related New Testament books. Some archaeological evidence is given to support the historicity of the some of the books. The wealth of manuscript evidence available for the New Testament is then described as well as internal evidence from the gospels. We ask how reliable and consistent are the documents, given that there are differences in the accounts. Finally we consider how we arrived at the so-called "canon" of New Testament books that we have today.

5.4.1 Authorship and Dating

According to New Testament scholar Craig Blomberg,[89] although the gospels are actually anonymous, the authorship of the gospels of Matthew, Mark, and Luke (the so-called synoptic gospels) are not generally in dispute. Mark carefully recorded Peter's observations without regard to the order of events as Peter was an eye witness; and Luke, the follower of Paul, were not even among the twelve disciples. Matthew, however, was a disciple but was originally a hated tax collector. Not a great choice of authors if one was trying to make up stories! Compared to the apocryphal gospels (discussed below) whose fictitious authors were given well-known names like Thomas and Mary to give more credibility, lesser names were associated with the gospels, which supports their authenticity.

The authorship of John's gospel has been in doubt as to whether it was John the disciple or a different John, the former being the more likely. John's gospel has a different linguistic style and a different theological perspective of Jesus, indicating independence from the other three synoptic gospels, especially with regard to the choice of stories and the claims of Jesus to divinity, though there are parallels in the synoptics.[90] Being the later writer, John would have wanted to add material not in the synoptics. I do not wish to enter into the debate about John's gospel and I refer the reader to Blomberg (2007: chapter 5) or Blomberg (2011) for more detail.

The church fathers provide supporting evidence for the gospels, and these include for example the epistles of Clement of Rome, Ignatius, Polycarp, and Barnabas. In particular Ignatius, the bishop of Antioch and martyred before

---

[89] Strobel (1998: 26–29).
[90] See in particular John 20:31.

AD 117, emphasized both the deity and humanity of Jesus, and certain basic facts about the life of Jesus. Even without the New Testament, we would still have a reasonable outline of the life of Jesus. Strong external evidence for the traditional authors is given by Papias around AD 125[91] and by Irenaeus around AD 180.[92]

It is important to recognize that the gospel writers vary in their writing styles and theological emphases. Blomberg writes: "Luke, probably a native Greek, writes much more artistically and fluently than Mark, whose prose is more awkward and more Semitic. Matthew and John's styles fall somewhere between . . . ."[93]

Dating the Gospels

The standard scholarship dating even amongst liberals is Mark in the AD 70s, Matthew and Luke in the 80s, and John in the 90s, all within the lifetimes of eye witnesses of Jesus, including hostile ones (who could correct any errors and embellishments).[94] This is too short a time for legends to build up. Blomberg[95] provides a strong case for an earlier dating of the gospels. Acts, also written by Luke, ends abruptly indicating it was unfinished or simply that the narrative was brought to a conclusion at the end of Paul's two-year stay. Either way suggests that Acts cannot be dated later than AD 62 before Paul was put to death. Also the destruction of Jerusalem in AD 70 is not mentioned, particularly as it was predicted by Jesus. Other facts supporting an earlier dating are: (1) there is no hint of the death of James at the hands of the Sanhedrin around AD 62, (2) the prominence of the Sadducees in Acts belongs to the pre-70 era, before the collapse of their political cooperation with Rome, and (3) there is no mention of the impending Jewish War in AD 66, or any particular deterioration of Christian relations with Rome involved in the Neronian persecution of the late 60s.

We note that Acts is part 2 of Luke's gospel, so Luke was written earlier. As Luke incorporates parts of Mark, Mark was written before AD 60, or even in the 50s. This puts Mark within 20 to 30 years of Jesus' death. Craig Evans,[96] a New Testament historian respected by both liberals and conservatives, suggested that there are good arguments for the synoptic gospels being written in the 50's and 60's with a preference for the first gospel Mark in the 60's close to the Jewish-Roman war of AD 66–70.[97] Clearly the New Testament accounts about Jesus were recorded by men who had either been eyewitnesses themselves or who related the accounts of eyewitnesses. We see

---

[91] He said that Mark "made no mistake" and "did not include any false statement."
[92] Strobel (1998: 28–29).
[93] Blomberg (2007: 71).
[94] McGrew and McGrew (2009) state that the evidence points to Matthew being written first, and therefore at an earlier time.
[95] Strobel (1998: 41–42) and Blomberg (2007: 26).
[96] Strobel (2008: 30).
[97] Mark 13:18.

this, for example, in the first five verses of Luke's gospel where Luke writes about following everything closely for some time to produce an accurate account. These accounts were being circulated within the lifetimes of those alive at the time of Jesus' life, and this included the critics as well. These people could certainly confirm or deny the accuracy of the accounts. For example, Peter in his speech when talking about Jesus added the words "as you yourselves know."[98] We also have the early letters by the eyewitnesses Peter[99] and James, with the latter showing recollections of Jesus' Sermon on the Mount. It is interesting to note that there is no mention of teaching by Jesus about issues that proved to be controversial in the early church such as circumcision, the role of women, food offered to idols, or the use of gifts of the Spirit. If the Church Fathers were inserting doctrinal stuff into the New Testament record, Jesus would have been depicted as making statements about these matters. The evidence, then, is that these were not issues during his time on earth.

Luke the Historian

Luke was not only a gospel writer and a physician, but is also regarded as an excellent historian as so much in his writing can be checked through archaeology.[100] For example, in Luke 3:1-2 we have 16 historical items:

> In the fifteenth year (1) of the reign of Tiberias Caesar (2) when Pontius Pilate (3) was governor (4) of Judea (5), and Herod (6) was Tetrarch (7) of Galilee (8), his brother Philip (9) Tetrarch (10) of Iturea (11) and Trachonitis (12), and Lysanias (13) was Tetrarch (14) of Abilene (15), in the priesthood of Annas and Caiaphas (16), ...

Right through the book of Acts Luke gives details that can be checked. If you are going to write a forgery you wouldn't bother to give such details! Although Luke was not an eye-witness to Jesus' life, we see in Acts chapter 16 that he starts using the first-person plural "we" that indicates that he joined Paul on his evangelistic tour of the Mediterranean cities. In chapter 21 he accompanied Paul back to Palestine and to Jerusalem. He was with Paul in Rome where Paul was imprisoned and anticipating martyrdom.[101] Clearly Luke was in direct contact with the eyewitnesses of Jesus' ministry in Jerusalem. Sir William Ramsey, a world famous archaeologist, began as a sceptic trying to discredit Lukes writings. Ramsey ended up becoming a Christian and concluded that

> Luke is a historian of the first rank; not merely are his statements of fact trustworthy ... [he] should be placed along with the very greatest of historians.[102]

Ancient historian Adrian Sherwin-White wrote: "For Acts the confirmation of historicity is overwhelming ... any attempt to reject its basic historicity even

---

[98] Acts 2:22.
[99] See 1 Peter 5:1 and 2 Peter 1:16.
[100] For extensive details see Geisler (1999: 4–8).
[101] 2 Timothy 4:6,11.
[102] Ramsey (1915: 222).

in matters of detail must now appear absurd."[103] The well-known archaeologist John McRay said: "The general consensus of both liberal and conservative scholars is that [the book of] Luke is very accurate as a history."[104] He also noted that Luke was an educated man and a writer of excellent Greek. The wealth of Luke's accurate historical and local knowledge is seen very clearly and in great detail by biblical scholar Colin Hemer.[105] Luke's aim, clearly stated in Luke at the beginning of chapter one, was to write an orderly and accurate account of what happened right from the beginning of Jesus' life. Although the other synoptic gospels do not give such a reason, John states that the reason for writing his gospel was to encourage the readers to believe that Jesus was the messiah and Son of God, and hence have salvation through Jesus.[106]

Paul's letters

It should be noted that the books of the New Testament are not in chronological order and the gospels were written after almost all of Paul's letters. Paul's writing probably began in the late 40s and early 50s, and incorporated some early creeds such as Philippians 2:6–11 and Colossians 1:15–20 that were formulated soon after the resurrection and reflected the beliefs of the early church about Jesus. For example, he called Jesus "the Son of God"[107] and the "image of God."[108] Of particular note is 1 Corinthians chapter 15 whose technical language indicates a fixed oral tradition,[109] indicating a belief in the resurrection going back to a few years after the event. Paul letters therefore contain some very early reports about Jesus. In Paul's conversion he said he encountered the risen Christ and such was its effect that he transformed from being a persecutor of Christians to being the foremost Christian missionary who suffered enormous deprivations because of his faith.[110] After that experience he met with some of the eye witnesses of Jesus to make sure he was preaching the same message as they were. His letters, although not written primarily to refer back to the teachings and deeds of Jesus, as these would be well known by the Christian churches, they do indicate that Paul had a basic outline of Jesus' life and was very familiar with Jesus' teachings.[111]

As far as the dating Paul's letters are concerned, it is generally agreed that 1 Corinthians was written by AD 55 or 56 and the letter repeatedly claims to be written by Paul because of personal references. Specifically mentioned are

---

[103] Sherwin-White (1963: 189).
[104] McRay (1991: 129).
[105] Hemer (1989: chapter 8).
[106] John 20:31.
[107] Romans 1:4; 2 Corinthians 1:19; Galatians 2:20.
[108] 2 Corinthians 4:4; Colossians 1:15.
[109] We refer to this passage again in Section 6.4.2.
[110] See 2 Corinthians 11:24–28.
[111] For details by Craig Blomberg see http://www.free-online.org/free-thinking/lifes-big-questions/bible/is-the-new-testament-historically-reliable.htm, accessed August, 2105.

the twelve disciples and James the brother of Jesus. Internal evidence strongly supports this early date and there are parallels with the book of Acts. There is also external evidence from early Christian writers including 600 quotations from 1 Corinthians in the writings from the church fathers Irenaeus, Clement of Alexandria, and Tertullian alone. Along with 1 Corinthians, 2 Corinthians and Galatians are well attested and early. They all show an interest in the events of Jesus' life and give facts that agree with the gospels. There are 12 epistles that have a greeting from Paul at the beginning of the epistle, indicating his authorship. Blomberg[112] notes that most, if not all, of Paul's letters predate the Gospels, so that Paul's knowledge of the life of Jesus and his teachings provide added confirmation of the early traditions that went into the formation of the Gospels.

James' Letter

Of particular interest is the letter of James. Of all the letters it is the one that contains the most passages that resemble the teaching of Jesus. Right from the beginning we see echoes of the gospels, especially Matthew, indicating that James was familiar with the Gospels, or at least with Matthew. Blomberg[113] suggests there are good reasons for dating James very early, in the mid-to-late 40s, making it one of the earliest New Testament writings. If the dating is correct, then we have convincing evidence that the gospel traditions were known and applied very soon, thus confirming their validity.

Peter's Letters

Some have queried whether 2 Peter was actually written by the apostle Peter, as the writing style is different from 1 Peter. In answer to this we note that:[114]

(1) In the first letter Peter had Silas as a secretary (1 Peter 5:2), but he seemed to have written the second letter himself.

(2) The circumstances of the letters were different. The first letter was written to encourage believers who are suffering, while the second contains warnings about false teachers; so different purposes can lead to different styles.

(3) There is good internal evidence for the second letter. For instance, verse 1 says it is from Peter, and in 1:14 the author remembers the words of Jesus concerning his death in John 21:18–19. In 1:16–18 the author was an eyewitness to what happened on the mount of transfiguration.[115]

(4) The author refers to the letter as his "second letter" (3:1).

---

[112] Blomberg (2007: 284).
[113] Blomberg (2007: 292–293).
[114] Geisler and Howe (1992a: 537).
[115] Matthew 17:1–8.

(5) The author is aware of the writings of Paul and calls him a beloved brother (3:15–16); see also Paul's special visit to see Peter (Galatians 1:18).

(6) There are some likenesses between the letters. They both place an emphasis on Christ; his suffering in the first letter and his glory in the second. He also mentions Noah and the flood in both letters (1 Peter 3:20; 2 Peter 2:5, 3:5–6).

(7) The discovery of the Bodmer papyrus (P^{72}, ca. AD 250) that contains both letters indicates that the second letter was highly respected in Egypt at an early date. The noted archaeologist William Albright dated 2 Peter before AD 80. Numerous early church fathers including Origen, Eusebius, Jerome, and Augustine cited the letter as authentic.

5.4.2 Archaeological Evidence for the New Testament

Although it is not the role of archaeology to verify the stories of Jesus, it does provide evidence for the authenticity of the New Testament writers. We have already mentioned above Luke's reliability and accuracy as a historian. There were a number of examples when Luke was thought to be mistaken but later excavations affirmed his accuracy. The gospel of John has been dismissed in the past as a historical source, yet we now find archaeological support for it. For example, excavations in 1956 have identified the pool of Bethesda and its five colonnades (porticos)[116] when it was originally believed that there were only four porches; the pool did not exist in the second century. The pool of Siloam was discovered in 2004.[117] Names of people mentioned in the gospels have also been confirmed from external sources. Blomberg[118] lists a number of findings including the synagogue in Capernaum,[119] with Simon Peter's home,[120] and that of the synagogue ruler, Jairus.[121] A first-century tiled mosaic of a fishing boat with the inscription of "Magdala" helped to identify the location of Mary Magdalene's home town;[122] and we have Jacob's well at Sychar, where Jesus was said to have met the Samaritan woman.[123] As Tim and Linda McGrew comment:

> Archaeology has not been kind to literary criticism of the gospels and Acts. The discovery in Caesarea Maritima in 1961 of an inscription bearing Pilates name and title, the discovery of a boundary stone of the emperor Claudius bearing the name of Sergius Paulus (cf. Acts 13:7), the very recent discovery of the Pool of Siloam (John 9) from the time

---

[116] John 5:2.
[117] John 9:7.
[118] Blomberg (2007: Appendix A).
[119] cf. Mark 1:21.
[120] Mark 1:29.
[121] Mark 5:22, 38.
[122] Luke 8:2.
[123] John 4:5–6.

of Jesus, and numerous other discoveries indicate a level of accuracy incompatible with the picture of the development of the gospels as an accretion of legend over the course of two or more generations.[124]

5.4.3 New Testament Manuscripts

There are no surviving originals of the New Testament so that all we have are copies.[125] Were they close to the original? Fortunately we have a large number of manuscript copies written within a couple of generations of the originals for cross checking, as well as early translations into other languages. We also have many quotations in commentaries, sermons, and letters of the early church fathers from which most of the original text could be constructed. In particular, more than 5700 New Testament Greek manuscripts from different places have been catalogued.[126] This is hugely more and more recent than those for other revered ancient writings that have been unquestionably regarded as authentic (e.g., Tacitus, Josephus, and Homer's Iliad). There are also another 18,000 or so manuscripts in other languages: 8,000-10,000 in Latin vulgate plus a total of 8,000 in Ethiopic, Slavic, and Armenian. A great majority of the manuscripts are complete for the purposes that the scribes intended, for example some were intended to include just the gospels while others just Paul's letters; 60 have the entire New Testament.

The earliest portion of the New Testament is a papyrus fragment of John's gospel containing five verses from chapter 18 dated from its writing style between AD 100 to AD 150 with the earlier date preferred. It was also written in Egypt very far away from where it was probably originally composed, pushing the gospel to an earlier date than originally proposed. Finally, if we compile the 36,289 quotations by the early church fathers of the second to fourth centuries we can construct the whole of the New Testament except for 11 verses.[127] The famous archaeologist Sir Frederick Kenyon concluded: "The last foundation for any doubt that the scriptures have come down to us substantially as they were written has now been removed."[128] This comment reflects the view of a number of other experts.

Of particular interest has been the differences in the gospels with regard to the resurrection stories because of the resurrection's centrality to the Christian faith. There are certainly quite a a number of differences in the details such as the time of events, the number of people present, the number of angels seen, and so forth. Theologian and historian Michael Licona[129] notes that the genre of biographies in those days allowed biographers a fair amount of

---

[124] McGrew and McGrew (2009).
[125] Much of this section is from Metzger, a leading authority on the biography of Jesus quoted in Strobel (1998: chapter 3).
[126] For some details see Geisler (1999: 531–537).
[127] Geisler (1999: 532).
[128] Kenyon (1940: 288).
[129] Licona (2010: 201 ff., 593).

license with how material was arranged including paraphrasing for example and what emphases were made, while being faithful to the teachings and beliefs of the subject. Furthermore, the differences support the idea that we have independent witness accounts, though many of the differences can be explained, as in Section 5.4 below. For example, consider the "three-day motif" that is discussed in detail by Licona.[130] We have the tradition of Jesus being raised on the third day with variants "after three days", "three days and three nights", "in three days", "today and tomorrow and on the third day", and "until the third day", which indicate that we have a figure of speech essentially meaning a short period of time, say within three days. This is also confirmed in Esther (4:16, 5:1). We use a similar type of colloquial phrase when we say "I will be with you in a second."

How does the historical information about Jesus compare with such information about any other founder of an ancient religion? The following summary speaks for itself:[131]

Zoroaster (approximately 630 to 550 BC though earlier dates have been given). His scriptures were not put into writing until the third century AD, and his most popular Parsi biography was written in AD 1278.

Buddha (6th century BC) The scriptures were not put into writing until after the time of Christ, and the first biography of Gautama Buddha was written in the first century AD.

Muhammad (AD 570 to 632) His writings were in the Qur'an compiled around AD 650, and his biography was not written until AD 767.

5.4.4 Reliability of the Gospel Writers

Can we be sure that the gospel writers were able to present accurate versions of Jesus' life, given the possibility of faulty memories, wishful thinking, and the possible growth of legends? It should be noted that in the absence of the printing press and the rarity of scrolls of papyrus, passing on material orally was the norm and part of the culture at the time. For example, rabbis were famous for memorizing the whole of the Old Testament, and from the earliest age children were taught to faithfully memorize sacred tradition. In first century Palestine "the ability to learn and remember large tracts of oral tradition was a highly prized and highly developed skill."[132] We can therefore expect that the disciples of Jesus would have been well able to remember more than the four gospels put together, especially as 80 to 90% of Jesus' words had a poetic flow using various figures of speech such as meter, parallelism, balance lines, and so forth, enabling memorization.[133] Also, in studies of cultures with

---

[130] Licona (2010: 324–329).
[131] Strobel (1998: 114).
[132] Craig (1998b).
[133] Strobel (1998: 54).

oral traditions, some flexibility was allowed as to what was included and what was left out, provided key points were unchanged. It should be noted that one of the hallmarks of a godly person was truthfulness, mentioned in both the Old Testament[134] and in the New Testament.[135]

As far as legend building is concerned, Sherwin-White,[136] a professional historian, showed from the writings of Herodotus that even two generations is too short a time span to allow legends to wipe out the hard core of historical facts. C. S. Lewis,[137] an expert literary critic, commented that the style of writing and the detail in the gospels indicates history, not fiction. Jesus was not a myth as his history can be located and dated.

As already noted, we cannot expect the synoptic gospels to be too similar as we would suspect collusion. In fact according to Bloomfield,[138] there is about 10 to 40 percent variation among the synoptics on any given passage, which might be expected from oral traditions when the emphasis was on preserving the words of Jesus. Finding minor differences in the recording of events, even by reliable historians and eye witnesses is not unusual as there will always be some selection and emphasis, and even contradictions. This however does not negate or undermine the credibility of the reports. For example, our principal sources for the life of Tiberius Caesar are Suetonius, Tacitus, and Dio Cassius and these authors contradict each other in a number of places. Philosophers Timothy and Lydia McGrew make the following comment:

> Almost no two authors agree regarding how many troops Xerxes marshaled for his invasion of Greece; but the invasion and its disastrous outcome are not in doubt. Floruss account of the number of troops at the battle of Pharsalia differs from Caesar's own account by 150,000 men; but no one doubts that there was such a battle, or that Caesar won it. According to Josephus, the embassy of the Jews to the Emperor Claudius took place in seed time, while Philo places it in harvest time; but that there was such an embassy is uncontroversial. Examples of this kind can be multiplied almost endlessly.[139]

A similar comment applies to legal evidence, as there is an expectation of there being some minor differences in testimony from witnesses, otherwise collusion is suspected.

5.4.5 Consistency of the Gospels

Given that the manuscripts were copied by hand we would expect some errors would creep in even if the scribes were extremely careful. I was surprised when a figure like 200,000 to 400,000 variations among manuscripts was quoted until

---

[134]Exodus 20:16; Numbers 22:18; Psalm 15:1–3.
[135]Ephesians 24:25.
[136]Sherwin-White (1963: 188–191).
[137]Lewis (1967).
[138]Strobel (1998: 55).
[139]McGrew and McGrew (2009: 6).

it was made clear that many of these could be easily explained. To begin with, we have a large number of manuscripts so that it would not be hard to generate a large number of trivial differences. The following are some sources of variation.[140]

(1.) *Spelling variations.* If a single letter in a single word of a single verse is misspelled in 2000 manuscripts it is counted as 2000 variations. If a single fourteenth-century manuscript misspells a word it counts as a single variant. Even if the misspelling in Greek makes no difference to the meaning of the word it still counts as a variant. Between seventy to eighty percent of all textual variants are spelling differences that cannot even be translated into English and have zero impact on meaning.

(2.) *Nonsense variants.* Here it is obvious we have a typo that is easy to correct.

(3.) *Synonyms* "When Jesus knew" or "When the Lord knew" (John 4:1) are both true irrespective of which version goes back to the original.

(4.) *Definite articles.* A lot of variants involve the Greek practice of using a definite article with a proper name, which is not the case in English, for example "the Mary" instead of "Mary" where there is no impact on the meaning.

(5.) *A change in word order that does not change the meaning of a sentence in Greek like it does in English.* For example "Jesus loves Paul" can be written sixteen different ways in Greek even though the English translation is the same in each case.

(6.) *The practice of paraphrase.* This was the most common difference in comparing parallel accounts, and unlike today it was a practice very common in the ancient world, though authors had to remain faithful to original meanings. Some of the "quotes" from the Old Testament by the gospel writers are paraphrases,[141] and Peter's confession is stated slightly differently in each of the synoptic gospels.[142]

One example is Jesus' teaching to his disciples about family relationships. Luke 14:26 reports that Jesus said that if someone comes to him and does not hate his family members and even life itself they cannot be his disciple. Matthew seems to greatly tone down the statement to saying that anyone loving parents or children more than Jesus is not worthy of him (Matthew10:37). The strong contrasting language of Luke reflects a Jewish way of speaking, especially as the word "hate"

---

[140] Strobel (1998: 82; 2008: 37–39); Blomberg (2007: chapter 4).
[141] For example, Matthew 2:6 and 4:14–16.
[142] Matthew 16:16; Mark 8:29; Luke 9:20.

to the Jew has a broader range of meaning than in English. Matthew's paraphrase is a reasonable interpretation.

Similar explanations are given for other "problem" passages by Blomberg[143] where a change by a gospel writer is often for theological clarification. Sometimes a change is made to suit the audience, whether Greek (as with Luke) or Jewish (as, say, Matthew). For example, Luke's version of the parable of the mustard seed has the plant growing in a garden rather than in a field as in the Graeco-Roman world, whereas Jewish purity forbade the seed being planted in a garden.[144] It should be noted that a paraphrase of a saying of Jesus can still reproduce its original meaning as faithfully as a literal translation. We see this in some modern translations of the Bible. Also the gospel writers would have believed they had access to accurate information as John claimed that Jesus promised the Holy Spirit that would teach them "all things", help them to remember his sayings,[145] and guide them into "all truth."[146]

(7.) *Partial reports of longer sayings, or different parts selected by different authors.* A good example is what the high priest and Jesus said to each other during his interrogation by the Sanhedrin on the night of his arrest.[147] Compression of a narrative also happens such as the raising of Jairus' daughter, where Matthew reduces the twenty-three verses used by Mark (5:21–43) to nine (Matthew 9: 18–26).

(8.) *Pairs of passages (doublets) either among the Gospels or within one Gospel.* Sometimes critics treat every pair of passages that are similar as variations of a single saying or event in Jesus life. Yet good teachers or preachers will regularly repeat themselves, and no doubt Jesus did the same.[148] However, when it comes to events it is not always clear when we have a single event or a pair of events. For example, Matthew and Mark describe how Jesus miraculously fed both five thousand and four thousand,[149] while Luke only includes the former.[150] Here it seems that the events are different given the different audiences (mostly Israelites or mostly Gentiles), and different words are used for "basket."

(9.) *Differences in chronology.* There are a number of examples where a gospel writer has arranged passages in a thematic or topical order rather than chronologically. Blomberg[151] gives a number of examples includ-

---

[143] Blomberg (2007: 160–162).
[144] Luke 13:19; cf. Matthew 13:31.
[145] John 14:26.
[146] John 16: 13.
[147] See Luke 22:67–70 with the shorter versions in Matthew 26:63–64 and Mark 14:61–62.
[148] I have done this in my own public speaking, which has been extensive in the past.
[149] Mark 6:32–44, 8:1–10; Matthew 14: 13–21, 15:32–39.
[150] Luke (9:10b–17).
[151] Blomberg (2007: 168–171).

ing the following. Luke moves his statement about John the Baptist's imprisonment from its place in the middle of Jesus' Galilean ministry[152] to its beginning so that it forms a natural conclusion to his section on John the Baptist's mission.[153] Matthew and Luke give the temptations of Jesus in a different order.[154]

(10.) *There are sometimes omission of entire passages or sections.* Given the extensive ministry of Jesus it is not surprising there are some omissions. The same applies to details within passages. Any differences in the gospel accounts only emphasizes the independence of the writers, especially as the scrolls used for writing would limit the amount of material recorded. What stands out is the amount of material repeated in more than one gospel. Being selective doesn't imply historical inaccuracy. John's gospel, being a later gospel, would be expected to include material not in the other gospels. We cannot argue from silence that something is in error.

(11.) *Variation in names.* The reason for some of the differences have already been mentioned in (1.) above.

(12.) *Variation in numbers.* This occurs more in the Old Testament and we consider just one example, namely 2 Kings 8:26 that gives the age of king Ahaziah as twenty-two, but 2 Chronicles says forty-two. The second number must be wrong and therefore a copying error, otherwise Ahaziah would be older than his father. With regard to the New Testament there are occasions when one Gospel refers to one person and a parallel Gospel refers to two people.[155] A natural explanation is that there were probably two present each time, but one acted as spokesman with nonessential details such as the other person omitted.

(13.) *There are two longer additions that are often referred to by skeptics.* These are the story of the woman taken in adultery[156] and the ending of Mark's gospel.[157] These have been known about for a long time and most modern translations mention that these passages are not found in the earliest manuscripts. The fact that they were picked up by cross-referencing different manuscripts shows the usefulness of having so many manuscripts to compare, and should not undermine the rest of the New Testament. Physicist and philosopher David Glass comments: "As a

---

[152] Mark 6:14–29.
[153] Luke 3:1–20.
[154] Matthew 4:1–11; Luke 4:1–13.
[155] For example, two blind men in Matthew 20:30 vs one in Mark 10:46; two demoniacs in Matthew 8:28 vs one in Mark 5:2; and two angels at the tomb in Luke 24:4 vs one in Mark 16:5.
[156] John 7:53–8:11.
[157] Mark 16:9–20.

result of textual criticism we know where there is some dispute about the text and, more importantly, where there is no dispute (i.e. almost all of it)."[158]

As with the above examples, the majority of the variations are not significant. Even the more significant ones do not contradict any doctrine of the church. Most good translations alert the reader to any variant readings, which are rare. Some that are apparent contradictions have alternative explanations such as the well-known difference in the genealogies of Jesus in Matthew and Luke. One answer is that Matthew reflects Joseph's legal lineage and Luke considers Joseph's human lineage with the divergence occurring when somebody in the line did not have direct offspring: legal heirs were required according to Old Testament practices.

There is biblical evidence to show that genealogies were sometimes deliberately abridged either to save space or perhaps emphasize the most important people (e.g., Matthew 1:1 where we have just Jesus, David, and Abraham). Matthew gives an abbreviated genealogy as a comparison of Matthew 1:8 and 1 Chronicles 3:11–12 indicates he missed three generations between Joram and Uzziah (Azariah). The short genealogy in Ezra omits six generations (cf. 1 Chronicles 6:3–14). There must be gaps in the Genesis 5 and 11 genealogies as numbers don't add up and there would be later inconsistencies in the growth of populations.[159] There are also differences in the numerical information given by the Hebrew Masoretic text, the Greek Septuagint, and the Hebrew Samaritan Pentateuch. The word "begat" can mean "became the ancestor of." The view that the genealogy is "open" and not "closed", that is has gaps, is supported historically by the existence of civilizations before 4000 BC such as in Egypt, for example. Peoples seem to have wandered to North America since 10,000 BC.

It is important to note that the New Testament deals quite honestly with disagreements when they occur such as the issue of circumcision and whether Christians can eat meat offered to idols. There is no attempt to cover up any embarrassing events such as the disciples arguing who would be first in God's kingdom, the flight of the disciples after Jesus arrest, Peter denying Jesus three times, the failure of Jesus to work miracles in Galilee, the reference of some of his critics to his possible insanity, and Paul disagreeing with Peter. Nor are some of the difficult sayings of Jesus left out (e.g., Jesus crying out "My God, my God, why have you forsaken me?"). The narratives have the ring of truth. Philosopher and historian Will Durant comments: "No one reading these scenes can doubt the reality of the figure behind them."[160] Take for example someone like Peter the disciple. When Mary told Peter and John about the empty tomb they raced off to the tomb with the younger John

---

[158] Glass (2012: 273).
[159] Geisler (1999: 268–270).
[160] Durant (1994).

outrunning Peter and getting there first. John paused outside the tomb and peered in, perhaps awe-stricken or sensitive to treading in a so hallowed spot. However Peter, true to form, is uninhibited, impetuous, and practical, and goes straight inside the tomb followed by John. They then see the empty grave clothes. Later, Peter and John are once again true to form when the disciples are fishing and they are told by a person on the beach where to cast their net. John recognizes the person as Jesus and whispers "It is the Lord." Peter the man of action jumps over the side and splashes ashore.

Summing up, Blomberg[161] provides a scholarly account of the gospels and shows that although there are some apparent inconsistencies, the similarities far outweigh the differences. Of the differences, many simply reflect varying theological interpretations or rewording of the same historical events without calling into question the fundamental historicity of the events themselves.

5.4.6 Choosing the Canon

Do the 27 books of the New Testament represent the most reliable information about Jesus, given that other so-called apocryphal gospels such as, for example, the gospels of Thomas, Mary, Judas, Peter, and the Nativity of Mary exist? These other writings endeavored to fill in details of Jesus' life and some of them are known as the "infancy gospels." In evaluating these other gospels, Craig Evans, a respected New Testament historian, says that we have to ask whether they were culturally accurate, supported historically, and written in a time and place that has proximity to Jesus' life.[162] He notes that the answer is generally no to these questions as they were written too late to be historically accurate with inaccuracies at key points. They were derived from earlier resources and written from other places, with rather strange contexts.

The gospel of Thomas, although mentioned several times in history, is not generally believed to have been written by Thomas the disciple (in fact by Judas Thomas), or that it is authentic or early. Since according to Evans, "Over half of the New Testament writings are quoted, paralleled, or alluded to in Thomas",[163] the gospel of Thomas appears to have been written later as it contains too much of the New Testament and reflects later developments in Luke or Matthew. It also has material from John's gospel, which was written later, as well as having distinctive Syrian sayings and Syriac catch words, pointing to a date after AD 175, and probably closer to AD 200.

The gospel of Mary would be dated between AD 150 to 200, and Evans says: "There's nothing in the gospel that we can trace back with any confidence to the first century, or to the historical Jesus, or to the historical Mary."[164]

---

[161] Blomberg (2007).
[162] Strobel (2008: 19).
[163] Strobel (2008: 21).
[164] Strobel (2008: 27).

These other gospels were therefore written much later than the four gospels and have prominent names unrelated to their real authorship to try and give some credibility. They present a distorted image of Jesus and conflict with the four gospels, which have more credibility. For example, the gospel of Peter has a very embellished version of the resurrection including a talking cross! The apocryphal gospels often refer to bizarre events like the Roman statues bowing down to Christ during his trial before Pilate (Acts of Pilate 5-6). For further information on these writings see Blomberg[165] who concludes :

> The probability that any of the extra-biblical sources preserve accurate information, otherwise unknown, about the life and teaching of Jesus is very slight, apart from the possibility of a few unparalleled sayings surviving.[166]

How did we arrive at the New Testament canon that we have today? It seems that the elders of congregations approved certain writings and rejected others as they became available. Some documents were hand-delivered by friends of the apostles to elders who also knew the apostles personally. Other documents promoted strange doctrines and could be ruled out. The elders received the writings of the apostles and their close companions as authoritative, as well as the writings the latter authors endorsed. For example, the gospels of Matthew and John were written by apostles, while the gospel of Mark was written by Peter's closest companion, and the gospel of Luke and the book of Acts were written by apostle Paul's close companion Luke. Thirteen letters were received from Paul. Similar comments apply to the other books. A few books took a while to be universally accepted.

By about AD 170 most churches were in general agreement having approved much the same books independently, especially the four gospels and the Pauline writings. It took some time to achieve consensus as transportation and communication were slow for the hand circulation of books to build up collections, especially between the Eastern and Western churches. However, gradually these two branches of the church arrived at a common agreement on the list of sacred books. Various bishops governing groups of churches as well as some regional Councils ratified these lists. The first official document that canonized just the twenty-seven books we have today was Athanasius's Easter letter for the year AD 367. The approved books were called the "canon" of scripture and modern scholarship has confirmed the choice of these books in terms of inspiration and authenticity. It could be said that God determined the canon and the church discovered it through the prompting of the Holy Spirit.

[165] Blomberg (2007: 112-115, 264-280).
[166] Blomberg (2007: 279).

5.5 PROPHECY IN THE BIBLE

An important aspect confirming God's hand in the source and existence of the Bible is the role of fulfilled prophecy. We shall look at some of these prophecies. Theologian John Barton Payne[167] has assembled a very comprehensive collection and analysis of nearly 600 topics of prediction in the Bible with about 200 in the Old Testament that he says gave rise to 8,352 predictive verses. Of the Old Testament, 70 percent are fulfilled in the Biblical narrative itself and the remaining unfulfilled 30 percent concern the future (e.g., second coming of Christ, the millennial kingdom, and so forth). As noted by theologian Gleason Archer,

> Virtually no prophetic utterance recorded in Scripture pertaining to any event through to the advent of the Church in New Testament narrative has failed to be fulfilled. Biblical prophesy is precise, explicit, and accurate with a record of proven fulfillment that stands as its own testimony of conclusive evidence as to the veracity of Scripture.[168]

The test of a true prophecy is that it is fulfilled.[169] False prophets were condemned such as Hananiah in Jeremiah chapter 28. In the past there have been number of psychics who have made prophecies such as Nostradamus[170] in the sixteenth century and Jeane Dixon in the twentieth century.[171] However, they are either mostly wrong and possibly no better than guessing, or their prophecies are so vague that they can fit a variety of situations. Some, like the two people mentioned, were involved in cultic practices, which might have helped them; Satan has some power. In contrast, biblical prophecies were specific, as seen in the following selection.

(1.) The fall of the Northern kingdom to Assyria in 722 BC.

(2.) The destruction of Babylon.[172]

(3.) The destruction of Tyre and its rubble swept into the sea.[173]

(4.) The destruction of Jericho.[174]

(5.) Nebuchadnezzar attacking and devastating Egypt.[175] The Greeks never mentioned this, and the account by the Jewish historian Josephus who did refer to it was regarded as a fabrication until a small fragment of a Babylonian chronicle from about 567 BC and an inscription on a statue confirmed the account.

---

[167] Payne (1973).
[168] Archer (1994: 564).
[169] Deuteronomy 18:20–22.
[170] See Geisler (1999: 544–546, 615) for a critique.
[171] For a critique by Geisler, see Strobel (2000: 189).
[172] Jeremiah 32:28ff, 50:13, 39 and chapter 51.
[173] Ezekial chapter 26; 1 Kings 16:34.
[174] Joshua 6:26.
[175] Jeremiah 43:8–13.

(6.) The fall of Jerusalem and deportation of the Jews under Nebuchadnezzar in 586 BC.

(7.) The identification of Cyrus, king of Persia,[176] who began the restoration of Israel. Daniel presented Cyrus with the writings of Isaiah which included a letter addressed to Cyrus by name, but written 150 years earlier!

(8.) The prophesy that Josiah, a descendent of David, would obliterate the idolatrous worship introduced by Jeroboam 1 in Bethel.[177] This was fulfilled explicitly 300 years later.[178]

(9.) The Book of Daniel is full of detailed predictions about the course of human history over many hundreds of years. Its dating is discussed in Section 5.2.2 above.

(10.) The doom of Edom (Petra) to desolation.[179] Although regarded as impregnable it was conquered by Muslims in AD 636 and is now deserted apart from travellors and tourists.

(11.) The desert flourishing in Palestine.[180] Today Israel has transformed the desert.

The Bible itself makes the fulfillment of prophecy the test for its own validity, and thus the reputation of the Bible stands or falls on prophecy. God invites others to do the same test.[181]

5.6 ALLEGED BIBLICAL CONTRADICTIONS

It has been my experience that on occasion someone will say that the Bible is full of contradictions, but is unable to produce one when challenged as they haven't read the Bible. There are however difficulties and contradictions, and philosopher Norman Geisler and biblical scholar Thomas Howe[182] have written an extensive book that provides suggested answers to some of the difficulties or so-called inconsistencies in the Bible. We now give a number of general principles that need to be applied in comparing documents, and include some of the introductory comments of Geisler and Howe.

(1.) We don't expect different eye-witness accounts to agree exactly otherwise we would suspect collusion. We referred to this above with regard to the gospel accounts.

---

[176] Isaiah 44:28–45:1–6.
[177] 1 Kings 13:2.
[178] 2 Kings 23:15–16.
[179] Jeremiah 49:16–17.
[180] Ezekial 36:33–35.
[181] See Isaiah 41:21–24 and 46:9–10.
[182] Geisler and Howe (1992a).

(2.) In the Old Testament we have many different authors writing from different perspectives and emphases so that their editing of events may focus on different details. For example, the books of 1 and 2 Kings recount the histories of the kings of Israel, while the books of 1 and 2 Chronicles do the same, but their narratives often include quite different details. This does not mean that the two accounts are contradictory. We see this with history writing today where different writers emphasize different historical details.

(3.) We need to take into consideration the cultural background of the writings. For example, in one instance Daniel uses the Hebrew calendar while Jeremiah uses the different Assyrian calendar.[183] It has been already noted above that the biblical languages had no quotation marks, but are added in English translations. Being totally precise with quotations is a modern phenomenon and in the ancient world it was acceptable for a writer to paraphrase someone else's words as long as the original meaning was not lost.[184]

(4.) A common mistake of critics is to take a passage out of context or neglect to interpret difficult passages in the light of clear ones. An example of the latter is Ezekial 18:20 which says that God does not punish sons for their fathers' sin and vice versa, but are each responsible for their own sins. In contrast, Exodus 20:5 says that God visits the iniquity of the fathers on the children to the third and fourth generations. The difference is that Ezekial refers to the guilt of the father's sin, while Moses refers to the consequences of the father's sin. As a counselor I regularly see the effects of parents' inappropriate behavior impinge on grown-up children and even grandchildren. In counseling terminology there is an inner child that needs healing. Also because of bad parental modeling, children may buy into the same inappropriate behavior as their parents (e.g., drunkenness, abuse, dishonesty) and come under the same condemnation as their parents. There is also the effect of epigenetics discussed in Section 2.6.3.

(5.) A passage can be unclear because a word is used that is not used elsewhere. We need to remember that almost all of the Old Testament was written in Hebrew (a language without vowels), except for a few Aramaic chapters in Ezra and Daniel, and later translated into Greek (the Septuagint). The New Testament was written in everyday Greek, although Aramaic was the common language spoken in Israel in Jesus' time, and it was probably the language he spoke on a daily basis. Trans-

---

[183] Daniel 1:1b and Jeremiah 46:2.
[184] For example compare Matthew 3:17 and Mark 1:11; Matthew 16:16, Mark 8:29, and Luke 9:20.

lating from these languages into English would no doubt lead to some translational difficulties.

(6.) We need to remember that the Bible uses a number of literary devices[185] and different writing styles. We also have different human perspectives, for example from a shepherd (Psalm 23), a prophet (Kings), a priest (Chronicles), a historian (Acts), and a pastor (2 Timothy).

(7.) One mistake made by biblical critics is to assume that the unexplained is unexplainable; a bit like assuming guilty until proved innocent! Scientists will often tell us that because we can't understand something right now, science will eventually provide a solution. There are a number of examples where critics have assumed the Bible to be wrong, but historical evidence has arisen to prove otherwise. For example, it was suggested that Moses couldn't have written the first five books as there was no writing at the time. However it has been found that writing existed a long time before Moses.

Another example where the Bible was regarded as wrong was when it mentioned the Hittite people, assumed to be nonexistent. However, we now know that they existed through their library found in Turkey. The destruction of Sodom and Gomorrah from fire and brimstone raining down was regarded as fictitious. However, the five cities of the plain have been located and the evidence is substantial. Round balls of brimstone, or nearly pure sulfur, have been found embedded in an ashen area near the Dead Sea, which shows clear signs of what were ancient building structures! The area is on top of a burnt out oil field and an earthquake (caused by God?) would have caused liquid fire in the skies from gas and oil coming up along the fault lines.

(8.) One misconception by critics is to assume that a partial report is a false report or that divergent accounts are false ones. As already noted above, Biblical authors, as in the gospels, often focus on different details without contradicting the essence of the story.

(9.) It was not uncommon for people to have two names, which can lead to some confusion. For example we have the more well-known ones Abram (Abraham) and Israel (Jacob) in the Old Testament and Cephas (Peter) and Saul (Paul) in the New Testament. One other example (Nehemiah 2:19, 6:6) is Geshem (Arabic) and Gashmu (Hebrew).

(10.) Round numbers, as used in the Old Testament, may differ as they are just that —round numbers. For example, in 1 Kings 7:23 the value of π, the ratio of the circumference to the diameter of a circle, is rounded to three when it is actually 3.14159.

---

[185]See for example http://carm.org/bible-literary-techniques, accessed August, 2015.

(11.) General statements should not be confused with universal ones. Finding an exception to a statement does not mean the statement is wrong. Rules of life give general guidance and will admit exceptions, as in the book of Proverbs, for example.

(12.) It is important to remember that later revelation supersedes previous revelation and that God does not reveal everything at once. A good example is that although God originally commanded that animals be sacrificed for people's sin, this is no longer necessary as Christ offered himself as the perfect sacrifice.[186] The sacrificial system looks forward to the sacrifice of Christ on the cross. A change of revelation is not a mistake as revelation is progressive, with each command being suited to the circumstance. Commands, however, that deal with God's unchangeable nature do not change.[187]

(13.) Words can have more than one meaning, as in the following selection of examples. In Genesis 15:16 the word translated "generation" means a 100 years, an amount of time rather than a number of descendants. In Genesis 22:2, Isaac is referred to as Abraham's only son even though there was another son Ishmael. The latter was the son of a concubine and therefore not regarded as an heir to the promised inheritance. Also the phrase "only son" may be equivalent to "beloved son."[188] The word for "all" can also mean "the majority."[189]

(14.) Copying errors do exist, but these in general do not cause problems. Geisler and Howe[190] show how easy it was for scribes to make copying errors when you see the similarity in appearance of many Hebrew characters. For instance, in Hebrew, the letters d (daleth) and r (resh) are almost identical and m (mem) and t (taw) are quite similar. When it comes to numbers, the ancient Hebrew language unlike English did not have a set of numerals but instead used words and letters of their alphabet to stand for numbers. Numbers were therefore very easy to miscopy. An example of number confusion is given in 1 Samuel 13:5 that says that the Philistines had an army of 30,000 chariots and 6000 horsemen, whereas a more appropriate number of chariots is 3000. This would give a better ratio of chariots to horsemen. Another type of copying error arises in 2 Samuel 21: 19 where it says that Elhanan killed Goliath when in fact Elhanan killed Goliath's brother.[191]

---

[186] Hebrews 10:11.
[187] Malachi 3:6; James 1:17.
[188] John 1:18, 3:16.
[189] Exodus 20:7, 22; 9:6,19.
[190] Geisler and Howe (1992a: 558–562).
[191] 1 Chronicles 20:5.

(15.) Sometimes two events are actually different and are not two versions of the same event (e.g., feeding of the five thousand and the feeding of the four thousand.[192]

5.6.1 Unpleasant Old Testament Stories or Practices

A fundamental objection raised against the Old Testament in particular is that, quoting philosopher and neuroscientist Sam Harris,[193] there are the instructions to stone people to death "for heresy, adultery, homosexuality, working on the Sabbath, worshipping graven images, practicing sorcery, and wide variety of other imaginary crimes." It is true that there are a number of unpleasant accounts such as David's adultery with Bathsheba and getting rid of her husband, and even horrific stories in the Old Testament like Judges 19:16–30. The fact that they are recorded does not mean that the Bible condones them. Matthew Flannagan, theologian and philosopher,[194] points out that while some of the laws reflect justice and equity for all people, the laws were for Israel only, as Israel had a covenant treaty with God that the Gentiles were not party to.[195] Gentile Christians and Christians today are also exempt, a fact clearly iterated in the New Testament but ignored by Sam Harris.

A comparison of Leviticus and Deuteronomy and other contemporary ancient Near Eastern law codes suggests that they have a similar genre. In particular they had harsh penalties like the old Babylonian law; for example where the hand that assaults is cut off. Quoting from a number of Near Eastern scholars, Flannagan points out that there was a big difference between the statement of laws and their actual practice of carrying out the punishment. The severity of the stated punishment was meant to emphasize the seriousness of breaking the law, but in practice there was a "ransom" that could be paid instead of the prescribed punishment. For example, instead of losing a limb or even life itself the transgressor could make a monetary payment and, or possibly, agree to a lesser penalty usually set by the courts.

In near Eastern practice the laws were written and read with the background assumption that penalties were often ransomed and not literally carried out. It appears that the Torah operated with the same assumption. For example Exodus 21:29–32 refers to the situation where an ox gores another person to death due to negligence of the owner of the ox. In this case the owner must be put to death also. But in the very next verse it says that if payment is demanded of the owner, the owner can ransom his life by paying whatever is demanded. We find a similar allusion to a payment in 1 Kings 20:39. There were some sixteen crimes that called for the death penalty in the

---

[192] Mark 9:19–21.
[193] Harris (2006: 8).
[194] Flannagan (2011).
[195] Psalm 147:19–20; the term "God of Israel" is mentioned 203 times.

Old Testament but in Numbers chapter 35 capital punishment was applied to a murderer without allowing a ransom (verses 31–32) when the murder was premeditated. This is widely interpreted to imply that the other fifteen cases could be covered by a ransom. Clearly a ransom was so prevalent that if the biblical law wanted it excluded it had to be specifically prohibited.

From the outset it must be recognized that the culture of the time (including the Jews) was cruel and bloodthirsty so that events need to be seen in this context. Also, as was the custom of the time, a certain amount of rhetoric and exaggeration were used in describing warfare and entrance into the "promised land." However, the God of the Old Testament has been described as vindictive and jealous as he exterminated large numbers of people through pestilence, famine, and warfare, sometimes for offenses that might seem unreasonable today. Without downplaying the significance of such events, it is important to remember it is easy to be very selective and forget the positive things in the Old Testament about God such as His concern for the orphans, the widows, the poor, and foreigners.[196] Frequent references are made to the fact that God is merciful, gracious, compassionate, slow to anger and abounding in steadfast love.[197] Such themes run right throughout the Old Testament and are conveniently ignored by critics.

Before discussing further some of the difficult stories and practices I want to set the Old Testament scene. God's aim was to create for Himself a race of people who would worship Him and eventually give rise to the Messiah, Jesus Christ. They would take over land occupied by nations who were idolatrous and cruel and create a God-space as a light to the surrounding nations. God formed a covenant with the Jews that required them to subject all their lives including worship and social, family, and individual aspects to Himself, and to guard the purity of both their worship and their land. This was done through regulations and a sacrificial system, as there was no Bible then to guide them. If the Jews obeyed God, they and their land would be blessed, but they would face dire consequences if they disobeyed; covenant conformity brought blessing but covenant rejection brought tragedy and sorrow. To maintain purity of worship, God was against intermarriage with surrounding nations as this typically led to idolatry and other inappropriate practices such as Baal worship. Keeping the race pure was very much on God's agenda for this reason as we see by the admonition in Joshua 23:6–13 not to intermarry or become involved with foreign gods.

In the rest of this section we shall look at some of the more prominent events and practices that have caused controversy.

(1.) *Destruction of the Amalekites.* God told the Israelites to totally wipe out the Amalekites and every living thing belonging to them, namely,

---

[196] For example Deuteronomy 10:18, 15:11; Exodus 23:6.
[197] Exodus 24:6; Numbers 14:18; Psalms 86:15, 103:8, 145:8; Jonah 4:2.

men, woman, children, and livestock.[198] The Amalekites were a horrible and depraved people, persistently warring, and bent on wiping out Israel from whom the Messiah Jesus would come. They chased after the Israelites killing off the vulnerable and the stragglers who could not keep up. They persistently attacked Israel when the latter was most vulnerable,[199] and they and any remnants would always be a problem. You may ask, "What about the innocent young children?" Clearly no one could look after the orphans, and children who had not reached the age of accountability would go to heaven and avoid later corruption.[200] Also most of the women, young children, and noncombatants would have escaped (there is in fact no mention of them being killed), leaving just some fighting men behind. That is why there were plenty of Canaanites around after the conquest of the land, as the biblical record attests. The purpose of the conquest was to drive the tribes out of the land to occupy it. Some allowance probably needs to be made for the cultural aspect of exaggeration used at the time for war stories. We sometimes have a coach tell his team to go out and kill the opposition![201]

We are also told that the Amorites, another nation under God's judgement, had a long time to repent (over four generations) before they were finally destroyed.[202] God told Abraham prior to Israel's bondage in Egypt that his offspring would sojourn in a land not theirs for four hundred years and be servants there, for the "iniquity of the Amorites is not yet complete." God was patient for all that time, but the offenses had become intolerable, and God did not want the Canaanites to corrupt His people.[203]

God is sovereign over life and God used Israel to carry out His judgement. At one stage he even used Assyria to carry out His judgement on Israel, calling Assyria His rod; but Assyria was later judged as well.[204] We see a similar command with regard to seven other nations in Deuteronomy chapter 7, where the emphasis is on breaking down their idols and avoiding intermarriage because of their corrupting influence. The Canaanites, a very cruel nation, practiced incest, bestiality, cultic prostitution, and child sacrifice by fire. That was why Jericho was totally wiped out, apart from Rahab who helped the spies (Joshua, chapter 6). It should be noted that God commanded the Israelites to first make an offer of peace whenever they went into an enemy city.[205]

---

[198] 1 Samuel 15:1–3.
[199] Numbers 14:45; Judges 3:13, 6:3-4.
[200] Deuteronomy 1:39; Jeremiah 19:4–5; Mark 10:14.
[201] For further discussion see Copan and Flannigan (2014).
[202] Genesis 15:13,16.
[203] Deuteronomy 20:18.
[204] Isaiah 10:5–6,12.
[205] Deuteronomy 20:10–13.

God saved those who repented, no matter how bad they were. For instance, God took pity on the inhabitants of Nineveh, a pagan Assyrian city, and sent Jonah to warn them to repent. They did, and were saved. Jonah was extremely upset as the Assyrians had a terrible reputation; for example they introduced crucifixion. We read that God does not take pleasure in the death of the wicked[206] and is described as being long suffering.[207] The themes love and judgement are found in both the New and Old Testaments about equally.

The above comments should be balanced by the fact that God loved the sojourner or foreigner[208] and commanded the Israelites to do the same. This is reiterated again by Solomon in his prayer of dedication of the temple.[209] Much of the book of Joshua is about how to live appropriately and peacefully in a pagan land.

(2.) *Mauling by Two Bears.* Forty-two children were apparently mauled at God's instigation by two bears because they slung off at God's prophet Elisha.[210] However, in the original Hebrew, the children were actually young men and could be regarded more as a large street gang threatening Elisha and putting a man chosen by God in danger. By telling him to "go up" they were saying that he should ascend into heaven like his predecessor Elijah, and prove who he was. They were attacking Elisha about his priestly role and mocking Elijah's ascent. As lepers shaved their heads, they were insulting Elisha as well when they called him baldy. In that society, what they did was a serious offense against religion and against God.

(3.) *The Bible seems to condone polygamy.* Solomon had 700 wives and 300 concubines, and other men that God praised such as Abraham and David had multiple wives and possibly concubines. Monogamy was God's standard for marriage[211] and polygamy was only tolerated. The Law of Moses prohibited polygamy, commanding not to multiply wives.[212] God also warned Israel not to marry worshippers of other gods as they would turn people away from God, which is what happened to Solomon.[213] His wives did a great deal of damage to the house of David and Israel. In fact every polygamist in the Bible, including David and Solomon, suffered for their actions. Even Abraham took a concubine, which turned out to be a disaster.

---

[206] Ezekial 33:1.
[207] Exodus 34:6.
[208] Deuteronomy 10:18–19.
[209] 1 Kings 8:41–43.
[210] 2 Kings 2:23–24.
[211] Genesis 1:27, 2:24; Matthew 19:4; 1 Corinthians 7:2; 1 Timothy 3:2, 12.
[212] Deuteronomy 17:17.
[213] 1 Kings 11:1–6.

(4.) *God seems to sanction rape.* In Numbers 31:17–18 we read that Moses encouraged his men to kill all the boy captives and female captives who are not virgins but keep virgins alive for themselves. God did not rebuke him but urged him to distribute the spoils (verses 25–40). We see earlier in the story that this was a compromise situation as there weren't supposed to be any captives. In fact the passage needs to be understood in terms of Deuteronomy 21:10–14 where Moses specifically stated what was to be done with female captives. God never condones sexual activity outside of a lawful marriage so an Israelite could only have sexual intercourse with a female captive if he first makes her his wife, thus giving her all the rights and privileges due to a wife. He must also allow her a period of mourning and cleansing first.

In another passage[214] it seems that a woman's rights and her welfare are ignored. If a man rapes a virgin who is not betrothed (engaged) and they are discovered, the rapist has to pay a bridal fee to the father and marry the woman, but no divorce is allowed in the future and he always has to support her (which could certainly be a deterrent!) If the woman is betrothed (which virtually means married in that culture) and didn't cry out, she and the rapist would be put to death as it was seen as consensual adultery. If the rape occurred in open country so that the victim's cry for help could not be heard, the rapist would be put to death. The passage does not sanction, condone or approve of rape. In fact one needs to look carefully at the original Hebrew words as they give a different story. The word translated as rape in the NIV translation of the Bible comes from two Hebrew words *taphas* and *shakab*, neither of which necessarily means rape (e.g. *taphas* can mean seize or grasp as in using a harp, sickle, bow etc., and does not necessarily mean the use of force, while shakab can mean lie down or consenting to sex).

Alternative words (e.g., *chazaq* indicating force) are used in other passages where rape had occurred. A weaker word *anah* meaning "to humble, afflict"[215] need not indicate sex at all. Usually Deuteronomy 22:28–29 is interpreted in terms of seduction. For example, Exodus 22:16–17 says that a man who seduces a virgin who is not betrothed, has to pay the father the dowry and marry the woman. If the father refuses the marriage (and who would want a daughter to marry a rapist), the man still has to pay the dowry. (The word *patah* is used here meaning coaxing or luring.) Before marrying, a man must seek the father's permission and negotiate a bride-price.

One story that shows the grief of a rape victim is given in 2 Samuel 13. Amnon, who raped his half-sister, was not forced to marry Tamar,

---

[214] Deuteronomy 22:22–29.
[215] For example, Exodus 1:11, 22:22.

though she apparently wanted it rather than live as an ashamed, unmarriageable, and single woman for the rest of her life, which then happened (verse 20). In that culture, a woman without a husband had a difficult time providing for herself.

(5) *God seems to sanction slavery.* When it comes to the topic of slavery we need to realize that the word today carries more modern emotional connotations such as the often brutal slavery in nineteenth-century America, as well as the more recent idea of sex-slaves. The picture we have in the Old Testament is very different from this as slavery was a very common feature of society in those days; it was a common and an accepted part of life. Also we need to realize that some things were not ideal but were allowed because of human weakness; divorce and slavery were in that category. What do you do with survivors of wicked nations who did things like sacrifice their children to the god Molech by burning them, or with people who could not pay their debts? For example a thief would have to make multiple restitution[216] or else be sold so that he could work off his debt.

Today such people would be put in prison and may be forced to do or not do certain things. In those days slavery was used instead, which in some cases can be regarded as a reasonable alternative. Many of the slavery regulations pertained to people who deserved far worse and perhaps deserved the death penalty. However, the slavery laws provided some social recognition and legal protection that was advanced for its time; slaves had rights.[217] Special rights applied to Jewish slaves because they were citizens; for example, they were to be freed after six years and not be sent off empty-handed.[218]

All slaves were not to be oppressed,[219] whether Jew or alien, as God is impartial,[220] and they were all entitled to take part in the Sabbath rest and other festivities.[221] Even the difficult slave was not to be beaten to death or the master would face punishment,[222] while special provision was made for those enslaved because of debt.[223] Such was the relationship of slave and master that some slaves did not want to leave their masters even when entitled to.[224] If in fact a master did abuse his slaves,

---

[216] Exodus 22:1–3.
[217] Exodus 21:20–27; Job 31:13–15 and, in particular, Leviticus 25:39–55.
[218] Deuteronomy 15:12–14; Exodus 21:2.
[219] Deuteronomy 24:14.
[220] Deuteronomy 10:17–19; Leviticus 19: 34.
[221] Exodus 20:10; Deuteronomy 16:9–17.
[222] Exodus 21:20.
[223] Leviticus 25:47–49.
[224] Deuteronomy 15:16–17.

and the slaves decided to run away, then God made it unlawful for a runaway slave to be returned to his or her own master.[225]

In the surrounding cultures of the time there was the practice of selling your own children, especially women, as slaves. This practice still existed among the Israelites so that regulations were introduced to ameliorate what was already ongoing, even if not condoned. Finally, if a man kidnapped another man and sold him as a slave, he faced capital punishment.[226] Clearly slavery was normally not oppressively harsh. Slavery in the New Testament is discussed in Section 9.5.4.

(6.) *The psalmist seems to condone the killing of babies.* In Psalm 137:9 we read that the psalmist calls blessed the one who dashes Babylonian babies against a rock. The context of the Psalm is the Babylonian captivity and the cruel treatment of Israel by the Babylonian tormentors. The Psalmist is not rejoicing over the death of babies but rather is referring to the retributive justice that God would eventually bring on the Babylonians. It does not mean that the Psalmist's request has God's approval, nor did God actually carry it out. God allows us to "sound off", and works within the culture and understanding of the time, as he does now.

5.6.2 New Testament Difficulties

The above difficulties have focussed on the Old Testament. However there is one in the New Testament that has raised a number of questions. A point of contention with critics is the length of time that Jesus and his followers thought would elapse before the end of the world. Did Jesus make a mistake about some of his disciples seeing the kingdom come in their lifetimes (Matthew 16:28; Mark 9:1)? Jesus also said that this generation would not end until all the signs and wonders associated with his second coming would occur in his era (Matthew 24:34 and Mark 13:30). In Matthew 10:23 he says that the disciples will not finish going through the towns of Israel before he returns.

There are several possible answers to these verses as a lot of Jesus' teaching that is accepted as authentic even by the most radical scholars presupposes the continuing existence of Jesus' followers teaching a Christian community about him. Jesus also took a stand on a number of ethical issues and, in particular, applied the Old Testament commandments in a new way to everyday life, as in the Sermon on the Mount. He said that no one, including angels and the son, knows the hour when he will return except God. (Matthew 24:36; Mark

---

[225] Deuteronomy 23:15–16.
[226] Exodus 21:16.

13:32). Some alternative explanations to the above questions are a follows.[227]

First, Matthew 16:28 could refer to Jesus' subsequent transfiguration (mentioned in the next verse) that foreshadowed his final coming. Second, in Mark 13:30, "generation" can mean "race" in the Greek indicating that the Jewish race would be preserved, or else it can refer to the people who will be alive when the things predicted happened and are still alive at the end. Also the phrase "all these things" refers to signs, persecution, and the destruction of Jerusalem, which happened soon afterwards. They would take place before the Son of man came. Third, in Matthew 10:23 Jesus is predicting the continually incomplete mission of preaching the gospel to all the Jews. In those days it was quite common for the Jews to expect that the end of the age would come quite soon, so it is perhaps not surprising that the disciples interpreted Jesus' teaching to mean he would return in their generation.

5.7 THE BIBLE AND SCIENCE

Does science contradict the Bible? In dealing with this question, another question has to be added, namely "Which science?" Science continually changes. For example in the 19th century science was deterministic (but what about Heisenberg's uncertainty principle?), reductionistic (with all laws seeming to be derivable from a few basic laws; not so), and realistic (as scientific theories were literal descriptions of nature and therefore independent of the experimenter; not so with the theory of relativity and quantum theory). We find that theology and its interpretation of the Bible also changes so it is not surprising if there are clashes. Clearly the Bible is not a book on science but describes, through various types of literature, God dealing with His people. God did not dictate the Bible to its writers but somehow inspired them to write it. This will mean that writers used thought forms of the day to express God's message, and something of the current culture will show through at times. Does this mean that some statements will be factually or scientifically wrong? There are some who say no and systematically provide alternative explanations to problem passages.

One of the problems is that it is not obvious when metaphors rather than facts are being used. An interesting example is how the Bible views the cosmos. For example, 1 Samuel 2:8 talks about the earth supported on pillars while Job 26:7 says that God hangs the earth on nothing. In Isaiah 40:22 we read about the "circle of the earth" in some translations, but this verse is highly controversial and is capable of other interpretations.[228] The verse also refers to stretching and spreading that could be related to modern cosmol-

---

[227] For example, Blomfield (2007: 64–65).
[228] e.g., http://www.bibleandscience.com/bible/books/genesis/genesis1_circleearth.htm, accessed August, 2015.

ogy, but the main image here is that of God's tent. However clearer verses about God stretching out the heavens are available.[229] The Bible also speaks about innumerable stars[230] when people believed they could count the stars, which we cannot do today. A detailed discussion on Genesis 1 and 2 and its relationship to science is given in Section 8.9.

What is particularly interesting are the hygiene laws spelt out in the Old Testament as they are so forward looking in many ways and different from beliefs in the ancient world. For example, the Israelites were forbidden to eat the flesh of any animal that had died a natural death,[231] those with contagious diseases had to be quarantined,[232] and sewage had to be be disposed of outside the camp of Israel.[233] Also the Israelites were forbidden to drink water from small or stagnant pools or from water that had been contaminated by coming into contact with animals or meat.[234] Rabbit and pork were off the menu, which makes sense as both are susceptible to infectious parasites, though we can eat them today when properly prepared. Two animals, the rock hyrax and the rabbit are described as chewing the cud,[235] although technically this is not the case, but they have a chewing action that resembles chewing the cud. Anything in the waters that doesn't have scales or fins was not allowed, and certain insects were forbidden as well as certain birds, particularly birds of prey.[236] We note that pest control was also introduced. Moses commanded Israel to set aside one year in seven when no crops were raised.[237] God promised sufficient harvest in the sixth year to provide for this period. This method served to eliminate insect cycles that plagued the ancient world.

Summing up, there are three points that can be made about the food laws: (1) As already mentioned in previously discussing Leviticus, the laws were part of God's covenant with Israel to separate them from other nations, and were not imposed on Gentiles; (2) the laws had health benefits; and (3) the laws were symbolic of correspondences between animals and humans. An example of the symbolism is shown in the story of Peter who, when praying, had a vision of a sheet being let down from heaven by four corners containing so-called unclean animals.[238] A voice told him to rise, kill, and eat, but he refused as they were unclean. The voice then said that what God had cleansed was not unclean. This happened three times. Then three Gentiles turned up at the house and God's spirit told Peter to go with them, thus making no distinction between Jew and Gentile. The message is: Once people of all nations could

---

[229] Isaiah 42:5, 44:24, 45:12, 48:13, 51:13; Jeremiah 10:12, 51:15.
[230] Jeremiah 33:22.
[231] Deuteronomy 14:21.
[232] Leviticus 13:45–46; the word translated "leprosy" referred to a number of skin diseases.
[233] Deuteronomy 23:12–13.
[234] Leviticus 11:29–36.
[235] Leviticus 11:5–6.
[236] Leviticus chapter 11.
[237] Leviticus 25:1–24.
[238] Acts 11:1–12.

belong to the people of God, those food laws that had symbolized Israel's election and separation from other nations became irrelevant too.

Does the Bible ever anticipate science? Some find references to wireless telegraph (Job 38:35), atomic theory of matter (Hebrews 11:3), motor cars (Joel 2:3–4), airplanes (Isaiah 31:5 and 60:8), and many other modern discoveries by suitably translating certain Biblical passages. But we must resist the temptation to choose the translation that suits us, and try to determine the most natural interpretation of each passage we consider. What is important about the Bible is that it does not buy into some of the errors in other writings of the time. For example we don't have alchemy, astrology, divination, bizarre medical prescriptions, magical control of disease, magic and omens, belief in a living earth, the deification of nature, the magical or demonological control of nature, and an incorrect understanding of earthquakes, storms, oceans, mountains, lightning, and other physical phenomenon.

5.8 SCRIPTURAL INSPIRATION

What does the Bible say about itself? We read in the Old Testament that God spoke through His holy prophets[239] using phrases such as, "Thus says the Lord" and "The Word of the Lord came to." Other examples are prophets speaking all the words of the Lord[240] and God putting words in a prophet's mouth.[241] Amos said that you are compelled to prophesy when God speaks to you,[242] and Jeremiah was commanded to stand in the courtyard of the Lord's house and tell the people everything God commanded him to tell without omitting anything.[243] Jeremiah did not want to be a prophet because he was only a youth who didn't know how to speak, but God touched his mouth and put words in his mouth (Jeremiah 1:4–11). He was beaten and put in the stocks (20:1–4), threatened with death (26:8,11,24), imprisoned (32: 2, 37:15), and dumped in a miry cistern to die because of his unwelcome prophecies, but which he escaped from (Chapter 38).

Some writers were moved to speak by the Holy Spirit such as David, as God's word was on his tongue.[244] This idea is well summed up by Peter when he said that scripture came from God through the Holy Spirit of God and not through the will of man.[245] Paul summed up the Old Testament by saying that all scripture and its prophecies are inspired by God,[246] where the Greek word for "inspired" means "God-breathed." Jesus believed the Old

---

[239] Luke 1:70; Acts 3:21.
[240] Exodus 4:27–30.
[241] Deuteronomy 18:18.
[242] Amos 3:8.
[243] Jeremiah 26:2.
[244] 2 Samuel 23:2.
[245] 2 Peter 1:20–21, 3:2.
[246] 2 Timothy 3:16–17.

Testament to be the indestructible Word of God[247] and used the same phrase, "The Word of God", when referring to certain Old Testament passages. In fact Jesus charged the Pharisees with replacing the commandments of God with their own tradition so that they were guilty of invalidating the Word of God.[248] For Jesus the scriptures were fully authoritative and therefore couldn't be broken.[249]

In the New Testament Paul referred to his message as the word of God[250] and it came through a revelation of Jesus Christ.[251] Peter refers to Paul's writing as "scripture."[252] Jesus promised inspiration to his disciples as the Holy Spirit would guide them into all truth and help them to remember what Jesus had told them.[253]

5.9 CONCLUSION

Many different topics have been covered in this chapter. However the evidence and arguments presented indicate that the Bible is a very special book and much of the criticism leveled against it is not supported. The Bible is very different from other religious books in that it is founded in history and stands up to historical investigation. Although the Old and New Testaments have many different authors, there is a common underling thread running throughout the whole Bible showing a progressive understanding of God and culminating in the life and teaching of Jesus. The Bible has also withstood the march of time in that new discoveries continue to support its reliability.

---

[247] Matthew 5:17–18.
[248] Mark 7:13; Matthew 15:6.
[249] John 10:35.
[250] 1 Corinthians 14:37; 1 Thessalonians 2:13.
[251] Galatians 1:11–12.
[252] 2 Peter 3:16.
[253] John 14:26, 16:12–13.

CHAPTER 6

WHO IS JESUS?

6.1 EVIDENCE FOR HIS EXISTENCE

No person in the whole of history has polarized people like Jesus Christ. People are happy to talk about God, but mention Jesus and the conversation may stop! Jesus asked his disciples "Who did men say that I am?" In a similar manner we get various responses to the same question today. For some people Jesus is just a swear word, others would describe him as a sort of guru or great teacher or (like the Muslims) a great prophet, some would see him as a purveyor of secret knowledge (the gnostic Jesus),[1] and others would describe him as a non-violent revolutionary who brought into the harsh Roman world ideas of humility, love, kindness, and self-sacrifice.

As noted by theologian Hans Küng,[2] Jesus was above all a non-legalistic Jew from whom a Jew could learn how to pray, fast, love God and neighbor, and understand the meanings of the Sabbath and God's kingdom. He was neither a priest nor theologian, nor a member of a political party such as

---

[1]Strobel (2008: chapter 1).
[2]Küng (1984: 178–188).

the Sadducees, for example, yet called into question the ecclesiastical establishment. Nor was he a social revolutionary who agitated against the Roman occupation (e.g., no call to refuse paying taxes to Caesar) yet stood up for justice and spoke against social abuses on behalf of the poor and the oppressed. Nor was he a religious recluse who advocated withdrawal from the world, but he socialized with all kinds of people including those marginalized such as lepers, tax collectors, and disreputable characters. Although not legalistic, he reinterpreted the law such as anger was like murder, adulterous desire was like adultery, untrue words were like perjury, and uncleanness came from within and not from the outside. He focused on the heart rather than on legality, and was against self-righteousness. Küung writes:

> Throughout all the gospels he appears as the unarmed, itinerant preacher and the charismatic physician who does not inflict wounds but heals them: one who relieves distress and does not exploit it for political ends, who proclaims not militant conflict but God's grace and forgiveness for all.

Also Küng concludes:

> Jesus apparently cannot be fitted in anywhere... He is neither a philosopher, nor a politician, neither a priest nor a social reformer.... Jesus in his wholeness turns out to be completely unique—in his own time and ours. ... the historical Jesus cannot be interchanged either with Moses or with Buddha, either with Confucius or with Mohammad.[3]

Although the New Testament is the major source about Jesus' life, his existence is also documented in pagan, Jewish, and Christian writings outside the New Testament. For example, Josephus, a Jewish priest and Pharisee who was regarded as a reliable historian, wrote in his "Antiquities of the Jews" around AD 93 a passage that corroborated important information about Jesus.[4] He referred to James, the brother and later follower of Jesus,[5] and also mentioned people like the high priests Annas (Ananias) and Caiaphas, the Roman governor Pontius pilate, King Herod, and John the Baptist. Tacitus, in AD 115, said that Nero persecuted Christians as scapegoats to divert suspicion away from himself for the great fire that devastated Rome in AD 64. He refers to the death of Christus under Pontius Pilate and the spread and success of Christianity.

From Josephus and Mara bar Serapion we learn that the Jewish leaders made a formal accusation against Jesus and took part in the events leading up to his Crucifixion. Pliny the Younger around AD 111 talks about the worship of Jesus as God, the rapid spread of Christianity, and the high ethical standards of the Christians who were not easily swayed from their beliefs. The Roman historian Suetonius, in his history "The Twelve Caesars", stated regarding the emperor Claudius: "Because the Jews at Rome caused continuous disturbances at the instigation of Chrestus, he expelled them from the

---

[3] Küng (1984: 212).
[4] Though with three of the comments apparently interpolated later (Strobel, 1998: 105).
[5] cf. 1 John 7:5 and 1 Corinthians 15:7.

city." This event occurred about the year AD 52,[6] and there is a strong case that Chrestus is an alternative spelling of Christus, or Christ. He also mentions the persecution of Christians by Nero (AD 54–68). Some other possible contributions to the life of Jesus are by Thallus[7] and Lucian of Samosata (AD 115–200), who describe Jesus as a man crucified in Palestine and was still worshipped, and the Babylonian Talmud "Sanhedrin 43a" that describes Jesus as a false teacher who practiced sorcery and was put to death on the eve of the Passover. The reference to sorcery might confirm the belief that he worked miracles. According to Luke Johnson, New Testament scholar and historian,

> Even the most critical historian can confidently assert that a Jew named Jesus worked as teacher and wonder-worker in Palestine during the reign of Tiberius, was executed by crucifixion under the Roman prefect Pontius Pilate and continued to have followers after his death.[8]

Together these sources located Jesus in Palestine, confirmed his execution and the continuing presence of a group called Christians who followed his teachings, worshipped him, and were willing to be tortured and even die for their beliefs. Also Jesus died under Pontius Pilate with Jewish leaders being involved, and during his life he was a teacher and worked miracles. Ed Sanders, a leading scholar in the field, wrote[9] that there are "no substantial doubts about the course of Jesus' life."[10] Although the number of external sources mentioned seems few in number it is actually a good number of resources compared to resources for other ancient events. For example there are just four Roman sources for the reign of Tiberius (AD 14–37).

6.2 THE ETHICS OF JESUS

I want to pause here and comment on the Judaeo-Christian ethical system. In particular, I want to quote from Paul Johnson, a famous journalist, author and speaker, who toured New Zealand in 1995. The following quote is from his published address to Canterbury university entitled "What are universities for?" He said:

> Knowledge cannot be divorced from ethics. Universities must operate within an ethical framework. My own strong belief is that the framework must spring from the Judaeo-Christian tradition, the only system of human behavior ever devised by man which makes consistent sense, which successfully promotes justice over the whole range of human endeavor, and which is capable of continuous improvement. If a secular system is devised, it is likely, if righteous, to vary only in detail from the Judaeo-Christian one.[11]

---

[6] See Acts 18:1, 2.
[7] http://christianthinktank.com/jrthal.html, accessed August, 2105.
[8] Johnson (1996: 123).
[9] Sanders (1993: 5).
[10] For further references see, for example, Habermas (1996) and Craig (1998b).
[11] Johnson (1995: 32).

What is suggested is that the Judaeo-Christian ethical system, in its purity, is about the best we can have. Unfortunately some of its practitioners have got away from the original principles and brought discredit to Christianity with the danger that people trying to throw out Christianity may throw the baby out with the bath water! What are some of the principles? The following are some of the teachings of Jesus:

(1.) Love your neighbor as yourself (without prejudice).

(2.) Love your enemies and pray for those who persecute you. Why, because God cares about everyone, even our enemies.

(3.) Jesus says a great deal about forgiveness. For example, in the Lord's prayer and his parables he says that we should forgive others if we want God to forgive us. When Jesus was asked how often should we forgive someone, his answer was always forgive, without limit. Jesus forgave all sorts of people, including those who crucified him.

(4.) We should not judge others if we don't want to be judged ourselves. I should take the log out of my own eye before I try to remove the speck from someone else's eye.

(5.) When we make a donation we should keep it a secret!

There is this and much more about peace, justice, righteousness, and so forth.

There have been many attempts to paint a portrait of Jesus. Many of these portraits have not provided much of lasting value. The first real break from the traditional portraits was given by that outstanding humanitarian, musician, and scholar Albert Einstein when he wrote his book "The Quest of the historical Jesus" in 1906. Since then there have many attempts at finding the historical Jesus and determining who he really is. Many of the sensational ones have been attempts at destroying Christianity. What is even more disconcerting is that some of these writers have come up with portraits that are totally contradictory.

Schonfield,[12] for example, suggested that Jesus engineered circumstances in order to "fulfill" prophecy so that he "proved" to be the Messiah and put forward a swoon theory to explain Jesus' resurrection, which is ruled out below. Jesus would then be a cunning deceiver, which is out of keeping with his character, and he would not have been able to fulfill all the prophecies. Clearly Schonfield's book is contrary to all the evidence and is a work of fiction.

Another portrait was given by Barbara Thiering called "Jesus the Man: A new interpretation from the Dead Sea Scrolls" published in 1993. Some problems with her work are as follows:

---

[12]Schonfield (1967).

(1.) She dates the Dead Sea Scrolls about 200 years later than other scholars.

(2.) The writers of the Scrolls claimed that the prophecies of the Old testament were written in code and required an interpretation ("pesher"). All scholars agree that this is how several of the scrolls "work." Thiering then claims that this community wrote its own story in code, namely the New Testament, which must be interpreted in the same way. No evidence or reason for this new use of the method is given.

(3.) In the pesher documents the word pesher itself is used all the time, usually preceded by a gap in the text. There is nothing of this in the gospels.

(4.) Her code leads to ridiculous conclusions about places and people.

(5.) She does not try to code away the crucifixion but endeavors to give her own unusual explanation of the resurrection, roughly akin to the swoon theory.

Thiering's theory on the dead sea scrolls and their relationship to the gospels presented in earlier books have not convinced any reputable scholar who studies the scrolls, whether they be Jewish, Christian, or agnostic. Her ideas are totally fanciful and the only scholar who seems to take her views seriously is Thiering herself. In contrast, the Dead Sea Scrolls actually confirm that the Essenes were not the source of of early Christianity as some writers in the past have suggested.[13] As also mentioned by philosopher and biblical scholar Norman Geisler:[14]

(1.) The Essenes emphasized hating one's enemies, but Jesus stressed love.

(2.) The Essenes were exclusive regarding women, sinners, and outsiders, while Jesus was inclusive. There was no salvation outside of Qumran.

(3.) The Essenes were legalistic sabbatarians, while Jesus was not. He showed astonishing freedom in regard to the law.

(4.) The Essenes stressed Jewish purification laws, but Jesus attacked them.

(5.) The Essenes did not stress the kingdom of God, but Jesus did.

(6.) The Essenes were Jewish nationalists, whereas Jesus made salvation available to the Gentiles.

(7.) The Essenes practiced asceticism but Jesus did not.

(8.) The Essenes were subject to monastic rule so that their day was strictly regulated. For Jesus there was none of this.

---
[13] See Charlesworth (1992) for details.
[14] Geisler (1999: 189, 216).

(9.) The Essenes believed that two Messiahs would come, but Christians held that Jesus was the only one.

The Essene leader, the so-called "teacher of righteousness", lived long before Jesus, was a priest, was not worshipped as God nor resurrected from the dead, and atoned for no one at his death.

Another recent attempt at removing all supernatural aspects of Jesus' life and reducing him to just another itinerant social critic was presented by the so-called Jesus Seminar founded by Robert Funk.[15] They treated the gospel of Thomas on an equal footing with the four gospels (see below) and showed disregard for New Testament mainstream scholarship. They have been heavily criticized by some respectable scholars.[16] According to Geisler: "Their conclusions are contrary to the overwhelming evidence of the New Testament and the reliability of the New Testament witnesses."[17] Theologian and historian Michael Licona, in his extensive historiographical volume, comments:

> It is no surprise that during the twentieth century somewhat of a proverb circulated and continues to this day that historical Jesus scholars end up reconstructing a Jesus that reflects their own convictions and preferences.[18]

This has led to a wide range of conflicting portraits of Jesus.

The most important aspect of the life of Jesus is that his life was too extraordinary to have been invented. Will Durant, former professor of the philosophy of history and agnostic said:

> "That a few simple men should in one generation have invented so powerful and appealing a personality, so lofty an ethic and so inspiring a vision of human brotherhood, would be a miracle far more incredible than any recorded in the Gospels."[19]

6.3 OUTSTANDING FEATURES OF JESUS' LIFE

In the following sections I want to look at the life of Jesus in more detail. His life has changed the world and I wish to highlight some of the special features of his life.

6.3.1 He Had an Extraordinary Introduction into History

The Old Testament spoke about a coming Messiah or anointed one from God. He is described in prophecies written at least 400 years BC (some more than

---

[15] See, for example, Funk (1985) and Funk et al. (1993).
[16] For a critique see Craig (1998b).
[17] Geisler (1999: 388).
[18] Licona (2010: 46).
[19] Durant (1994: 555).

1000 BC). We are told that the Messiah would fulfill the following prophecies, which are listed with the prophecy and biblical references in brackets:

- He would be born of a virgin (young woman) (Isaiah 7:14 and Matthew 1:21–23, Luke 1:26).

- He would be a descendant of Jesse of the line of king David (Isaiah 11:1–5, 10).

- He would be born in the rather obscure town of Bethlehem Eprathah (Micah 5:2 and Matthew 2). The town apparently had less than 1000 people there.[20]

- He would be brought up in Nazareth, another small place (referred to as "the branch" in Zechariah 3:8, 6:12, Jeremiah 33:15 and Matthew 2:23).

- His ministry would be heralded by a messenger like Elijah, who turned out to be John the Baptist (Malachi 3:1, 4:5 and Matthew 11:10).

- He was to bring a message of healing and liberation to the gentiles (Isaiah 61:1–2 and Luke 4:18–21).

- He would use parables (Psalm 78:2 and Matthew 13:35).

- He would make a triumphant entry in to Jerusalem sitting on a donkey (Zechariah 9:9 and Matthew 21:1–11).

- He would be betrayed for thirty pieces of silver that would be cast on the floor of the temple and used to buy a potter's field (Zechariah 11: 12–13 NIV translation, where "treasury" is translated as "potter", and Matthew 26: 14–15.)

- We have details of his death, particularly from the Psalms. He would be silent before his accusers, his hands and feet would be pierced though not a bone would be broken, and would be given vinegar to drink. He would be numbered with the transgressors, mocked by his enemies, and lots cast for his clothing (Psalm 22, Isaiah 53, Psalm 34:20, and 69:21).

- He would be despised and rejected, and suffer terribly for the sins of all (Isaiah 52:13–53).

- His words on the cross were predicted (Psalm 22:1–2 and Matthew 27:46).

- He would be buried in a rich man's grave (Isaiah 53:9 and Matthew 27:57–60).

---

[20] It took a decree by the mighty Caesar Augustus to bring this event to pass.

There are 60 prophecies and 270 ramifications given by different prophets at different times. All of these appear to be fulfilled by Jesus of Nazareth. Taking a very liberal view, the prophetic books were completed at least 400 years before Christ and the Book of Daniel no later than 165 BC., well before the events. It is true that a number of prophecies are tucked away in obscure places in the Old Testament and it may seem surprising that the unlearned disciples knew where to look. Suggested answers to this are: (1) many of the Messianic prophecies were well-known among the Jews and would have been drummed into them, and (2) Jesus enlightened them (Luke 24:44–45). Also some of the prophecies are yet to be fulfilled (e.g., Isaiah 9:6–7; 11:3–5).

Let's put these prophecies in a modern context. I am going to make a prediction about a future leader of a country (e.g., a prime minister or president) who will take up his office in 400 years time. He will satisfy the following:

- be a descendent of ABC

- be born in DEF, an obscure town

- spend his early life in GHI

- be brought to the public's notice by a somewhat unconventional friend who takes a back seat from then on

- receive great adulation when driven through the city in an open vehicle

- his message will be to the working class and he will be particularly engaged in medical and prison work

- he will introduce a number of completely revolutionary reforms

- he will be quietly spoken and upright

- surprisingly he will be forced to give a up his position through being wrongly accused in his leadership role

- his death will be tragic one in that he will be murdered, but shortly before his death his final words will be predicted

- he will be buried in an unusual place

Let us make it more incredible. Suppose I told you that the prophecies were not all mine but came from different people, who prophesied in the name of God. If such prophecies came true you would wonder what manner of man this leader would be who could make such an impact even four centuries before he was born. What then are the chances of all the prophecies relating to the Messiah happening by accident? I will not try and estimate a probability, but those who have end up with a vanishingly small value—virtually zero.

If Jesus fulfilled so many prophecies, why did the Jews reject him?[21] Abraham was promised that through his descendants, namely the nation of Israel, the whole world would be blessed.[22] The Jews were looking for a conquering hero who would establish a new kingdom of Israel, so they rejected Jesus as he did not measure up to their criterion. Jesus spoke a lot about the kingdom of God as being a kingdom that was not of this world, and made the same comment to Pilate;[23] the kingdom was within us.[24] Even Jesus' disciples, who were Jews, still did not fully understand him as they asked him after his resurrection whether he would restore the kingdom to Israel at the time.[25] The Jews did not understand a number of prophecies.

First, they did not realize that the Messiah would be disfigured and would suffer and die for the sins of the world as described in the prophetic passage Isaiah 52:13–53:12 (see also John 12:38). They thought he suffered for his own sin. Second, some of the prophecies that spoke about the Messiah coming in glory such as Daniel 7:13 were in the future and would be fulfilled when Jesus came again. Third, Jesus fulfilled several prophecies[26] that there would be a second temple that the Lord would come to, which would have had to be before AD 70 when that temple was destroyed. Finally, the Jewish thinking about the resurrection to glory and immortality was that it referred to after the end of the world and did not appreciate the idea of it occurring within history. We see this confusion in comments made after the disciples came down from the mountain after the transfiguration of Jesus.[27] They asked Jesus why the scribes said that Elijah must come first because the scribes believed that Elijah would come first before the judgement day and the universal resurrection.

Before leaving this section about predictions concerning Jesus, I wish to refer to a somewhat controversial topic already mentioned in Section 5.3 about the possible existence of hidden codes in the Old Testament. What is embarrassing to our Jewish friends is the encryption of the word *Jeshua* in Equidistant Letter Sequences (ELS's) behind every major Messianic prophesy. Engineer and Bible scholar Chuck Missler[28] mentions that counting the letter intervals up to 100, *Yeshua* occurs in over 5,538 times in the Old Testament with 2919 going from right to left, with 136 having no intervals, and 2619 going from left to right. He does point out that there is an inherent difficulty in that the name *Yeshua* has only four letters in Hebrew, of which two are the most common in Hebrew and the other two are quite frequent. This makes it easier for the word to appear by chance. However, what is more startling is

---

[21] For further comments and a Jewish insight on this question see Michael Brown in Strobel (2008: 50ff).
[22] Genesis 12:2–3, 16:18.
[23] John 18:36.
[24] Luke 17:21.
[25] Acts 1:6.
[26] e.g., Malachi 3:1–5.
[27] Mark 9:9–11.
[28] Missler (1999: 149).

the number of codes found in Isaiah 52:13–53:12 where we have the description of the Messiah of Israel described as the Suffering Servant.

As already mentioned in Section 5.2.1, the accuracy of the book of Isaiah in relation to the original version has been confirmed by the Dead Sea Scrolls written six centuries prior to the later Masoretic text. A number of books have been written on the subject of these codes.[29] What these authors found encoded within this passage was Yeshua Shmi or "Yeshua (Jesus) is My Name", as well as over 40 names and key places in 15 sentences including the Hebrew names of the disciples (except Judas), Messiah, Galilee, Caiaphas, and so forth. What is interesting is not just their occurrence, but also their proximity, links, and relevance to the text. There are those who will debate the relevance of such findings, given the nature of Hebrew, and in some cases a statistical analysis is needed to determine the odds of the codes existing other than by chance.

6.3.2 He Had Extraordinary Abilities

We have a tantalizing glimpse of the boy Jesus at twelve years of age debating in the temple with learned rabbis.[30] Right through his ministry people were astonished at his wisdom and understanding, and he confounded those who tried to trick him.[31] The essence of what he taught was not only new, but the way he taught it was unique. He used parables very effectively and the ones like the good Samaritan (Luke 10), the prodigal son (Luke 15), the lost sheep (Luke 15:4ff.), and the sower (Matthew 13:18–23) particularly stand out. Even in his own home town people wondered where he got his wisdom from and his mighty works, given he was just the carpenter's son.[32]

Much later in his ministry he was in Jerusalem during the feast of tabernacles and we read in John 7 that he went up into the temple and taught. The Jews marveled at his learning, given that he had supposedly never studied. Jesus answered them that his teaching was from He who sent him. On the last day of the feast the chief priest and the Pharisees sent some soldiers to bring Jesus to them because Jesus had claimed to be the Messiah. However they returned empty-handed saying that nobody ever spoke like this man.[33] Soldiers are not easily persuaded to neglect their duty!

Jesus as a Jewish boy would have had basic training. To become a Rabbi, for example, involved long and intense training and consisted of three stages of formal Jewish education. The first, Bet Sepher, began at age six which involved studying and memorizing the Torah. Those boys who showed promise moved on to Bet Talmud around 10–13 years old when they memorized the

---

[29] For example Rabbi Yakov Rambsel (1996, 1997) and Grant Jeffrey (1996, 1997).
[30] Luke 2:46–47.
[31] Matthew 22:46.
[32] Matthew 13:54–55.
[33] John 7:45–47.

Talmud (the Old Testament). The top 2% would move on to Bet Midrash when they would spend the next two years immersing themselves in the teachings of the previous great Rabbis, finally graduating at age 15 when they would look for a Rabbi to study under. If the Rabbi accepted the student he would say, "Come follow me", reminiscent of the words Jesus used.[34] We don't know anything about Jesus' early years except for the previously mentioned reference in Luke 3:41–57 where Jesus is described as listening to and asking questions of the teachers in the temple at the age of 12; they were amazed at his answers and understanding.

Jesus was referred to as "Rabbi" in several places, for example by Nicodemus a member of the Sanhedrin. Given his Jewish background, we can understand some of his strong and colorful language, typical of Jewish discourse. For example, he said that unless we hate father, mother, wife, children, and even our own life we cannot be his disciple.[35] We see it again in how Jesus interpreted the law in the so-called beatitudes.[36] Lusting after a woman is likened to actually committing adultery with her in his heart, while being angry with someone is likened to committing murder. If your eye causes you to sin, pluck it out and throw it away, or if your right hand sins cut it off and throw it away. Obviously such language is not to be taken literally!

We find that Jesus had special insight with people. We watch him talking to different people, for example, the woman of Samaria (who was into her fifth de facto relationship), Nathaniel, Nicodemus, and Zaccheus. Jesus knew all about them from a distance and what they were doing.[37] He knew when his friend Lazarus had died.[38] In addition to his ability to work miracles and bring about healing, he had the gift of prophecy. For example, he foretold his own death and resurrection a number of times,[39] he foretold Peter's thrice denial mentioned in all four gospels, and foretold the destruction of Jerusalem and the temple, which followed in AD 70.[40]

As we watch him working with his disciples for three years and sharing a common purse, we find that the disciples got on each other nerves occasionally and there was some friction. However they never found in Jesus the sins they found in themselves. Two of his closest friends Peter and John described him in their letters as without sin—almost as an aside.[41] Peter on one occasion said "Depart from me Lord for I am a sinful man."[42] These men were brought up as Jews and would have had it drummed into them the universality of human sin. Yet Jesus was different. Even the enemies of Jesus conceded that he was

---

[34] Matthew 4:18–22.
[35] Luke 14:26.
[36] Matthew 5:1–7:28.
[37] John 4:17–19, 1:48, 3:9; Luke 19:1–10.
[38] John 11:11–15.
[39] Mark 8:31–32, 9:31, and 10:32–34.
[40] Mark 13:1–4.
[41] 1 Peter 2:21–23 and 1 John 3:5.
[42] Luke 5:1–8.

innocent, for example Herod, Pilate (and his wife), the penitent thief on the cross, the centurion at the foot of the cross, and even Judas when he threw down the 30 pieces of silver and proclaimed he had betrayed innocent blood. Paul expressed the belief of the early church that Christ had no sin,[43] which was also echoed by the writer of Hebrews (4:15). There are many aspects to the character of Jesus I could look at but there are five that particularly appeal to me.

First, Jesus was a man of prayer. We read how he went into the hills to pray and on one occasion he prayed all night. I often wonder what he prayed about. We know he prayed for the disciples and for those who would believe in him through the testimony of the disciples. Why did he need to pray? Being human he had our physical limitations and needed to focus spiritually just as we need to. He also set us an example.

Second, Jesus was a man of obedience. In Philippians 2:8 we read that Jesus became obedient to death, even death on a cross. At one stage he prayed that if possible God would not let him pass through all the suffering that would happen, yet all that mattered to him was to do the will of God rather than what he wanted. In John 4:34 when he was talking to the Samaritan woman at the well he said to his disciples that his food was to do the will of Him who sent him, and to accomplish His work. This is the standard of obedience. God's will must come before all else, sometimes even before our daily needs.

Third, Jesus was a man of forgiveness. In his Sermon on the Mount he mentioned forgiveness a number of times.[44] We need to forgive others so that God forgives us, the one requirement in the Lord's prayer.[45] How often should we forgive someone? Jesus said "seventy-times seven", in other words, always.[46] Jesus practiced what he preached. He forgave people. He forgave the woman caught in adultery.[47] In a special way he forgave his disciple Peter for denying him three times. In John 13 he showed both a humble and forgiving spirit when he washed his disciples feet including those of Judas whom he knew would betray him. And most of all, he forgave those who crucified him.[48]

Fourth, Jesus was a person for all people. It is very easy to be prejudiced these days and stereotype other people. Jesus accepted people in spite of prejudice, and he crossed social barriers, for example he touched a leper. He also spoke with a Samaritan woman to the surprise of his disciples as the Jews hated the Samaritans.[49] We see this reflected in John 8:48 when the Jewish leaders accused Jesus of being a Samaritan and having a demon, the

---

[43] 2 Corinthians 5:21.
[44] Matthew 5:23–24.
[45] Mark 11:25; Luke 6:37.
[46] Matthew 18:21–22.
[47] Although this story is not in all manuscripts it is in keeping with Jesus' character.
[48] Luke 23:24.
[49] See John 4.

two obviously going together in their minds. However, Jesus did not speak to just a Samaritan, but a Samaritan woman. It was not the thing to speak to a woman on the street, especially one of dubious reputation. She had had five husbands and was currently living in a de facto relationship. Yet Jesus saw her worth, even revealed to her that he was the Messiah, and many Samaritans from the nearby town came to believe Jesus because of her testimony.

The parable of the good Samaritan is a wonderful story. The priest did nothing for the unfortunate man who was beaten up, and the Levite (also a religious man) did nothing. It was the despised Samaritan who helped the man, and he did it generously, leaving money with innkeeper. It is a picture of Jesus and his compassion. He stressed that what we do for others we do for him.[50] Jesus stepped across another barrier when he made friends with Zaccheus, a hated tax collector. When Jesus went to his home they all marveled that Jesus had gone to be the guest of a man who is a sinner.[51] Yet Jesus recognized the man's worth and consequently Zaccheus completely changed his practices giving half of his goods to the poor and restoring to anyone he had defrauded four times the amount.

6.3.3 He Made Extraordinary Claims

It is not just the extraordinary powers of Jesus or his moral character that catch our attention, but we also have his extraordinary claims. He called himself the Son of Man 78 times (a Messianic title) and the Son of God over 40 times. The title of the Son of Man is actually a title of deity, as it refers to Daniel 7:13-14. It is mentioned only once outside the gospels in the rest of the New Testament[52] suggesting that it was not a title added later and written back into the traditions about Jesus. There are verses where Jesus identified himself completely with God: to know him was to know God; to see him was to see God; to believe in him was to believe in God; and to receive him was to receive God.

In Matthew 11:27 he said that no one knows the Son except the Father and no one knows the Father except the Son and anyone to whom the Son chooses to reveal Him. This can be regarded as an authentic saying for two reasons.[53] First, it is drawn from an old source shared by Matthew and Luke (called the Q document). Second, it is unlikely to have been invented by the early church because it says that the Son is unknowable; not a product of later church theology, because after the resurrection we can know the Son in spirit. Another early authentic saying is Mark 13:32 when Jesus says that no one, including the Son, knows when the second coming will occur, only the Father. Although this expresses the humanity of Jesus, the later church would

---

[50] Matthew 25:40.
[51] Luke 19:7.
[52] Acts 7:56.
[53] Craig (1998).

not have ascribed limited knowledge to Jesus. We also have the testimonies of the Godly people Simeon and Anna who recognized the baby Jesus as the Messiah in Luke 2. Even the demons that Jesus exorcised recognized that he was the Messiah.[54]

Jesus himself acknowledged that he was the Messiah to Peter,[55] to John the Baptist,[56] to the woman of Samaria at a well,[57] to the high priest,[58] to the two disciples on the road to Emmaus,[59] and to the disciples after his resurrection.[60] The soldiers were sent to arrest him because he claimed to be the Messiah in John 7. The gospel of John was written so that people would believe that Jesus was the Messiah and the son of God, and have life in his name.[61]

Jesus also claimed to do things that only God could do:

(1.) He claimed to forgive sin (Matthew 9:1-8).

(2.) He claimed that he would judge the world (John 5:22, 32).

(3.) He claimed to be Lord of the Sabbath (Luke 6:5; Matthew 12:8).

(4.) He claimed to be pre-existent (John 8:58 and 17:24).

(5.) He claimed to answer prayer (John 14:13–14).

(6.) He claimed to be the truth (John 14:6).

(7.) He claimed to have all authority (Matthew 28:18).

(8.) He claimed to be one in essence with God (John 10:30, 14:9–11).

(9.) He claimed to give eternal life (John 10:28).

(10.) He claimed that he was the only one who knew who the Father was (Luke 10:22).

(11.) He claimed to be from heaven (John 6:38).

With regard to Jesus' authority, he reinterpreted the Jewish law on his own authority and gave his own teaching on a number of matters. He used the phrase, "Truly I say to you"[62] to express his authority. People were astonished

---

[54] Luke 4:33, 41; 8:28.
[55] When Peter said he was the Christ in Matthew 16: 13–20; Mark 8:28–30: see also John 6:69 and Acts 2:36, when Peter affirmed this publicly.
[56] Matthew 11:2–6; Luke 7:19–23.
[57] John 4:25–26.
[58] Matthew 14:60–64.
[59] Luke 24:25–27.
[60] Luke 24:44–47.
[61] John 20:31.
[62] For example Mark 8:12.

at his teaching as he taught them as one who had authority.[63] He focussed on what was in people's hearts and not on the trappings of religion.

We note that Jesus pointed people to himself and demanded exclusive loyalty above all family relationships. He made claims like "I am the way, the truth, and the life", " I am the bread of life", "I am the light of the world", "I am the good shepherd", "I am the door", "I am the resurrection and the life", "I am the first and the last", and in John 3:16 we have that whoever believes in him has eternal life. Some of his claims are also claims made by God in the Old Testament such as God will be judge,[64] God is the first and the last,[65] and God is a shepherd,[66] for example. Yet when we match up the claims of Jesus with his life we see that his portrait in the New Testament is a balanced one. His teaching may have been unpopular at times but there is no trace of the eccentric or fanatic. We have the paradox of his self-centered teaching and his non-selfcentered behavior. In thought he put himself first but in deed last.

Some might argue that he didn't really make the above claims, but they were simply made up by the New Testament writers. This does not make sense as monotheistic Jews would not ascribe divinity to a man they had known and who did not directly claim to be divine, as this would be blasphemy. Such a claim must have come from Jesus himself. If we don't believe him or his disciples, what about his enemies? It is eye-opening to read the gospels just through the eyes of the opposition and the critics of Jesus. He was accused of blasphemy several times, such as claiming God was his father, which made him equal with God.[67] On one occasion onlookers said that only God can forgive sins.[68] On another occasion the Jews took up stones to stone Jesus[69] as they said they were going to stone him for blasphemy when he claimed to be God when he was only a man.

Then we have the trial of Jesus. Why did the High Priest condemn Jesus? Because of blasphemy. The fact that the High Priest rent his garment signifies this. He would be required by custom to do this if any blasphemy was uttered in his presence.[70] They had tried to condemn Jesus through false testimony,[71] but that failed. In the end Jesus was not tried for what he had done, but for who he claimed to be. He was condemned by the Jewish high court on a charge of blasphemy and then delivered to the Roman authority to be executed for the treasonous act of setting himself up as the King of the Jews. I believe there is no doubt that Jesus claimed to be one with God.

---

[63] Matthew 7:28–29.
[64] Joel 3:12.
[65] Isaiah 41:4, 44:6.
[66] Psalm 23:1.
[67] John 5:18.
[68] Matthew 9:1–8.
[69] John 10:31.
[70] Matthew 26:65.
[71] Matthew 26:59–60.

There has always been some controversy by what is meant by the trinity: God the Father, God the Son, and God the Holy Spirit, but only one God; Jesus had both a divine and a human nature. The disciples, who were monotheists, wrestled with what they experienced as they felt the divine power within them that they variously described as "the Spirit of God", "the Spirit of Christ", or simply "the Spirit."[72] We are faced with mystery here and the best we can do is to suggest metaphors to help our understanding. One idea is that God is a multidimensional being with three projections (like house plans) or self-disclosures of Himself, each revealed at different times. We have three "Persons" with the same essence of deity. It reminds me of a house: the architect or creator designs it, the builder constructs it, and the owner lives in it, with all three having some kind of ownership. Another weak metaphor is that H_2O can be water, steam, and ice. We can therefore think of God the Creator, Jesus the human face of God, and the Holy Spirit indwelling a person.

The Old and New Testaments state very clearly that there is only one God.[73] However, Jesus was conceived by the Holy Spirit,[74] and sometimes we see all three persons of the trinity referred to at the same time.[75] We have personal testimonies: Paul and John referred to Jesus as creator,[76] while the disciple Thomas confessed Jesus to be his God.[77] The disciples in a boat after Jesus walked on the water worshipped him as the Son of God. There is also evidence for the trinity throughout the Old Testament in terms of the use of many plural forms associated with God, though there is some debate as to whether the "royal we" is being used or not. The New Testament writers, although monotheists, tried to convey what they experienced. Jewish and Muslim writers would certainly deny the idea of a trinity. However this book is not a debate about theology but about evidence.

A Muslim argument that does not stand up on historical grounds is that someone else and not Jesus died on the cross, as the gospels denigrated Jesus on several occasions. We consider just two such times. First, Jesus was afraid of what was ahead of him, yet he wanted to do his Father's will,[78] and second he called out "My god, my God why have you forsaken me" on the cross.[79] This was not the Muslim idea of a martyr. In answer to this, William Lane Craig first points out that although the early church believed in the divinity of Christ, they recorded a number of incidents that were embarrassing or dif-

---

[72] See Romans 8:9 where all three terms are used together.
[73] Deuteronomy 6:4; Isaiah 45:5; 1 Corinthians 8:4.
[74] Matthew 1:20 and Luke 1:35.
[75] For example, Matthew 3:16–17, 28:19; John 14:16–17, 23; Romans 8:17; 1 Corinthians 12:4–7.
[76] John 1:2–4 and Colossians 1:15–20, 2:9; see also Philippians 2:6.
[77] John 20:28.
[78] Matthew 26:8–39.
[79] Matthew 27:46.

ficult for them, therefore strongly supporting that they actually happened.[80] Second, Jesus was also human (contrary to the Muslim viewpoint) and any human weakness supported the humanity of Jesus and his ability to identify with human suffering. Third, some Christians believe that God turned his back on Jesus when he was bearing the sins of the world, but Jesus had not died for our sins just yet. Craig therefore suggests that Jesus was speaking out the identical prophetic words of Psalm 22:1 as he prayed Psalm 22 out loud to God, a Psalm he would have known, having been steeped in the Old Testament. There was no sign of abandonment by God when Jesus spoke the final words "Father, into your hands I commend my spirit."

What do we make then of Jesus' claims? Was he crazy? Is it consistent to say that Jesus was a good moral man who said profound things, but was still a lunatic? In the words of C. S Lewis, was he on the same level as a man who claimed to be a poached egg? I think not. We don't have those options. When you match up his claims with his life we see that his portrait in the New Testament is a balanced one. His teaching may have been unpopular at times, but there is no trace of the eccentric or fanatic, and there is no evidence of paranoia or schizophrenia. His life and works supported his claims, and we have the paradox of his self-centered teaching and his non-selfcentered behavior. In thought he put himself first but in deed last, like when he washed his disciples feet.[81]

Jesus also crossed social barriers. We saw above that he talked to a Samaritan woman, he touched a leper (an untouchable), and he went to dinner at a (hated) tax collector's home, and so forth. Alright, if he was not crazy, was he a liar and a deceiver? There is no evidence for this, and this contradicts his outstanding moral teaching. For example, he told people to be honest, whatever the cost. Was he then prepared to die for the lie that he was divine? His disciples and family, who knew him best of all, eventually believed in Him, and his brother James became a pillar of the church. Our friends and family know us pretty well don't they! The only alternative is that he was what he said he was.

6.3.4 He Had extraordinary Powers

Jesus was a miracle worker. Even skeptical critics accept that there is evidence that Jesus had a ministry of miracle working and exorcism, unless you take a stance that completely rules out supernaturalism. Craig Blomberg[82] compares Jesus' exorcisms and healings with those described in his day. He mentions some common features with ancient accounts such as (1) the at-

---

[80] http://www.reasonablefaith.org/do-the-gospels-support-a-muslim-view-of-jesus, accessed August, 2015.
[81] John 13:5.
[82] Blomberg (2007: 122–123).

tempt to discover the demon's name in order to gain mastery over him,[83] (2) the use of touch or laying on of hands,[84] and (3) the application of spittle.[85] However, he emphasizes that the differences far outweigh the similarities. For instance point (1) occurs only once in the four main exorcisms in Mark, while (2) and (3) never occur in exorcisms, only in healings, and were often practiced as an appropriate treatment. When touch is involved, either by Jesus or by the person wanting to be healed, Jesus makes it clear that it is the person's faith and not magic that brings healing.[86] Blomberg said that healing by mere touch is generally not in pre-Christian accounts of miraculous healing, and although there were common practices such as the use of spells and magical objects with regard to healing and exorcism, Jesus used none of them. Furthermore, subtle variations in the gospel accounts suggest that they weren't invented by the gospel writers. Blomberg[87] provides a number of other arguments, concluding that "the accounts of Jesus' miracles differ from other miracle stories of the ancient Mediterranean world" and

> if one accept's Jesus teaching about his ushering in the kingdom of God, then one ought to accept the reality of his miracles. The two go hand in hand, since Jesus uses the miracles to authenticate his teaching. And it is fair to say that modern scholarship, even of the most skeptical variety, has agreed that if anything in the Gospels is reliable it is the teaching of Jesus about the kingdom of God.

Miracles are discussed in more detail in Chapter 7, where the so-called nature miracles are also mentioned.

6.3.5 He Had an Extraordinary Birth

The virgin birth of Jesus is a source of contention as it defies science. The priest Zacharias and his wife Elizabeth, parents of John the Baptist, would not want their son to be of lesser importance than Jesus, a younger cousin. Nor would they buy into the story of a virgin birth, which would be cultural suicide if it were not true, given the heartache that followed (John was killed by Herod, and Mary was told by the angel that a sword would pierce her heart). It has been claimed that the virgin birth was a borrowing from pagan religions as stories of supernatural births were common to pagan gods. For example, we find miracles associated with the births of Buddha, Confucius, Zarathustra, and Muhammad, and there are other parallels with the life of Jesus. However the story of the virgin birth doesn't show any signs of being mythical but reads like an ordinary narrative, without borrowed elements from pagan mythology.

---

[83] Mark 5:9.
[84] Matthew 9:29; Mark 6:59; Luke 4:40.
[85] Mark 7:33; 8:23; John 9:6.
[86] Mark 5:30–34.
[87] Blomberg (2007: 128).

As noted by Norman Geisler,[88] the New Testament was written when eyewitnesses were still alive thereby ruling out later myths and any legendary development. Furthermore, we see that names, places, and events connected with Christ's birth are accurate historically. Greek myths never referred to the literal incarnation of a monotheistic God into human form; gods were only disguised as humans and invariably mated sexually with humans. Greek myths of gods who became human postdate the time of Christ and therefore the gospel writers could not have borrowed from them.

6.4 HE HAD AN EXTRAORDINARY DEATH AND RESURRECTION

Paul stated clearly that the resurrection of Jesus was fundamental to Christianity as without it our faith was in vain.[89] The story of the resurrection has always been a subject of fierce debate because if it happened, people must take note of what Jesus said and consider his claims. Consequently all sorts of theories have been thrown up to try an avoid this, some of these theories being very far-fetched. As William Lane Craig noted in a debate,[90] it is important to distinguish between the evidence and the best explanation of the evidence.

6.4.1 What Happened?

The Jews were looking for a Messiah who would lead them to victory against the Romans and establish a new kingdom. They didn't realize that the new kingdom was to be a spiritual one that was not just for Jews but for a new Israel encompassing both Jew and Gentile, male and female, bond and free, and entered by spiritual birth.[91] When Jesus claimed to be the Messiah and Son of God[92] they accused him of blasphemy and pressured Pilate into condemning him to the cross on the grounds that he was perverting the nation and setting himself up as Messiah and King. Clearly the disciples did not make up his claims to be the Son of God as Jesus was crucified in part for these claims. There were also other reasons why Jesus was crucified. As theologian Hans Küng pointed out, Jesus

> had offended against almost everything that was sacred to this people, this society, and its representatives: that without bothering about the hierarchy, he set himself in word and deed above the cultic taboos, the fasting customs and particularly the Sabbath precept...[93]

---

[88] Geisler (1999: 201–202).
[89] 1 Corinthians 15:12–20.
[90] http://academics.holycross.edu/files/crec/resurrection-debate-transcript.pdf, accessed August, 2015.
[91] Galatians 3:28.
[92] Matthew 27:63–65; Mark 14:61–64. See Peter's acknowledgement in Matthew 16:15–17.
[93] Küng (1984: 291–292).

Jesus said that the Sabbath was made for man not man for the Sabbath.[94] [

Jesus not only questioned the law and the social system, but interpreted the law differently and changed it, and set himself above it. He claimed to be greater than Solomon who built the temple. The Pharisees never did understand Jesus because they looked on religion outwardly through the eyes of narrow nationalism and legalism. Jesus, on the other hand, taught that God looks on the heart and not on outward appearances, and condemned the hypocrisy of the Pharisees. They were looking for a Messiah who would establish a kingdom just for the Jews. They did not realize that the Messiah would establish a new Israel, a new Jewish nation, made up from all peoples of the earth. To become a member of this new Israel you would have to be born into it as with the old Israel, but with the difference—this must be a spiritual and not a physical birth.[95] No one can read the life of Jesus without realizing how different he was from the current thinking and perceptions of the time.

Some readers might find it difficult to shake off the idea that we are already assuming the reliability of the New Testament when we examine what these documents have to say about Jesus, thus suggesting reasoning in a circle. However it is important to realize that New Testament historians treat the New Testament documents in a neutral fashion just like any other ancient documents and try and find out what historical facts can be obtained from them. When viewed in this manner, there are good grounds for their reliability in how they portray Jesus with regard to his life and what happened to him after his death, including his burial, the empty tomb, the postmortem appearances, and the transformation of the first disciples.

Although Jesus was crucified, his death was a voluntary death. He said that he had the power to lay down his life on his own accord that he might take it up again.[96] However, the events leading up to the crucifixion of Jesus, when looked at from a medical point of view, make for horrendous reading.[97] The suffering began after the Last Supper when Jesus went to the garden of Gethsemane to pray all night. Knowing what he had to face he was so stressed psychologically that he began to sweat drops of blood.[98] This uncommon medical condition is called hematidrosis. Severe anxiety can release chemicals that break down the capillaries in sweat glands releasing a very small amount of blood into the sweat leaving the skin in an extremely fragile condition.

The next day Jesus was brutally flogged (scourged) by a soldier for at least thirty-nine lashes using a whip of braided leather thongs with metal balls and pieces of sharp bone attached. The whipping would lay open the flesh on back, buttocks and legs down to muscle and sinew, and this would often kill

---

[94] Mark 2:27–28.
[95] John 3:3–7.
[96] John 10:18.
[97] The following is based on Strobel (1998: 260–267).
[98] Luke 22:44.

recipients or at least put them into hypovolemic shock. This means suffering from losing a large amount of blood, thus causing a drop in blood pressure that led to fainting or collapse and a loss of fluids leading to extreme thirst. Jesus was in this state when carrying the horizontal beam of the cross up to the crucifixion site. When he collapsed they conscripted Simon of Cyrene to carry the beam; later on the cross when Jesus said he was thirsty they offered him a sip of vinegar.

Once at the crucifixion site, Jesus was laid down and attached to the crosspiece by nailing long metal spikes through his wrists crushing the median nerves and producing excruciating pain.[99] Jesus was then lifted up as the crossbar was attached to the vertical post that had been set permanently in the ground and his feet were nailed to the post. His arms would have been stretched so that his shoulders were dislocated. As he hung there the stresses on muscles and the diaphragm put him in an inhaled position so he could only exhale by pushing himself up on his feet. While on the cross the soldiers gambled for his tunic. Eventually he was completely exhausted, and excess carbon dioxide caused his heart to beat erratically indicating to him that he was near death and able to say, "Lord into your hands I commit my spirit" before dying of heart failure. With such a death there would have been a collection of fluid around the heart and lungs.

Now Jesus was crucified between two thieves, and because the Sabbath and the Passover were due the Jewish leaders wanted the whole business over by sundown. The soldiers then shattered the lower legs of the two thieves so they would die by asphyxiation, but found that Jesus was already dead. A soldier confirmed this by thrusting a spear into Jesus' side piercing heart and lung; clear fluid and blood flowed out, as observed by the eyewitness John.[100] As the soldiers were expert in killing people and that they would forfeit their lives if a prisoner escaped, there was no doubt that Jesus was dead.[101]

After his death, Jesus' body would have been wrapped in a linen cloth and about a 100 pounds of aromatic spices mixed with a gummy substance applied to the wrappings in accordance with Jewish burial customs. The body was then placed in a solid rock tomb and a large stone weighing about one or two tons was rolled by means of levers against the tomb entrance. The tomb was guarded by a well-disciplined group of Roman soldiers who put a Roman seal on the tomb to stop any attempt at vandalization.

6.4.2 Paul's Early Writing

It is generally agreed that Paul wrote First Corinthians in AD 55 based on information about his missionary journeys and local history. He was the first

---

[99] A pain so bad that a new word was invented for it, namely "excruciating" meaning "out of the cross."
[100] See John 19:34.
[101] Mark 15:44–45.

writer to refer to the resurrection in 1 Corinthians 15: 3–8, and declared what he had received, namely that Jesus died, was buried, and rose again on the third day according to the scriptures. He then said that Jesus successively appeared to Peter, to the twelve disciples, to more than 500 people at one time (most of whom were still alive), to James, to all the apostles, and finally to Paul himself.[102] If the resurrection had not occurred, why would Paul give such a list of supposed eyewitnesses who could be questioned? Here Paul refers to testimony that goes back to being very close to the actual event (say within 25 years or even earlier) before citing what appears to be an early pre-Pauline oral tradition common to all Christians that predates the gospels.

We are reminded by Michael Licona[103] of the major role of oral traditions in the Greco-Roman world at the time. He provides evidence to support a number of further pre-Pauline traditions that refer to the resurrection of Jesus.[104] He also lists 37 texts that many scholars believe are short formulas reflecting a confessional tradition that refers to the resurrection. Paul, as a former Pharisee prior to his conversion, was zealous about tradition[105] so that apostolic traditions were important to him in his new faith. Licona refers to a number of components in 1 Corinthians 15:3–7 that are consistent with Paul's asserting that he was imparting tradition. We also note that Paul refers to himself as "one untimely born",[106] a violent birthing image relating to the idea of a Cesarean section, indicating that what happened on the Damascus road was different to what happened to the others who had seen the resurrected Jesus. This is endorsed by the words "last of all" in verse 8 indicating that he wasn't referring to the ordinary conversion experience of later believers.

The conversion event is well attested. For example, Paul testifies within approximately twenty to thirty years after the event to his role as a persecutor of the church in a number of letters.[107] In addition we have Luke's five references in Acts to Paul's persecution of the Christians written thirty to sixty years after the event.[108] Also the story of his conversion was circulating within three to a little more than ten years after the event.[109] Paul, a fanatical persecutor of Christians, became the greatest missionary of the gospel.

---

[102] Some evidence that he appeared to many people other than just the disciples is given by Peter where they wanted to pick a replacement for Judas from those who knew Jesus after his resurrection (Acts 1:21–22). There were apparently several candidates with two finalists.
[103] Licona (2010: 220 ff.).
[104] For example, Romans1:3b-4a and Luke 24:34 (Jesus appeared to Peter).
[105] Galatians 1:14.
[106] 1 Corinthians 15:9.
[107] Galatians 1:12–16, 22–23; 1 Corinthians 15: 9–10; Philippians 3:6–7; 1 Timothy 1:13.
[108] Acts 7:58, 8:1–3, 22:1–5, 26: 4–5, 9–11.
[109] Galatians 1:22–23; cf. Acts 9:21.

6.4.3 The Evidence

We shall now look at some of the evidence for the resurrection. We note first that the actual resurrection is not described, and that the narrative is straightforward without the embellishment and imaginative additions that might be expected from a later legend. There was not enough time for there to be legendary influences. The burial is part of very old source material used by Mark in his gospel and it is interesting that there is no other competing burial story, which would be the case if Mark's account was basically legendary fiction. Although there are some differences in the gospel accounts, as might be expected from independent observers, the following observations generally hold with most of the accounts.

(1.) *Jesus predicted his death and resurrection.* Jesus told his disciples what would happen to him on a number of occasions[110] but it didn't register with them; not something they would be proud of later. Mark 8:27–33 is of particular interest as it is linked with Peter rebuking his master Jesus, and then Jesus rebuking Peter who would later have a leadership role in the church. Because these statements are embarrassing, the passage is unlikely to have been invented by the early church. Jesus referred to his impending death in his introduction of the Last Supper; a reminder to his disciples that his body would be broken and his blood poured out for them.[111] The close similarity of 1 Corinthians 11:25 and Luke 22:20 in the Greek indicates a pre-Pauline tradition independent of Mark.[112] One final source refers to what happened in the Garden of Gethsemane when Jesus anticipated a violent death.[113] Again not a made up story as, in contrast to brave martyrs, he was very troubled and distressed, and wanted to avoid what would follow if at all possible.

(2.) *Jesus died.* Some writers have said that Jesus did not die, but only swooned and revived later. However, Jesus had received a brutal flogging which would have left severe injuries, and he was pierced by a sword to ensure death. If he survived, which was virtually impossible, it would mean that someone who was half dead and in need of serious medical attention somehow got out of the grave clothes and out of the tomb by rolling back the stone and then appeared to the disciples as a conquering hero. He also walked some distance on the road to Emmaus on wounded feet with two disciples. Such a theory is no longer regarded as acceptable, though it has been resurrected more recently by some skeptics.

---

[110] For example, Matthew 16:21; Luke 9:22; Mark 8:31–32, 9:31, and 10:32–34.
[111] Luke 22:15–20; Mark 14:22–240.
[112] Licona (2010: 286).
[113] Mark 14:32–41; Matthew 26:36-45; Luke 22:39–46.

(3.) *The tomb was well known.* As mentioned in all four gospels, it belonged to Joseph of Arimathea, a prominent member of the Sanhedrin that had condemned Jesus. Also the early church was hostile toward the Jewish leaders, so that Joseph of Arimathea was unlikely to be a Christian invention, given the care he showed. Then why mention Arimathea, an obscure town of no particular significance? The location of the tomb therefore could not have been mistaken, being known to Jew and Christian alike. The women noted the location of the tomb to visit later and if they got it wrong they would have been corrected by the disciples or the Jewish authorities. Also according to John, Nicodemus, who was another member of the Sanhedrin, attended and brought an embalming mixture of myrrh and aloes.

The Jewish authorities went to Pilate because Jesus prophesied that he would rise again in three days[114] and they didn't want the disciples to steal the body so they could claim falsely that Jesus had risen. Pilate then gave them permission to have a guard set that could have been Jewish or Roman.[115] If the claim that Jesus had risen was due to a mistaken tomb, then the Jewish authorities would have produced the body from the proper tomb to quell the story. The body could not be produced. In fact the soldiers were bribed by the chief priests to say that the disciples stole the body while they were asleep.[116] If the soldiers were Roman, there would have been a severe penalty for soldiers falling asleep on the job.

(4.) *The Roman seal was broken.* The consequences of breaking such a seal were extremely severe. Anyone caught doing this faced execution by crucifixion upside down so that people would be afraid of this. Would Jesus' disciples have had the courage to break the seal given they hid themselves when Jesus was captured,[117] and later Peter denied Jesus three times?

(5.) *The large stone was moved.* All the gospel writers mention the fact that the stone was rolled away. This was such a large stone that three women could not move it.[118] Matthew called it a great stone.[119] Who moved the stone?[120]

(6.) *The tomb became empty.* The grave clothes were still there in place along with the head covering, which was folded up by itself and separated

---

[114] Matthew 27:62–66.
[115] For a discussion of this see Craig (1984).
[116] Matthew 28:11–15.
[117] Matthew 26:56.
[118] Mark 16:1–8.
[119] Matthew 27:60.
[120] This is the title of an excellent well-reasoned book written by Frank Morison,[121] a freelance writer who set out to disprove the resurrection but ended up becoming a Christian.

from the strips of linen. (If the body was stolen why unwrap it?) When Peter looked inside the empty tomb and saw the grave clothes there he was puzzled.[122] The empty tomb was also attested to by the Jewish historian Josephus and other external writings; both hostile sources. On the Sunday morning after the crucifixion a group of women followers found the tomb empty. The disciples had gone back to Jerusalem where any false teaching would have soon been disproved.[123] There were six independent accounts of the empty tomb, with one in each gospel and two in Acts. If the tomb had not been empty, it probably would have been treated as a shrine, which was common with the graves of holy men in first century Judaism. The disciples never repeatedly returned to the grave nor treated it with any special reverence; there was nothing there.

(7.) *Jesus' resurrection was bodily.* Some critics claim that the resurrected body was a non-physical body, but rather a spirit. When Jesus first appeared to the disciples they thought he was a ghost, but he soon dispelled that idea. Support for a bodily resurrection is seen in Luke 24: 39–43 where Jesus showed his disciples his hands and feet (which would have been scarred) and said that a spirit does not have flesh and bones. He invited them to touch him as his scars were visible, and he also ate a piece of fish in front of them. In John 20:26–28 Jesus invited doubting Thomas to touch his crucifixion wounds; skeptical Thomas saw and believed and called Jesus Lord and God. Jesus' resurrection was clearly a bodily one! The critics say that Paul apparently did not understand that Jesus' appearance to him (cf. 1 Corinthians 15: 8) was a bodily one, and argue from verses 47–50 that Jesus' body is a spiritual one (whatever that means). Arguments for and against are given by William Lane Craig[124] and I will give just a few comments from his discussion.

There is evidence that the appearance of Jesus to Paul[125] was not just visionary because in all three accounts there was light and voice and physical phenomena, also experienced by Paul's companions. Also Paul equated the appearance of Jesus to him with the appearances of Jesus to the disciples in 1 Corinthians 15:8, though it doesn't necessarily mean that all the other appearances were of the same mode. Paul is anxious to include himself with the other apostles (those commissioned by Christ), and having seen Jesus would give confirmation to his claim.

---

[122] Luke 24:12.
[123] For a detailed discussion of the historicity of the empty tomb and the disciples' inspection of the empty tomb see Craig (1985, 1992, 2001a).
[124] Craig (1980).
[125] Described in Acts 9:1–9, 22:3–16, and 26:9–23.

Another critical argument that is put forward is that Paul equated Jesus' resurrection body with our future resurrection bodies that will be spiritual, so that Jesus' resurrection body was spiritual. While the first part of the the argument is true,[126] Craig[127] carried out an extensive analysis of the Greek words used in the verses about the resurrection to disprove the conclusion of the argument. For example, the Greek word *soma* refers to the physical body itself and not the person separate from his or her body. He also maintains that the resurrection was a transformation of the body and not just the resuscitation of a corpse, as with Lazarus who would die again someday. Jesus rose to eternal life[128] with a body that possessed special powers such as suddenly appearing behind closed doors, suddenly vanishing, and moving from one place to another. There is no reason why God could not provide a new form of matter no longer bound by the second law of thermodynamics. We are told that Christians in the afterlife will be resurrected to new bodies.

(8.) *Jesus appeared to many different individuals.* Some have argued that all these people hallucinated. This idea is psychologically untenable for the following reasons.

   (a) They occurred with different people, places, and conditions; for example the appearances to the women, to James, to Cleopas and friend on the road to Emmaus, to the disciples behind locked doors on two occasions and on the shore of Galilee, and to Paul on the road to Damascus. The diversity of the appearances cannot be adequately explained by hallucinations. As noted by Craig:[129]

   > The appearance to Peter is independently attested by Paul and Luke (1 Corinthians 15:5; Luke 24:34), the appearance to the Twelve by Paul, Luke, and John (1 Corinthians 15:5; Luke 24:36–43; John 20:19–20), the appearance to the women disciples by Matthew and John (Matthew 28:9–10; John 20:11–17), and appearances to the disciples in Galilee by Mark, Matthew, and John (Mark 16:7; Matthew 28:16–17; John 21). Taken sequentially, the appearances follow the pattern of Jerusalem-Galilee-Jerusalem, matching the festival pilgrimages of the disciples as they returned to Galilee following the Passover/Feast of Unleavened Bread and travelled again to Jerusalem two months later for Pentecost.

   (b) A visual hallucination is a private event. This makes it hard to explain the appearances to groups of people.

---

[126] As noted by Paul in Philippians 3:21.
[127] Craig (1980).
[128] Matthew 28:18–20; Luke 24:26; John 20:17.
[129] http://www.leaderu.com/offices/billcraig/docs/visions.html, accessed August, 2015.

(c) On three separate occasions the "hallucination" was not recognized as Jesus.[130] This is a very embarrassing detail for those who want to dismiss the stories as fictional. Anybody inventing the story would have almost certainly described instant and certain recognition. Also, the disciples weren't expecting to meet him again after his crucifixion. Jesus' resurrected body was different in some way as he could suddenly appear in a locked room and disappear again. There was, however, a continuity of personality in the same way that every cell in our body is replaced every seven years, but there is both a conscious and unconscious continuity of personality.

(d) The disciples were thoroughly dejected at the death of Jesus and despite Jesus' predictions were not expecting a resurrection. Mary Magdalene thought that Jesus' body had been stolen before actually seeing him alive again.[131] There was no longing by the disciples for something that had failed to occur. In fact they did not believe the women when they said that Jesus was risen,[132] nor did they believe the two who reported their encounter with Jesus on the road to Emmaus,[133] and Thomas was obstinate in his doubt.[134] Some of the eleven even doubted at the great commission in Matthew 28:17, much like those today who struggle with the miraculous. The idea of a dying, much less risen, Messiah was completely foreign to the Jew.[135] The Messiah was expected to be a conquering hero who would liberate Israel, and not someone condemned as a criminal! After all, Old testament law dictated that anyone executed by hanging on a tree was under God's curse.[136] Yet the disciples did not go and find another Messiah.

(e) The appearances ceased after a certain period of time and did not reoccur. This would be in contradiction to any hallucination theory.

(f) Some critics claim that Matthew's mountain-top experience mentioned in Matthew 28 was some kind of kind of vision or illusion. However, it is clear that Matthew considered Jesus' appearance to be physical, as we see from his appearance to the women in verses 9 and 10 and his commissioning of the disciples. The women and the disciples worshipped him, with the women taking hold of his feet. Also the Greek for Jesus coming toward the disciples seems to definitely indicate a physical appearance.

---

[130] See Luke 24:13–31; John 20:14–16 and 21:4.
[131] John 20:2,15–18.
[132] Luke 24:11.
[133] Mark 16:13.
[134] John 20:24–31.
[135] For example Mark 8:32 and 9:10.
[136] Deuteronomy 21:23.

(9.) *The disciples were transformed in spite of initial unbelief.*[137] As already noted above, the Jews had certain expectations of the Messiah and would not envisage a dying and rising Messiah. Also the Jews had certain beliefs about the afterlife that generally precluded the idea of a special individual resurrection from the dead. The disciples were fearful and depressed[138] and Thomas was openly skeptical. However, according to church history and tradition,[139] ten or eleven of the twelve disciples died a martyr's death, mostly in horrible ways, and they died on the basis of two beliefs: the resurrection of Christ and their belief in him as the Son of God. This is very different from people who are prepared to die for an ideology. People may die for a lie they believe to be true, but who would die for a lie they know to be a lie! We read that despondent Peter who lied that he was a follower of Jesus was transformed into a fearless preacher at Pentecost in Acts 2, and before a crowd in Caesarea proclaimed why he and the other disciples were so convinced that Jesus was alive.[140]

On the day of Pentecost all the disciples were present and they almost invited martyrdom by their boldness and the commotion they caused! Acts states that about three thousand joined the new movement that day, which would imply a big crowd listening to Peter. Later Peter and John, in spite of threats, told the leaders that they could not stop speaking about Jesus because of what they had seen and heard.[141] However they took the threats very seriously.[142] Arrests followed later and even after a flogging they did not stop teaching and proclaiming that Jesus was the Christ.[143] We also note that James the oldest brother of Jesus (named James the Just to distinguish him from the disciple, James the son of Zebedee) apparently died a martyr's death, as reported by Josephus, Hegesippus, and Clement of Alexandria.[144] Paul included James among those he listed as having seen the resurrected Jesus.

(10.) *Skeptical James, the oldest brother of Jesus, became spiritual leader of the church at Jerusalem.* Jesus' mother, and at least some of his brothers that originally did not believe in him,[145] were among his followers after the resurrection[146]

---

[137] For a good discussion of this topic see the online version of McGrew and McGrew (2009).
[138] John 20:19.
[139] See http://bibleprobe.com/apostles.htm or http://www.apostles.com/apostlesdied.html, both accessed August, 2015.
[140] Acts 10:39–41.
[141] Acts 4:18–20.
[142] 4:23–30.
[143] Acts 5:17–19, 41–42.
[144] For details see Licona (2010: 455–460).
[145] Mark 3:21, 31–35; 6:3; John 7:1–10.
[146] Acts 1:14; 1 Corinthians 9:5.

(11.) *There was a forty-day delay before the ascension of Jesus and forty-nine days before the first public proclamation.* Anyone prefabricating the resurrection story would not have included this embarrassing delay. Luke tells us that in those forty days Jesus appeared to his disciples on many occasions, ate and drank with them,[147] informed them about the kingdom of God, and told them to wait in Jerusalem until they received the Holy Spirit.[148] According to Luke they therefore waited there just nine days and they were baptized by the Holy Spirit on the Jewish day of Pentecost.

(12.) *An individual bodily resurrection was almost unbelievable by world views at the time.* New Testament scholar Tom Wright,[149] after an extensive survey of first century Mediterranean thought of non-Jews, showed that the bodily resurrection was regarded as impossible. The Greeks and Romans believed that the material body was corrupt so that putting a spirit that was good back into the body, which was a prison, was completely unthinkable and undesirable. The Jews, on the other hand, thought that the body was good and death was a tragedy. He argued[150] that the idea of an individual resurrection was completely contrary to the belief of an eventual general resurrection of all the righteous, where a new order would be established. However it should be noted that Herod[151] believed that Jesus was John the Baptist raised from the dead and mentioned others who believed it was Elijah or a prophet.

(13.) *After the resurrection, Jesus first appeared to Mary Magdalene and other women.* If the story had been made up, Jesus' first appearance would not have been to women and especially to someone with Mary Magdalene's reputation! According to Jewish principles of evidence, women witnesses were regarded as invalid witnesses. Jewish society was patriarchal and we see this when the women who had found the tomb empty were not believed by the disciples. Licona comments:

> Thus there is a double-embarrassment factor present as the women serve as both witnesses and as the recipients of divine revelation while the men are presented as thickheaded. These are not the kind of reports one invents to boost confidence in church leadership.[152]

(14.) *Jesus appeared to Paul.* As a result Paul's life was completely turned around. He suffered many hardships including martyrdom for his faith.

(15.) *There were other witnesses to Jesus' resurrection.* Paul refers to more than five hundred in 1 Corinthians 15:8, and the existence of other

---

[147] Acts 10:41.
[148] Acts 1:3–5.
[149] Wright (2006: 113).
[150] Wright (2006: 200–206).
[151] Mark 6:14–16.
[152] Licona (2010: 355).

witnesses is attested by the selection process leading to the appointing of Matthias, a replacement for Judas.[153] In fact at the time there were about 120 people present. Why appoint another disciple if Jesus didn't rise from the dead?

(16.) *The early church grew rapidly in a very unsympathetic world.* They were prepared to die for their faith, and their upfront behavior was almost inviting martyrdom! We read about the commotion on the day of Pentecost in Acts 2, their refusal to stop speaking about Jesus in Acts 4, their being thrown into prison in Acts 5:17–19, 33, the stoning of Stephen in Acts 7, and the death of James the brother of John in Acts 12. It was not the early church making up the story of the resurrection, but rather the fact of the resurrection that created the early church.

6.4.4 Probability Arguments

The reader is not allowed to escape from this chapter without meeting some probability arguments! In Section 2.9 we briefly considered the nature of probability and showed how we can obtain a (posterior) probability of the existence of God given the evidence in Chapter 2 based on a (prior) probability of God's existence without using the evidence. We now apply similar arguments to find a subjective posterior probability that Jesus rose from the dead given the evidence, namely $P(R \mid E)$, given a prior probability $P(R)$ of the resurrection. Although a number of arguments in support of the resurrection have been presented, Licona[154] endorsed three facts put forward by Habermas as being regarded as historical by an almost universal consensus of scholars writing on the subject since 1975, namely:

(1.) Jesus died by crucifixion.

(2.) Soon after Jesus' death, the disciples had experiences that led them to believe and proclaim that a resurrected Jesus had appeared to them.

(3.) A few years after the death of Jesus, Paul experienced what he interpreted as a resurrection experience of Jesus that led to his conversion.

Also 75% agreed that the tomb was found empty. Taking a very conservative view we will assume just the statements (2) and (3) to be true as well as the testimony of some of the women with regard to the empty tomb and apply some probabilities below. We begin by introducing the concepts of statistical independence and conditional independence and then go on to show the validity of the resurrection has a high probability given the evidence.

---

[153] Acts 1:21–22.
[154] Licona (2010: chapter 4).

Statistical Independence

If two events A and B are *physically* independent, then the occurrence or non-occurrence of one of the events will not affect the occurrence or non-occurrence of the other. If I toss a coin in the kitchen and my wife tosses a coin in the lounge, the outcome of my toss will not affect the outcome of her toss. Mathematically this means that if A is the event of me getting a head and B is the event of her getting a head, then $Pr(A \mid B) = Pr(A)$ and $Pr(B \mid A) = Pr(B)$, that is the probability of A occurring given B occurring will not affect the probability of A occurring, and vice versa. Using the laws of probability, it is readily shown that both expressions lead to the symmetrical expression (which we would expect) $Pr(A\&B) = Pr(A)Pr(B)$, that is the probability that both A and B occur is the product of the individual probabilities.

Conditional Independence

The above idea also extends to conditional probabilities, namely if A and B do not affect each other *given* that C has occurred, then we have conditional independence defined by $Pr(A \mid B\&C) = Pr(A \mid C)$ and $Pr(B \mid A\&C) = Pr(B \mid C)$, which both lead to the symmetrical expression

$$Pr(A\&B \mid C) = Pr(A \mid C)Pr(B \mid C).$$

It should be noted that the (unconditional) independence of A and B does not necessarily imply any kind of conditional independence, and vice versa, without further assumptions. For example, even if

$$Pr(A\&B \mid \sim C) = Pr(A \mid \sim C)Pr(B \mid \sim C),$$

as well as the previous equation, it doesn't follow that $Pr(A\&B) = Pr(A)P(B)$.

As conditional independence is a tricky concept to explain in a real situation, I give two examples. Suppose Jack and Jill live in the same large city but in places some distance apart, and they don't know each other. Let A be the event that Jack drives to the movies on a particular evening and B be the event that Jill drives to the movies (different theatre) on the same evening. Given the event C that driving conditions are normal in the city, the events A and B will be conditionally independent, *given* C. That is, given normal driving conditions, Jack and Jill will be independent of each other with respect to going to the movies. However, if there is a snow storm over the city on that day so that we have the event $\sim C$ (or not C), then if we know that A did not occur, that is Jack did not drive to the movies, it is likely that B did not occur, that is Jill did not drive to the movies either. This means that A and B are not independent given $\sim C$.

Another example showing conditional independence, which may or may not help, involves the rolling of a blue die and a red die, each by a different person.

Then the event A that the outcome of the blue die is a three and the event B that the outcome of the red die is a five will be independent events. Suppose we now have the event C that the blue die does not give a one and the red die does not give a two, then given C, A and B will be conditionally independent as given C the outcome of the blue die will not affect the outcome of the red die. Now we have made that all very clear (!) we apply these ideas to the resurrection; the following is based on the chapter of Timothy and Lydia McGrew.[155]

Probabilities Given the Evidence

We note first of all that sixteen pieces of evidence E_i ($i = 1, 2, \ldots, 16$) for the resurrection are given above in the previous section. However, to simplify matters we shall consider just three pieces of evidence that are generally accepted, as not all sixteen will be accepted by some. These are the events assumed to have occurred: W is the event of the women's reports of an empty tomb and the risen Christ, D is the event of the disciples' testimony, and P is the event of the testimony and conversion of Paul. We assume that we have reliable New Testament records (e.g., the four gospels, Acts, and some undisputed Pauline letters such as Galatians and 1 Corinthians). If the resurrection actually happened, then the conditional probability that all three events occurred, $Pr(W\&D\&P \mid R)$, will be much greater than $Pr(W\&D\&P \mid \sim R)$, the conditional probability that they occur given $\sim R$, that is "no resurrection." This means that the existence of the resurrection will explain the evidences very much better than no resurrection. Using the laws of probability, a formula for the ratio of the two probabilities and the three events is

$$\frac{Pr(R \mid W\&D\&P)}{Pr(\sim R \mid W\&D\&P)} = \frac{Pr(R)}{Pr(\sim R)} \times \frac{Pr(W \mid R)}{Pr(W \mid \sim R)} \times \frac{Pr(D \mid R\&W)}{Pr(D \mid \sim R\&W)}$$
$$\times \frac{Pr(P \mid R\&W\&D)}{Pr(P \mid \sim R\&W\&D)}.$$

The last three ratios on the right-hand side are called Bayes' factors, and the above expression is a special case of a more general formula for n events E_i ($i = 1, 2, \ldots, n$).[156] We now ask whether or not the three events W, D, and P are independent. First, given R is true, W and D will be conditionally independent as the women were not believed by the disciples; the same is true given $\sim R$ is true. Also, being later, P will be at least conditionally independent of W and D, given R or $\sim R$, or possibly (unconditionally)

---

[155] McGrew and McGrew (2009).

[156]
$$\frac{Pr(R \mid E_1 \& \ldots \& E_n)}{Pr(\sim R \mid E_1 \& \ldots \& E_n)} = \frac{Pr(R)}{Pr(\sim R)} \times \frac{Pr(E_1 \mid R)}{Pr(E_1 \mid \sim R)} \times \frac{Pr(E_2 \mid R\&E_1)}{Pr(E_2 \mid \sim R\&E_1)} \times \cdots$$
$$\times \frac{Pr(E_n \mid R\&E_1\& \ldots \&E_{n-1})}{Pr(E_n \mid \sim R\&E_1\& \ldots \&E_{n-1})},$$

independent. This means that the probabilities $Pr(D \mid W\&R) = Pr(D \mid R)$ and $Pr(P \mid W\&D\&R) = Pr(P \mid R)$ hold for R and also for R replaced by $\sim R$. Hence

$$\frac{Pr(R \mid W\&D\&P)}{Pr(\sim R \mid W\&D\&P)} = \frac{Pr(R)}{Pr(\sim R)} \times \frac{Pr(W \mid R)}{Pr(W \mid \sim R)} \times \frac{Pr(D \mid R)}{Pr(D \mid \sim R)}$$
$$\times \frac{Pr(P \mid R)}{Pr(P \mid \sim R)}$$
$$= r \times r_1 \times r_2 \times r_3, \qquad (6.1)$$

where r_1, r_2, and r_3 are the Bayes' factors. We now consider each of the three sets of testimonies.

Women's Testimony

If the resurrection occurred, then the fact of the women's testimony is perfectly believable and $P(W \mid R)$, the conditional probability of getting the testimony, would be a reasonable value. The women weren't expecting what happened and were consequently bewildered. Furthermore, their story of an empty tomb was definitely not believed by the disciples,[157] and as they were the first witnesses they would not have been affected by the disciples' testimony. The meeting of Jesus with Mary Magdalene[158] led to Mary rushing off to tell the disciples and, although skeptical, Peter and John considered it worthwhile to run and see for themselves. In Matthew 28:9-10 a brief account is given of Jesus meeting the women who had been to the tomb in a group. The fact that there are slight differences in the accounts and some loose ends indicate a lack of collusion. If Jesus wasn't resurrected it is hard to explain the women's testimony because of the various discredited theories given above such as Jesus just swooned, the story of Jesus' appearance was made up, the women had hallucinations, the tomb was mistaken, and so forth. Therefore $Pr(W \mid \sim R)$, the probability of getting the women's testimony given the resurrection did not occur (i.e., life carried on as usual), would be small, giving a large value to the first Bayes' factor $r_1 = Pr(W \mid R)/Pr(W \mid \sim R)$.

Disciples' Testimony

We begin by considering the testimony of the disciples as a group, then later as individuals. The big supporting factor for their joint testimony is the sudden change in the disciples and their being prepared not only to die for their faith but almost all actually dying for their faith. It is therefore not credible that the disciples stole the body, and the other alternative theories mentioned for the women don't hold here either. The theory that they had just a vision of Jesus does not fit with the story mentioned above that Jesus ate with them and invited them to touch him.[159] We see then that their testimony makes sense if

---
[157] Luke 24:11.
[158] John 20:11–18.
[159] Luke 24:36–43.

Jesus was actually resurrected, but is most unlikely to have occurred if Jesus wasn't resurrected. The second Bayes' factor $r_2 = Pr(D \mid R)/Pr(D \mid \sim R)$ will therefore be large.

Paul's Testimony

What is surprising about Paul's testimony is his complete transformation from persecuting the Christian church[160] to being one of its greatest missionaries who was prepared to put up with all kinds of severe suffering and even die to further the gospel and spread the message of a risen Christ.[161] He was such an ardent persecutor of Christians that people were loathe to approach him after his conversion to Christianity.[162] Clearly he wasn't affected by the testimony of the disciples. The only alternative explanation is that Paul had an hallucination and for some reason it was followed by blindness and Paul's total transformation. This does not make sense psychologically. The third Bayes' factor $r_3 = Pr(P \mid R)/Pr(P \mid \sim R)$ will therefore be large.

Combining the Probabilities

Referring to the equation (6.1) we have

$$\frac{Pr(R \mid W\&D\&P)}{Pr(\sim R \mid W\&D\&P)} = r \times r_1 r_2 r_3,$$

where $r_1 r_2 r_3 = r_1 \times r_2 \times r_3$, will be very large. However, the other multiplying factor $r = Pr(R)/Pr(\sim R)$ will be very small as the prior probability that the resurrection occurred will be a lot smaller than the prior probability of the resurrection not occurring. If you don't believe in miracles (see Chapter 7) then you will have $Pr(R) = 0$, which is " end of story" as none of the above arguments will convince you. Related to a belief in miracles, or at least the possibility of one miracle, namely the resurrection, is whether you believe in God or not. If you believe in God then your $Pr(R)$ may not be zero but could still be very small.

The McGrews show that the odds in favor of R can be increased substantially by considering the disciples individually, so that we now look at the testimony of, say, thirteen disciples (including James). Let D_i be the testimony of the disciple number i ($i = 1, 2, \ldots, 13$) and, to simplify the argument, we initially restrict ourselves to just three disciples so that $D = D_1 \& D_2 \& D_3$. Then using the laws of conditional probability

$$\frac{Pr(D_1 \& D_2 \& D_3 \mid R)}{Pr(D_1 \& D_2 \& D_3 \mid \sim R)} = \frac{Pr(D_1 \mid R)}{Pr(D_1 \mid \sim R)} \times \frac{Pr(D_2 \mid D_1 \& R)}{Pr(D_2 \mid D_1 \& \sim R)} \times \frac{Pr(D_3 \mid D_1 \& D_2 \& R)}{Pr(D_3 \mid D_1 \& D_2 \& \sim R)}. \quad (6.2)$$

---

[160] Acts 22:1–22.
[161] 2 Corinthians 11:21–28.
[162] Acts 9:10–19.

The first question to ask is what sort of collusion might have occurred among the disciples. Assuming $\sim R$, then the most likely alternatives seem to be either a group hallucination or agreeing to lie about the resurrection, neither of which is credible. In fact it would then be very unlikely that there would have been any testimony from the disciples. This means $Pr(D_1 \mid \sim R)$ will be small and $Pr(D_1 \mid R)/Pr(D_1 \mid \sim R)$ will be large; the McGrews suggest a value like 10^3. Also given $\sim R$ and the fact that the disciples were familiar with one another, the martyrdom of one of them for their beliefs would no doubt put the others off and reduce the probability of them holding to their testimonies so that we might reasonably assume that

$$Pr(D_2 \mid D_1\& \sim R) < Pr(D_2 \mid \sim R)$$

and

$$Pr(D_3 \mid D_1\& D_2\& \sim R) < Pr(D_3 \mid \sim R),$$

particularly if the martyrdom of D_1 occurred while D_2 and D_3 were alive, for example. If the resurrection occurred, then there could be either reinforcement of their beliefs and mutual encouragement (because of the powerful impact of the the resurrection and what happened on the day of Pentecost) or else be put off from giving testimony, though not put off as much as when R is not true. Assuming the former, it would be reasonable to assume

$$Pr(D_2 \mid D_1\& R) > Pr(D_2 \mid R) \text{ and } Pr(D_3 \mid D_1\& D_2\& R) > Pr(D_3 \mid R).$$

Combining the two results, we get from (6.2)

$$\frac{Pr(D_1\& D_2\& D_3 \mid R)}{Pr(D_1\& D_2\& D_3 \mid \sim R)} > \frac{Pr(D_1 \mid R)}{Pr(D_1 \mid \sim R)} \times \frac{Pr(D_2 \mid R)}{Pr(D_2 \mid \sim R)} \times \frac{Pr(D_3 \mid R)}{Pr(D_3 \mid \sim R)},$$

which shows that assuming independence of the D_i underestimates the left-hand side. Extending these ideas to the thirteen disciples we now have

$$\frac{Pr(D \mid R)}{Pr(D \mid \sim R)} = \frac{Pr(D_1\& D_2\& \cdots \& D_{13} \mid R)}{Pr((D_1\& D_2\& \cdots \& D_{13} \mid \sim R)}$$
$$> \frac{Pr(D_1 \mid R)}{Pr(D_1 \mid \sim R)} \times \frac{Pr(D_2 \mid R)}{Pr(D_2 \mid \sim R)} \times \cdots \times \frac{Pr(D_{13} \mid R)}{Pr(D_{13} \mid \sim R)}.$$

In this case, assuming the McGrews' value of 10^3 for each ratio, the probability on the left-hand side is greater than that obtained by multiplying the individual ratios on the right-hand side together, namely $(10^3)^{13} = 10^{39}$. The McGrews give conservative values of 10^2 for the ratio $Pr(W \mid R)/Pr(W \mid \sim R)$ and 10^3 for the ratio $Pr(P \mid R)/Pr(P \mid \sim R)$, giving a value of the left-hand side of (6.1) of about 10^{44}, which should be sufficient to overcome a prior probability (or rather improbability) of, say, 10^{-40} for R and giving us a posterior probability, the left-hand side of (6.1), greater than 0.9999. Of course these values are all contestable but the method is at least rigorous if going to be used at all. Also, a more detailed and more complicated analysis taking into account other evidences could be developed. The above analysis, however, does demonstrate that the evidence for the resurrection is strong from a probabilistic point of view.

6.4.5 Some Alternative Views of the Resurrection

Most of the arguments against the resurrection have been dealt with previously. However some have argued that the resurrection of Christ is a myth patterned on the dying and rising of fertility gods. Such myths do in fact go back a considerable time before Christianity so we would expect them to be known to the Jews. The New Testament documents, however, rule out any mythical ideas as we have straight-forward narratives, and any parallels are too superficial, as the resurrection narrative, like the gospels in general, are to be interpreted from a Jewish perspective.[163] There were many other so-called messiahs in the first century that were executed, but there is no mention of them being resurrected, nor the sudden explosion of a new world view, as with Christianity. Why should Jews worship Jesus as Divine when Jews believed in one God?

Bishop Spong explains away the empty tomb by claiming that there was no tomb, that Jesus was buried in a common grave, and Paul mentioned no tomb at all. He also claimed that the gospels are midrash, a Jewish tradition of reinterpreting and retelling the stories. This doesn't fit as (a) midrash is a commentary on an actual biblical text, (b) midrash is tightly controlled and argued, and (c) midrash never included the invention of stories that were clearly seen as non-literal in intent. No, the gospels are biographies.[165]

6.5 IS JESUS THE ONLY WAY?

Is Jesus the only way to reach God? Some assert that claiming one religion to be the true religion is the sort of intolerance that has caused much bloodshed. Others believe that all religions lead to God, which is referred to as pluralism. However this cannot be the case as different religions tend to contradict one another. For example, Hindus have millions of gods and hope to strive through reincarnations to reach Nirvana, where they lose their personal identities. Hinayana Buddhists are atheists while Mahayana Buddhists worship images or idols. Some Eastern religions have an impersonal deity. Non-Christian religions teach salvation by works, but they often completely disagree over which works. The three largest monotheistic religions Judaism, Islam, and Christianity are vastly different, though they have a common source, the

---

[163] For discussions of this point see for example the articles by Craig (2001b), Yamauchi,[164] and the book by N. T.Wright (2003).
[165] For a criticism of Spong see Bott and Sarfati (1995).

Jewish nation. Because of the differences, some would argue that no religion possesses the entire truth, only some of it. Alister McGrath comments: "There is a growing consensus that it is seriously misleading to regard the various religious traditions of the world as variations on a single theme."[166]

Jesus declared that he is the way, the truth, and the life, and no one comes to the Father except through him.[167] We also have Peter's declaration that salvation only comes through the name of Christ,[168] and the Bible says that if we reject Jesus we reject God.[169] This exclusiveness of Christianity claiming that Jesus is the only way is a point of contention with many people. Religious pluralism maintains that every religion is true and adequate, though one may be better than others. Some believe that all religions are simply different paths to God, but unfortunately they can't all be true as they have contradictory views about fundamentals. Which bits are right? Also, unless God accepted every religion He would still be described as intolerant and unfair. We now consider this issue of exclusiveness.

6.5.1 All the Major Religions are Exclusive

It should be noted that all the major religions are in fact exclusive. We consider just Judaism, Islam, Hinduism, Buddhism, and Christianity. Confucianism, although a way of life, is not regarded by some as a religion. The Chinese religion Taoism or Daoism is a mixture of religious and philosophical beliefs, while Shintoism, the main Japanese religion, is a mixture of beliefs.

The Jews were described as God's chosen people,[170] were entrusted with the words of God,[171] were loved on account of the patriarchs,[172] were heirs of the prophets and of God's covenant,[173] provided Christ's ancestry, and were the people that Jesus was sent to first to lead them to repentance.[174] In this sense salvation is from the Jews.[175] Unfortunately the Jews' advantages were in vain if they did not trust their Messiah Jesus as their Savior and obey him as Lord.[176] Jesus told them that they would die in their sins if they did not acknowledge him as the Messiah.[177] As a religion, Judaism is exclusive as it requires one to follow the very many commands and prohibitions of the Torah to have a chance at salvation, so that salvation is by

---

[166] McGrath (1997: 155).
[167] John 14:6; see also Matthew 11:27, John 3:16–18 and 1 John 5:13.
[168] Acts 4:12.
[169] Luke 10:16.
[170] For example Deuteronomy 7:6, 14:1; Kings 3:8; Isaiah 3:1.
[171] Romans 3:2.
[172] Romans 11:28.
[173] Acts 3:25.
[174] Romans 1:16 and 15:8.
[175] John 4:22.
[176] Luke 13:28–30; John 3:36, 8:24.
[177] John 8:24.

doing the right deeds (or "works") using the Torah as a guide. The Jews reject Jesus as the Messiah and do not believe in his resurrection. For an interesting debate between a Rabbi and Messianic Jew (a believer in Jesus) see https://www.youtube.com/watch?v=Jg3Aq5qsY8Q, accessed August, 2015.

Islam is seen as exclusive both theologically and linguistically, since the Qur'an their sacred book must be read in Arabic, as any translation lowers its sacredness. However, Sura or Surah (chapter) 2:112–113 in the Qur'an says that although Jews and Christians each believe that they are the only ones to enter Paradise, whoever surrenders to God [Allah] and does good deeds as well shall find his reward with his Lord, and shall have no fear or grief. This seems to cut across any exclusiveness of Islam as long as you believe that in the Last Day you will go on to Paradise (as the Abrahamic faiths all do). However, Muslims believe that Judaism and Christianity have become perverted, and reject the divinity of Christ, treating the doctrine of the Trinity as blasphemy (or *Shjrk*).[178] However they misunderstand the trinity and link it to God, Jesus, and Mary.[179] Islam is also salvation by works, but with a different set of works. In fact this is a problem with non-Christian religions as they all have different good works to practice.

Hindus do not compromise on the ultimate authority of their scriptures (the Vedas), the caste system, the law of karma, and reincarnation. It may appear that Hinduism is the most inclusive of religions as all their gods are manifestations of the ultimate God (Brahman, the Ultimate Reality) and that all things are God (pantheism). However, unless you have faith in the teaching of Krishna, one of the manifestations of God, and if you don't hold the view that you do not exist and that you are actually God-Brahman, then you won't achieve Hindu salvation (enlightenment).

Buddhism rejects the ultimate authority of the Vedas and the caste system, and although it has a has a similar worldview to Hinduism it is basically a non-theistic religion (from a Western viewpoint) so that you can be an atheist Buddhist (e.g., Hinayana Buddhists). However, Mahayana Buddhists deified Buddha even though he denied his godhood. Enlightenment comes by learning to lose one's attachment to self and one's independent existence. We note that there is no forgiveness for personal sin with reincarnation, only a pay back in the next life, so what are you paying for in the very first life?[180] In other words, how did inequalities begin in the first place? If you don't know what you are paying for in this life from sins in past lives, how can reincarnation help you? In contrast, Christianity teaches that we die once and then are judged.[181]

---

[178] cf. Surah 4:172.
[179] See Surah 4:171, 5:73, and 5:116.
[180] Buddhism has a slightly different view of reincarnation from Hinduism and prefers the term "rebirth."
[181] Hebrews 9:27.

There is of course a lot more to the above religions and my brief comments do not do full justice to them, but it is clear from a careful study of them that Christianity, Judaism, Islam, Hinduism, and Buddhism are all mutually exclusive of one another. Apologist Ravi Zacharias said that:

> All religions do not point to God.... Anyone who claims that all religions are the same betrays not only an ignorance of all religions but also a caricatured view of even the best-known ones. Every religion at its core is exclusive.[182]

It should be noted that the truth is always exclusive; it excludes the false.

With regard to pagans, religious ones (and not just atheists) are without Christ, excluded from God's promises, without hope, and without God.[183] In the Old Testament we read that God was particularly hard on those who pursued other gods, and continually tried to prevent the corruption of the Jewish nation by forbidding intermarriage with pagans. The exclusiveness of God with respect to other gods is spelt out clearly in Isaiah[184] and in the Ten Commandments.[185]

Although Christianity is exclusive, it does respect people of other faiths. In fact there will be some overlap with other faiths and cultures as Christians believe that we are made in the image of God and therefore share something of the nature of God within us. Christians also recognize that they are also sinners, which doesn't leave room for pride. The question is why should the exclusiveness of Christianity lead to Christians caring for the sick and dying at great personal risk in different parts of the world and, for example, during the terrible plagues of the first two centuries. Pastor Timothy Keller answers:

> It was because Christians had within their belief system the strongest resource for practicing sacrificial service, generosity, and peace-making. At the very heart of their view of reality was a man who died for his enemies, praying for their forgiveness.[186]

In the end only God knows the future of an individual, and therefore it is not up to Christians to make judgments.

6.5.2 New Age

For completeness we shall look briefly at a form of religion called New Age that endeavors to promote a person's own power. New Age is difficult to pin down because it is a collection of many metaphysical thought systems, focussing on "correct thinking" and "correct knowledge" (whatever they might be), with a theology of "feel-goodism", "universal tolerance", and "moral relativism." For such followers, God is a higher consciousness within themselves so that

---

[182] Zacharias (2000: 7).
[183] Ephesians 2:11ff.
[184] Isaiah 42:8, 43:10–12, 44:6–8, 46:9–11, 45:5, 21.
[185] Deuteronomy 5:7–8.
[186] Keller (2008: 20).

pantheism is the underlying belief, i.e., "I am God." Many gods and goddesses are acknowledged, as in Hinduism, and New Age incorporates a wide range of ancient spiritual traditions. It teaches, for example, a wide array of eastern metaphysical and psychic techniques including breathing exercises, chanting, drumming, use of crystals, astral projection, and meditating to help a person develop an altered consciousness and one's own divinity. These techniques are used to speed up our evolution toward a new spiritual level, the aim being to establish a global unity of human with human and human with nature, where we are on an equal footing with the animals. New Age also teaches that the Earth is like a living being with its own intelligence, emotions, and deity, and is the source of all spirituality.[187] However, self supersedes all and is the source of reality. New Age opens people up to believing almost anything!

6.5.3 Mystery Religions

It has been suggested that the stories about Jesus were borrowed from the so-called Mystery Religions. In particular there was Mithraism, already existing in the region, that actually flowered after Christianity, so it is not surprising that there are similarities between Mithras and Christ. But there is also mythology. The god Mithras was born out of a rock, and he battled with and slew a primeval bull that became the ground of life for the human race. The so-called superficial parallels between the Mystery Religions and Christianity are not taken seriously by scholars any more, even by those rejecting Christianity, for a number of reasons. For instance, proponents of the cross-fertilization tend to take a bit from each of the various Religions to make the comparison, when the Religions are all very different and individually have little in common. Christian terms are used to describe pagan beliefs, which is totally misleading. There is no archaeological evidence that mystery religions were in Palestine in the first century A.D. and the Jews and early Christians were totally against syncretism with other religions.

There are a number of doctrines such as one God, universal sinfulness, blood sacrifice for sin, salvation by grace, and the bodily resurrection, for example, that are based in Judaism and have nothing to do with Greek mythology. The key difference is that the Mystery Religions are not centered on historical claims, but are based on the annual cycle of birth and death in nature in contrast to Christianity's linear view of history. Any borrowing would be from Christianity to the Mystery Religions, and not the other way round. Licona[188] argues that there is no causal connection of Jesus' resurrection to the Mystery Religions. He refers to the early disputes in the church reported by Paul concerning whether Christians should eat meat offered to idols, whether they should be circumcised and observe certain religious days,

---

[187] It is referred to as Gaia.
[188] Licona (2010: 149).

and whether Jewish believers were allowed to eat with Gentile believers.[189] Since such care about outside influences was exercised, it is unlikely for the same Christians to borrow from outside sources to create the resurrection story, especially when they stated that the truth of Christianity rested on the resurrection.[190]

6.5.4 Relativism

The next question to consider is which religion is the right one. Relativism (pluralism), which many hold to today implies that there is no way of knowing which religion is true or best, or which bits of any religion are true. Clearly you cannot put them all together to form a universal religion as there is really very little of substance in common. There is no universal agreement among the worlds religions as to who or what God really is. The Bahai faith, supposedly a universal religion that excludes all exclusivists, has no clergy, sacraments or rituals, but has a mixture of religious practices (reflecting its early ties to Shiite Islam), and endeavors to provide a world-wide religion with its belief in immortality and in one God. This God is transcendent and unknowable, but has revealed himself down through the ages through various prophets such as Adam, Abraham, Moses, Krishna, Zoroaster, Buddha, Jesus Christ, Muhammed, The Bab, and Baháu'lláh. Unfortunately these prophets do not all speak with the one voice! Baháu'lláh described God's purpose for man in the following way: "The purpose of God in creating man hath been, and will ever be, to enable him to know his Creator and to attain His Presence."[191] This makes good sense, but the question is how to do it! If all religions lead to God why start another?

Relativism has another problem that is self-refuting. If you state that there can't be one true religion and that you can't determine which beliefs are right or wrong, then how can we believe your belief statement? If you say that social conditions, like the home and society you were brought up in produced your beliefs, the same holds for your belief about your beliefs!

6.6 COMPARING RELIGIONS

Continuing on the theme of the previous section, I have heard it said, "Does it matter what sort of God we believe in? Surely all paths lead to the same destination." Those supporting this view would argue that there is an ethical core which all religions share in common. For example, we have some form of the Golden Rule about doing to others what you would have them do to you in many religions such as Buddhism, Christianity, Hinduism, Islam, Judaism,

---

[189]Galatians 2; 1 Corinthians 8, 10:18–33; Romans 14; Revelation 2:14, 20.
[190]1 Corinthians 15:17; Matthew 12:39–40; Luke 11:29–30; John 2:18–22.
[191]Baháu'lláh (1976: 70).

and Taoism. Furthermore, each of these traditions produces an ethical or moral transformation in the lives of its followers. The argument then is that is therefore difficult to prove that one religious tradition is more effective than others in transforming the lives of its followers. As we have seen above, some of the religious views cannot be true as some views totally contradict one another. Although we have compared several religions with regard to exclusiveness, we now compare them further on a number of basic ideas. This is not meant to be a detailed comparison, but some comments on this topic are relevant. Attitudes to suffering are considered separately in Section 8.9.6.

6.6.1 Their Founders

To compare the major religions we can begin first with their founders. We have already looked at the life of Jesus and the documentary evidence about his life in this and the previous chapter. His life began with the virgin birth that was prophesied a long time before the event. This is confirmed even in the Qur'an written six hundred years after Jesus' birth in Surah (19:18–23).[192] It seems that people of all faiths acknowledge that Jesus' life was the purest ever lived, being described in the Bible as sinless.[193] In contrast, leaders like Muhammad (also written as Mohammad or Mohammed), Krishna, and Buddha have their lives and struggles recorded within their own scriptures. Muhammad spent the last ten years of his life at war and was involved in retaliation.

As noted by Ravi Zacharias,[194] Muslims today believe that all the prophets were sinless even though this is not evident in their own scriptures, nor is it a view of the Old Testament prophets. However, in the Qur'an,[195] Moses said that he erred in slaying the Egyptian, and Abraham asked for forgiveness on the Day of Judgement.[196] Muhammad was told he would be forgiven for his sins in Surah 48:3.[197] Muhammad had eleven wives instead of the four prescribed for his religion, and heaven is described in the Qur'an as "wine and women",[198] which Muslims dismiss as metaphorical. He also was entitled to marry, under certain conditions, almost any woman.[199] Krishna's exploits with the milkmaids in the Bhagavad-Gita causes some problems with many Hindu scholars. Buddha said that he endured rebirths indicating a

---

[192]The numbering of verses may differ slightly (such as subtract one) depending on the translation used, as some introductory comments or titles may be listed as verses, e.g. Qur'an (2009); see also https://www.alislam.org/quran/search2/, accessed August, 2105.
[193]See Section 6.3.2.
[194]Zacharias (2000: 40–41).
[195]Surah 26:21.
[196]Surah 26:83.
[197]Muslims get round this by translating the word as faults or shortcomings rather than "sin", although it is translated as "sin" elsewhere, as for example in Surah 12.30.
[198]Surah 78:32ff.
[199]Surah 33:51–53.

series of imperfect lives. Christianity, more than any other religion, has made the doctrine of sin very clear. We are told that the heart is desperately wicked[200] and gives rise to to all kinds of sinful practices.[201] People refuse to acknowledge God even though the universe speaks of God as creator, and turn to sin worshipping the creature rather than the creator.[202]

Jesus also made claims that no other religious founder made, and the documentary evidence about the lives of other founders does not compare with that for Jesus. The militancy of Muhammad contrasts strongly with Christ's teaching in the Beatitudes and in his exhortation to his listeners to love their enemies and pray for those who persecute them.[203] Muslims are commanded not to make friends with Jews or Christians[204] and can kill someone if there is a just cause.[205] However the God of the New Testament loves unbelievers with a love that is unconditional and universal,[206] and we are told to do the same, whereas the God of the Qur'an has no love for unbelievers and loves only those who are faithful Muslims.

In contrast, Jesus showed love for the Samaritans, by teaching and healing them[207] even though they had corrupted Judaism with paganism and were hated by the Jews. He told the powerful parable of the good Samaritan[208] who was pictured as a better man than two Jewish religious leaders to explain what it meant to love your neighbor. Jesus went against cultural norms by evangelizing a Samaritan woman of low character as well as people from her village. He also rebuked James and John for requesting judgment on the Samaritans[209] and commanded his disciples to evangelize Samaria.[210] As already mentioned, Jesus said a lot about forgiveness, and even asked God to forgive his tormentors when he was crucified.[211]

It is of interest to note that the Qur'an refers to Jesus as Messiah,[212] the Word from and of God, and the Spirit of Allah, as well as mentioning his virgin birth, sinlessness, resurrection, ascension, miracles, and Second Coming. However Muslims don't believe that God could have a son.[213] Also, Muslims don't believe that Jesus was crucified, but that he was made to appear like one

---

[200] Jeremiah 17:9.
[201] Matthew 15:19).
[202] Romans 1:28–32.
[203] Matthew 5:1–11, 44; Luke 6:27–36.
[204] Surah 5:52.
[205] Surah 17:34.
[206] Matthew 5:43-48.
[207] Luke 17:11–19.
[208] Luke 10:25–37.
[209] Luke 9: 51–5).
[210] Acts 1:8, 8:1–25, 15:3.
[211] Luke 23:34.
[212] Surah 3:46.
[213] For further comments by William Lane Craig comparing Jesus with Muhammad see www.reasonablefaith.org/who-is-the-real-jesus-the-jesus-of-the-bible-or-the-jesus-of-the-quran, accessed August, 2015.

crucified.[214] Jesus claimed to be God incarnate; Muhammad never claimed to be God (only a messenger);[215] Buddha remained silent on the question of God; Confucius refused to discuss the idea of God; and Moses merely claimed to be a prophet of God. Finally, in contrast to Muhammad and Buddha, Jesus knew what his mission was right from the beginning. Even at the age of twelve he referred to the temple as his father's house.[216]

6.6.2 Universality

For a second comparison, we look at how universal is a particular religion. Christianity is open to people of all nations or languages, irrespective of their social or intellectual status, or whether they are male or female, or Jew or Gentile.[217] This universality was God's intention right from the beginning, with all nations being blessed through Abraham, Isaac, Jacob, and their descendants;[218] this theme runs throughout the book of Isaiah.[219] The good news of Christ's redemption is for all nations[220] and Christians inherit the promises made to Abraham's descendants.[221] Of course Islam is open to all people but men and women are not treated equally, especially with regard to worship. We don't find gender equality in Hinduism or Buddhism either.

6.6.3 Nature of God

A fourth comparison is when we consider how God is described by a religion. There is a big gulf between monotheistic religions (such as Christianity, Islam, and Judaism) that believe that there is only one God who created the universe from nothing, and pantheistic religions (such as Hinduism, Buddhism and New Age) in which God is described as being in everything including ourselves. Some Hindus do not believe in a personal creator but rather believe in Brahman, an impersonal absolute reality that permeates all things. Other Hindus believe that there are millions of deities (such as Brahma, Vishnu, Shiva and Krishna) that are manifestations of Brahman. As mentioned above, Buddhists don't necessarily believe in God. They don't worship Buddha, and generally reject the notion of any supernatural power. Meditation is more of a self-discipline. The Christian God is described as loving and personal in stark contrast to other gods, and can be known through Christ. We have the many Hindu gods that have similar character defects to those of the Greek pantheon. Also we are not identical with God as claimed by Hinduism but

---

[214] Surah 4:158.
[215] Surah 3:145.
[216] Luke 2:41–50.
[217] Galatians 3:28.
[218] Genesis 12:3, 18:18, 22:18, 26:4, 28:14 cf. Matthew 28:19.
[219] For example, Isaiah: 2:2–4.
[220] Matthew 24:14: Acts 1:8; Revelations 7:9.
[221] Galatians 3:28.

are created in God's image, thus providing a reason for our moral framework. Christians believe that God is all-loving, while it seems that Muslims believe that God loves only Muslims, and has no love for unbelievers so that they can be killed indiscriminately. Allah, the God of Islam seems to be distant, aloof, and capricious. For the Buddhist, God is unknowable.

6.6.4 Nature of Sin

A fifth comparison is how each religion considers sin. In Hinduism sin is understood as spiritual blindness. A person is regarded as a sinner when he or she breaks the ritual and caste laws or the rules of life according to Hindu laws. As the result of such actions karma is the evil that becomes attached to the personality. There are some Hindu theists who approach nearer to the Christian view of sin as seeing the individual asserting his or her autonomy from God.

The greatest sin in Islam is unbelief in Allah and in His prophet Muhammad. For the Muslim, sin generally means a failure to fulfill the commands of God so that any breach of the Shari'ah (law) is sin, though distinctions are made between moral and ritual sins. All sins require repentance, but God's reaction to repentance is unknown as he may forgive grave sin and punish the venial. In Islam, forgiveness is based on a combination of Allah's grace and the Muslim's works. On the Day of Judgment, if a Muslim's good works outweigh his bad ones, and if Allah so wills it, he may be forgiven of all his sins and then enter into Paradise. Sin is not such a serious matter as it is for Christians as it is a lapse due basically to ignorance; it is not seen as defying God or due to a corruption of the will as in Christianity.

For the Buddhist, suffering is the dominant element in human life and it results from the sinful desire to hang on to insubstantial things. Both Judaism and Christianity hold that sin arises from a misuse by people of their free wills, where there is disobedience to God's will and rebellion against Him. It becomes an attitude that is strengthened and hardened by repeated acts. The distinction between a state of sin and sinful acts, as in Christianity, does not seem to be in other religions.

6.6.5 Scriptures

A sixth comparison relates to the attitude towards scriptures which varies with the religion. It is important to have a reliable translation before making any comparisons. In contrast to other scriptures, the Bible is grounded in history so that it is supported for example by archaeology and history. Although the Bible is regarded as being inspired by God, even by Islam,[222] it is not seen by Christians as being directly dictated by God. It is certainly important to

---

[222] Surah 3:4–5.

Christianity, but in the last resort a person, namely Jesus Christ, rather than a book is the foundation of Christianity. A Christian does not need to wash his or her hands before reading scripture.[223]

On the other hand, Islam is a religion of the sacred book, the Qur'an, which is believed by Muslims to have been dictated to Muhammad by the archangel Gabriel from a book kept in heaven, and is accepted as the foundation of Islamic law, religion, culture, and politics. It must be read in Arabic, as any translation lowers its sacredness. The Qur'an is believed to be perfect,[224] totally infallible, and inerrant[225] with respect to every sentence. The Qur'an is ambivalent toward other religions such as Jews, Christians and Sabians.[226] Also it states that there should be no compulsion in religion.[227] The Qur'an refers to some Old Testament stories and people (prophets) such as Adam, Noah, Moses, Abraham, Enoch, Job, Jonah, and Joseph several times, but adding to, reinterpreting, and changing the stories. It also says that Jesus referred to a messenger who would come after him called Ahmad (Muhammad).[228]

6.6.6 Nature of the Afterlife

For a seventh comparison, there is a big difference in what various religions claim happens to individuals when they die. Islam, like Christianity, claims that we are judged by God (Allah) who decides where we will spend eternity. In contrast, many Hindus claim we shall live (and have already lived) many lives on earth. Hindus believe that the conditions of our past future existence are determined by the cosmic laws of karma. After death each of us is reincarnated into a different form (human, animal etc.) The Buddha rejected the Hindu belief regarding the cause of the endless cycle of birth, death and rebirth. Reincarnation is discussed further in Section 8.10.3.

6.6.7 Obtaining Salvation

For an eighth comparison, we look at how we obtain "salvation." Christianity is the only religion that teaches salvation solely by grace through faith alone. This simple fact makes it stand entirely apart from other religions. For nonchristian religions, salvation is by charitable works, good deeds, or performing certain rituals. But even with good works, adherents cannot be sure about their salvation. For Muslims, salvation requires faith as well as good works.[229]

---

[223] cf. Surah 56:80.
[224] Surah 2:3.
[225] Surah 29:39.
[226] Surah 2:63.
[227] Surah 2:257.
[228] Surah 61:7.
[229] Surah 5:10, 33:71–72, 49:15.

The Qur'an admits that all Muslims go to hell,[230] though Islamic scholars would say it is only temporary. Ultimately the destination of the Muslim is Paradise, the Garden of Bliss.[231] Those who die as martyrs defending the faith go straight to Paradise. Since works are involved, there can be no assurance of salvation as there is no criterion; one's outcome rests with Allah and whether the good deeds outweigh the bad. Muhammed could not offer any assurance of salvation[232] as he had none himself and didn't know what would happen to him or other Muslims after death.[233]

In Hinduism three ways of salvation are recognized: the way of good deeds and conduct, the way of knowledge, and the way of devotion and a form of grace, as in Bakhti sects. There is a fourth, typically only followed by wandering monks that uses meditation and yoga techniques. The idea is to seek release from the endless round of births and re-births, but there is no assurance as to when you "arrive."

In Buddhism, salvation (Nirvana) is by works from following the Eight-fold Path of right understanding: accepting the "Four Noble Truths" along with right aspirations, right speech, right conduct, right modes of livelihood, right effort, right mindfulness or contemplation, and right meditation. The four truths are: (1) Life means suffering; (2) The origin of suffering is the attachment to transient things (including objects, ideas, and self); (3) The cessation of suffering is attainable through extinguishing all forms of craving and clinging to things; and (4) The path to the cessation of suffering is by gradual self-improvement through following the Eightfold Path (which may take many lifetimes). To attain these, the life of the monk is the surest road.

In Christianity, people are "saved" and inherit eternal life when they are converted, that is when they believe in Jesus[234] and choose to make him Lord of their life. Consequently their life is turned around (the meaning of the word conversion in Greek) and they become in-dwelt by the Holy Spirit of God. When we repent of our sins and accept Christ and his representative death on the cross for our sins, God forgives us by His divine grace and our faith in Christ, and not through any works of ours.[235] We can have assurance of salvation,[236] but we cannot earn our salvation as we have all sinned and fallen short of God's standard of perfection;[237] Jesus becomes our righteousness. He didn't come into the world to make us good but rather to connect us with God and to make us spiritually alive.

---

[230] Surah 19:68–69, 72–73.
[231] Surah 4:14, 7:43–44, 57:22.
[232] Surah 16:94, 95, 17:14.
[233] Surah 46:10.
[234] John 3:16.
[235] Ephesians 2:8–9; Romans 4:5, 13–24, 5:1–2.
[236] John 3:16–18, 1 John 5:11–12.
[237] Romans 3:23–26.

There are two aspects of God's grace I want to address. First, some people may feel it is unfair if a person who has done some terrible things in his or her life can become a Christian and have it all forgiven. In answer to this, sin can be forgiven, but its consequences still follow. We see this in King David's life when he married Bathsheba and arranged the death of her husband; problems followed that are described as the sword not departing from his household.[238] Second, although good works won't save us, it does not mean that we should avoid doing good works. James 2:14–16 makes it clear that as people cannot see our faith and only see our works, we can justify ourselves before them only by our works, while it is faith that justifies us before God. Below I refer to the fact that one day we have to give account of ourselves.

6.6.8 Tolerance and Compulsion

In a final comparison I want to say something about two words, tolerance and compulsion, as regard to religions. In countries where there has been a foundation of Christianity, we generally find religious tolerance that allows people of different faiths to freely practice their religion. Although some religions may mention love as an aspect somewhere in their beliefs, love is the *central* theme of Christianity,[239] and it includes loving our enemies.[240] Muslims are authorized to slay infidels wherever they find them.[241] Of course in the past there have been times (and even now) where there has been some religious intolerance within Christianity, but this is not in keeping with the teaching of Jesus. In non-Christian countries intolerance is often alive and well. For example, there are some Islamic countries where it is illegal to proclaim Jesus Christ as Lord, and rejecting Islam can lead to serious persecution and even death. There is also persecution of Christians by militant Hindus in India. Compelling someone to become a Christian is contrary to the teaching of Jesus, though he did teach his disciples to go into all the world and preach the good news.[242]

6.7 THOSE THAT HAVEN'T HEARD

There remains a question that all religions as well as Christianity have to face: "What happens to people who have not heard about their particular religion?" In the case of Christianity, there are a number of views as to what happens to people who have not heard of Jesus. The Bible tells us that God will judge all people righteously and not arbitrarily,[243] so we can trust God. Those

---

[238] 2 Samuel 12:10.
[239] John 13:34–35.
[240] Matthew 5:44.
[241] Surah 9:5.
[242] Matthew 28:19–20.
[243] As for example in Psalm 9:8, 98:9, and 1 Peter 1:17.

Jews who died before Jesus was born (e.g., the patriarchs such as Abraham) were accepted as righteous (that is justified before God) because they were part of God's chosen people and had faith in God.[244] Even those who were not Jews before Christ came but who repented were accepted.[245] It could be inferred that God may be tougher on those who reject Jesus than on those who have never heard of Jesus.[246] Paul tells us that those who rely on the law (e.g., practicing Jews) will be judged by the law,[247] while those who have never heard of the law are judged by the law of God that he has placed in their hearts[248] and by the light they have received.[249] God has also revealed Himself in the things that he has made.[250] Unfortunately people may not fare too well under either law as all have sinned.[251]

Paul said that God did not leave Himself without a witness to the nations.[252] When he was debating with Greek philosophers in Athens he mentioned two things about God in Acts 17:26–27. First, God determines where people live and makes it possible for people to seek Him. Second, God is not far from any of us and, in the words of Hebrew 11:6, those who draw near to God and diligently seek Him will be rewarded. This implies that God will somehow break through to the genuine seeker of Himself either in a vision or dream, or by some other agency.

A good example of God interacting with us is the story of the centurion Cornelius who was a good man who regularly prayed to God and God supernaturally connected him with Peter so that he would learn of Jesus and receive the gift of the Holy Spirit in Acts 10. Ananias was sent by God to minister to Paul after his conversion to recover Paul's sight.[253] I have read a number of stories from non-Christian countries where Jesus has appeared to various people in visions or dreams that have transformed their lives.[254] William Lane Craig discusses the question of those who never hear about Jesus but are still not saved.[255] He argues that God providentially knows that such people would have rejected Jesus even if they knew about him. Those who never heard of Jesus will be judged on their response to God's revelation in nature and conscience. He said that a possible answer is that anyone who wants or even would want to be saved will be saved.

---

[244] Romans chapter 4 and Hebrews chapter 11.
[245] For example, the people of Nineveh in the book of Jonah who repented; they are also mentioned in Matthew 12:40–41.
[246] John 3:36, 12:48.
[247] Romans 2:11–12.
[248] Romans 2:14–15.
[249] See for example Matthew 11:20–24.
[250] Romans 1:20.
[251] Romans 3:10, 23.
[252] Acts 14:17.
[253] Acts 9:10–19
[254] I cannot personally vouch for these stories, but I have no reason to disbelieve them.
[255] Craig (2010: 279).

A related question I want to consider is what happens to babies and young children who die. Since they are considered innocent and do not fully understand the difference between good and evil,[256] we would expect them to be excluded from judgement and go to heaven. Psalm 139 talks about the unborn baby as a creation of God whose name is in God's "book" in heaven (verses 14–16). Jesus spells out their special relationship with God in a number of verses.[257] This raises the question as to what age a child becomes accountable, as this age will vary with the child. Another question is what happens to those who are mentally disabled. Clearly God can be trusted to do what is right in both cases.

6.8 IS GOD UNFAIR?

Another question commonly asked is: "How can a loving God assign people to hell?" To answer this we first begin by noting that God takes no pleasure in the death of the wicked but would rather they turn from their ways.[258] People choose where they want to end up. Second, we note that people misunderstand the nature of hell. It can be regarded not as physical torture as such, although described in strong and colorful language, but rather a separation from God whereby people carry on without God as they chose to when they were alive. It will be punishment because of the regret and anguish experienced over losing the good life that God designed for them, and the loss of potential to receive so many good things like love, joy, and wisdom. Third, we only get a glimpse of what will happen after death in the Bible so that our understanding of the afterlife is in a sense metaphorical as we cannot conceive what it will really be like.[259]

We all get angry if we see someone we love get hurt by others. In a sense God is no different. He is angry at evil and injustice because it is destructive of what is good and can hurt people who love Him.[260] Could we really worship a God who was complacent about evil? Some may argue that a belief in a God of judgement will lead to a more brutal society. This is not the teaching of Christianity that teaches loving one's enemies and praying for those who persecute us.[261] We are to leave vengeance to God.[262]

Some descriptions of hell are: being cast into outer darkness, experiencing weeping and gnashing of teeth;[263] being thrown into a fiery furnace and facing

---

[256] Deuteronomy 1:39; Isaiah 7:15–16.
[257] Matthew 18:1–5, 10–14, 19:13–14 and Mark 10:14.
[258] Ezekiel 18:23, 32; 33:11; see also 2 Peter 3:9; 1 Timothy 2:4
[259] 1 Corinthians 2:9.
[260] Psalms 145:17–20.
[261] Matthew 5:44–47; Romans 12:14–18, 20–21.
[262] Romans 12: 19.
[263] Matthew 8:12, 22:13, 24:51, 25:30, Luke 13:28.

eternal fire;[264] being in a place of torment and flames;[265] being cast into Gehenna (sometimes translated as Hades), an eternal type of city rubbish dump where there are maggots and continuous fires;[266] and being cast into a lake of fire.[267] Clearly the images of burning in hell are metaphorical as we have conflicting images such as darkness and flames together, and hell is a rubbish dump but also a bottomless pit.[268] We also note that the word "eternal" does not refer to endless time but rather a new state of being. We therefore find that the actual nature of hell is a controversial topic and there are different viewpoints.[269]

Some maintain that the wicked will cease to exist instead of being sent to an everlasting hell, so called *annihilationism*, often referred to as the "second death", as mentioned in Revelation 20:14. For example, in the Old Testament we read that the wicked will perish,[270] while in the New Testament we read that the wicked will be shut out of God's presence and face everlasting destruction.[271] However, the Hebrew word *abad* used to describe the wicked perishing in the Old Testament is also used to describe the righteous perishing,[272] which is certainly not everlasting. The word is also used to describe things that are lost but are later found.[273] As far as the New Testament is concerned, everlasting destruction cannot mean annihilation as the latter would be instantaneous, not everlasting. Other texts support the idea of eternal conscious punishment;[274] we have the repeated use of the phrase "weeping and gnashing of teeth" which clearly has some metaphorical significance. Any torment is self-inflicted brought about by separating oneself from God: we punish ourselves. We continue down the self-centered path leading to isolation and delusion. Whatever our theological view of Hell, it is not the most exciting place if we continue to exist there.

In the same way the Bible has metaphors for hell it also has them for heaven. We are not able to comprehend what God has in store for those who love Him.[275] After death we read that there will eventually be a final resurrection when we will all need to give account of ourselves before the judgment seat of Christ,[276] and this includes Christians although ultimately they are saved by faith and God's gift of grace.[277] Christians will also be

---

[264] Matthew 13:42, 25:41.
[265] Luke 16:23-24, 28.
[266] Mark 9:43–48.
[267] Revelation 20:14.
[268] Revelation 20:3.
[269] For four views see Crockett, Walvoord, Hayes, and Pinnock (1992).
[270] Psalms 37:20, 68:2, and 112:10.
[271] 2 Thessalonians 1:9; 2 Peter 3:7–10.
[272] Isaiah 57:1; Micah 7:2.
[273] Deuteronomy 22:3.
[274] Matthew 8:12, 22;13, 24:5, 25:30.
[275] 1 Corinthians 2:9.
[276] John 5:25–29; 2 Corinthians 5:10; Revelation 20:11–15.
[277] Romans 14:10–12; Ephesians 2:8–9.

judged and rewarded on the genuineness of their faith and how they lived out that faith. God will judge both our deeds[278] and our inner thoughts.[279] After judgement, one's destiny is either being with God or being dismissed from God's presence, labelled respectively as heaven or hell.[280] Some argue that we will get a second chance after death.[281] However, the story of the beggar Lazarus in Luke chapter 16, the lessons from Noah and the flood, and the destruction of Sodom in chapter 17 do not support this. We are also told we die once and then comes the judgement;[282] we read that people are without excuse when they ignore the evidence for God in the world.[283]

Another belief is that every one will eventually be saved, the so-called doctrine of *Universalism*, which is a bit similar to the idea that all religions lead to God, except the nonreligious are now also included. Christian universalism believes that eventually all people are saved through Jesus Christ.[284] Universalism has been strongly put forward by philosopher Thomas Talbott[285] who argues that a loving God would not consign anyone to eternal hell, and any time in hell would only be temporary (a form of purgatory). The latter smacks of torture on behalf of God to achieve His will, which is contrary to free will. Belief in a form of purgatory is supported by the Roman Catholic church through their tradition and the past practice of praying for the dead.[286] Purgatory is described as a state or place in which the souls of those who have died in a state of grace (i.e., Christians) are believed to undergo a limited amount of suffering to expiate their venial sins and become purified of the remaining effects of mortal sins making them fit for heaven. Sins are forgiven, but there is not a release from punishment. There are a few New Testament verses that some suggest refer to an intermediate state where the soul is purified.[287] Talbott argues that God could arrange the circumstances of a particular person so that he or she would freely accept God's grace and be saved. However Bill Craig[288] points that the circumstances required for one person may clash with the required circumstances for another so that it is not possible that there is total set of circumstances in which everybody can be saved. For example, he says: "It may be a tragic fact of the matter, for example, that Joe, Jr. will freely respond to God's grace and be saved only if his father Joe, Sr. failed to do so."

---

[278] Romans 2:6; 1 Peter 1:17.
[279] Romans 2:16; 1 Corinthians 4:5; Hebrews 4:12–13.
[280] The sequence of events is a bit more complicated than this but I don't wish to go into further details. See for example Prince (1993: part 6).
[281] See, for example, 1 Peter 3:18–20, 4:6.
[282] Hebrews 9:27.
[283] Romans 1:19–23.
[284] See for example Philippians 2:9–11.
[285] Talbott (1990a, 1990b, 1999).
[286] 2 Maccabees 12:43–46 in the Apocrypha.
[287] Matthew 12:32;1 Corinthians 3:11–15.
[288] Craig (1991).

With regard to universalism, there are a number of verses that could be interpreted as supporting universalism[289] where the emphasis is on the word "all."[290] One major plank of Talbott's argument is that Christians could not be happy in heaven if they were aware of loved ones in hell, and could not love and worship such a God. Talbott argues that Romans 5: 18–19 refers to all men being condemned because of Adam's sin but all are acquitted through Jesus, where "all" refers to everyone and not just to some. He also interprets other writings of Paul in a similar manner. However 2 Thessalonians 1:6–9, for example, speaks about those who don't know God and reject Jesus as facing eternal destruction and exclusion from the presence of God. Some believe that these verses may refer to the permanent annihilation of unbelievers.

With regard to loved ones in hell, Craig argues that we may be shielded from the knowledge about such loved ones, or the overwhelming presence of God may drive such knowledge from our consciousness; God endures the pain, not us.[291] The Bible tells us that there will be a new heaven and a new earth[292] in which former troubles are forgotten and the old earth is no longer remembered.[293] This raises a question about free will. If there will be no pain, suffering, or evil in heaven, why can't we all go straight to heaven and dispense with earth altogether. It would seem that God can create a world in which there is both free will and no suffering or evil. This world, however, would be different as it is for those who choose to go there (if they have the power of choice), and it is the place where God will step out of the shadows and be revealed in all His glory.

It is clear from the above discussion that there are a number of theological issues that can be debated, which will lead us off in a different direction away from the central theme of who is Jesus. How God will actually administer both justice and mercy is not clear to us, but we cannot talk about a God of love and forget about God's holiness and righteousness.

6.9 CONCLUSION

We have looked in detail at the extraordinary life of Jesus Christ. His resurrection is a key event, and substantial evidence is given supporting this event. We have also compared Christianity with other religions, and showed that exclusiveness applies to the main religions. In our comparison we saw that there are large differences in many important areas such as the founders of

---

[289] John 12:32; Romans 11:32; 1 Corinthians 15:22; 1 Timothy 4:10; Luke 2:10–11; 1 John 2:2.
[290] However, for an interesting discussion on these verses that do not support universalism see Philip Brown at http://www.newwine.org/Articles/Universalism.htm, accessed August, 2015.
[291] For further logical debate on this topic see Craig (1991).
[292] Revelation 21:1-5; 2 Peter 3:13.
[293] Isaiah 65:16–17.

religions, religious universality, and the religion's understanding of God, sin, scripture, the afterlife, salvation, and tolerance. Heaven and hell are briefly considered and some questions about those that haven't "heard" and those who will be "saved" are considered. Admittedly the latter part this chapter has been somewhat theological, but some of the questions raised have been used against Christianity so that they needed to be considered even if we don't know all the answers. We are dealing with almighty God so there is bound to be mysteries.

CHAPTER 7

DO MIRACLES OCCUR?

7.1 DEFINITIONS AND THEORIES

What do you regard as a miracle? For some people it is getting to bed early! In order to answer this question we have to define what we mean by a miracle. This is not straightforward as the word "miracle" is often used lightly (e.g., a miracle drug) rather than in a more technical sense. Thomas Aquinas defined a miracle as an event that exceeds the productive power of nature, where "nature" also includes us and any other creatures like us. We also have several other possible definitions. For example, a miracle is an event: "that appears inexplicable by the laws of nature and so is held to be supernatural in origin or an act of God"; or "that is of extraordinary nature manifesting divine intervention in human affairs."

Michael Licona[1] adds a further criterion, namely "it occurs in an environment or context charged with religious significance" so that "the event occurs in a context where we might expect a god to act. The stronger the context

---

[1] Licona (2010: 163).

is charged in this direction, the stronger the evidence becomes that we have a miracle on our hands." Most definitions of a miracle involve two components: (i) the fact that it defies what we understand as the laws of nature, and (ii) it is of supernatural origin where, by supernatural, we mean of God or some spirit being. Evil powers can also work magic, or at least mislead, as for example the magicians were able to reproduce some of Moses' plagues,[2] but not all of them or reverse them.[3] C.S. Lewis described a miracle as an interference with nature by a supernatural power. Obviously, to interfere with a natural law may not necessarily mean to break the natural law (e.g., the parting of the waters of the Red Sea). As noted by Lewis,[4] how you deal with what appears to be a miraculous event will depend on whether you believe that miracles are possible or not.

Often people don't believe in miracles as they don't believe there is a God who does miracles. If you don't believe miracles exist and you are confronted with one you will explain it away. You will probably do the same if you think they are possible, but unlikely. Without an open mind, any historical evidence that is obtained for the confirmation of miracles will not be accepted. Clearly identifying a miracle is problematic. However, without God miracles do not make sense, but we can't use miracles as evidence of God's existence as the latter needs to be established first.

At this stage it is important to distinguish between God working through miracles and God working through providence, though the distinction is blurry. By providence I mean an event that has a natural explanation, but the timing is special in some way such as the stopping of the River Jordan so that the Israelites could cross the river at a particular instant.[5] William Lane Craig[6] suggests that:

> By means of His middle knowledge, God can providentially order the world so that its natural causes of such events are, as it were, ready and waiting to produce such events at the propitious time, perhaps in answer to prayers which God knew would be offered.[7]

God is continually sustaining and preserving our universe according to predictable patterns that we usually describe as natural laws, and although He operates from the shadows He is in some mysterious way able to be involved with time and human affairs. From time to time He may step out of the shadows and directly intervene in the world to provide what appears to us to be a miracle. God is actively at work in both kinds of situations and is described as being both transcendent and immanent.

---

[2] Exodus 7:8–13.
[3] Exodus 8:18–19.
[4] Lewis (2001c: chapter 1).
[5] Joshua 3:14–17.
[6] Craig (1998a).
[7] Middle knowledge is considered in Section 4.3.1.

There has been considerable debate about miracles, especially those belonging to the Bible and their role in the life of Jesus. Some philosophers simply ruled them out for various reasons. For example, Spinoza held that the laws of nature flow from God so that a violation of nature was a violation of God. This is very much a deist view that sees God creating and setting the world in motion and then leaving it alone. Some have argued like Spinoza that biblical miracles are simply natural events, but described in metaphorical language so that they appear miraculous. They argue that since our knowledge of nature's laws is limited it doesn't mean that God is the cause of a particular event supernaturally. It is of course true that our understanding of so called "natural laws" is never final and in the next section we shall revisit this.[8]

Hume's Criteria

Hume says that any testimony to a miracle must surely be outweighed by the uniform experience of mankind of the unchangeable laws of nature. He sets the bar impossibly high for attesting to a miracle with the following comments (which again reflect deist objections to miracles):[9]

(1.) No miracle in history is attested by a sufficient number of men of good sense and education, of impeachable integrity so as to preclude deceit, of such standing and reputation so that they would have a good deal to lose by lying, and in sufficiently public manner.

(2.) People crave the miraculous and will believe absurd stories, as the multitude of false miracles shows.

(3.) Miracles only occur among barbarous peoples.

(4.) All religions have their own miracles and therefore cancel each other out in that they support irreconcilable doctrines.

As already mentioned, Hume effectively argues that the uniformity of human experience rules out miracles, a view held today by those who rule out the biblical miracles out of hand. However, according to research carried out by Craig Keener, New Testament scholar,[10] a former atheist and now a Christian, hundreds of millions of people today claim to have experienced miracles, and he argues that not all those people can be false witnesses. Some of the records of the miraculous events that are in places where there is adequate medical access have been verified medically. He explains on page 1 that,

> The book's primary thesis is simply that eyewitnesses do offer miracle claims... The secondary thesis is that supernatural explanations, while not suitable in every case, should be welcome on the scholarly table along with other explanations often discussed.

[8] See also Sections 1.1.1 and 1.3 about the laws of science.
[9] From Craig (1986).
[10] Keener (2011).

He argues that there is good solid historical evidence that the apostles and earliest followers of Jesus thought they were witnessing miracles rather than such reports being legendary or mythical additions. He shows that this is similar to what happens today when we have reports from people who have witnessed miracles, and cites hundreds of contemporary miracle accounts. Keener points out that Hume's argument is circular since he rejects miracles based on the uniformity of human experience but in order to do so rejects all credible eye-witness claims to miracles (which is extensive human experience). Keith Ward does not find Hume's argument at all convincing and concludes that "the laws of nature, as modern science understands them, do not exclude the occurrence of miracles."[11]

7.2 THE RISE AND FALL OF SCIENTIFIC LAWS

I now wish to look at "miracles" from the point of view of science. We saw in Chapter 1 that scientific laws come in all shapes and sizes. Many of our physical laws are derived mathematically from certain assumptions where, as we saw there, the assumptions refer to an ideal world. This is a world with planes that are perfectly flat, lines that are perfectly straight, circles that are perfectly circular, and some objects that are perfectly symmetrical or can be simply regarded as points in a large scale system. How well the laws work depends on how good our ideal world approximates to reality. We saw that such laws are on probation; they are accepted until a new experiment shows they are wanting in some way. They are then either discarded or refined. Our formulation of natural law is therefore never final so that we cannot empirically verify a scientific law. No matter how many observations we take we cannot come to a general conclusion that we can defend logically as the very next observation may contravene the law in some way. For example, the speed of light might change tomorrow. Just because it has been the same when it has been measured before, does not mean that it will always be the same. Can we prove that the sun will rise tomorrow?

Another problem pointed out by the philosopher David Hume is that we cannot prove a logical connection between cause and effect. All we can show is that there is a correlation between the two so that we cannot derive a scientific law from a logical connection. For example, we cannot prove that the mind is just a function of the brain and nothing else, as the materialist would argue, although the two are strongly correlated. No matter how many brains are examined there may be one for which the mind is not just the brain.

We saw in Section 1.3 that a law only qualifies as a genuine theory if it is testable through one being able to carry out an appropriate experiment that has the potential to disprove the law. We also referred to Newton's laws of motion there, which were verified by experiment and observation for over 200

---

[11] Ward (2008b: 106).

years, and are excellent approximations at the scales and speeds of everyday life. However, they break down under certain circumstances such as at very small scales (e.g., microscopic), very high speeds (e.g., near the speed of light), or very strong gravitational fields, and have been refined in the development of Einstein's theories of relativity and later quantum field theory. Eddington's eclipse observations in 1919 brought the first important confirmation of Einstein's theory of gravitation. New experimental phenomena such as the discovery of quasars, microwave background radiation, and black holes, provided a stimulus to general relativity theory.

Other laws are formulated from observations that give rise to so-called empirical laws. For example, a data plot of pairs of certain observations on a graph may suggest that the points are very close to lying on a straight line. This empirical relationship could then be described mathematically by the equation of the fitted line. Another example that is not mathematical arises in the following situation. Suppose we have a bus stop close to our home and we observed the arrival times of buses for say a year. The picture is not as clear as we would like it to be because of random variations in the system that affect bus travel. However, because of a general regularity observed in bus arrival times, we would suspect that a timetable actually exists designed by someone. We might not be aware that there are some unforeseen aspects of the timetable that we haven't allowed for (e.g., Christmas day, a non-repeatable event in our year's observation, or a bus breakdown). Our concept of the timetable might therefore have to be modified to allow for the occasional "miracle", that is a one-off event. There is no reason why we cannot construct our own timetable or empirical law and use it to catch buses most of the time! If we construct such a timetable what does it tell us about bus behavior? It tells us about average behavior! This is true in general. Most laws refer to what happens on average.

Another example relates to a gas law where the pressure of a gas is about the average rate of impact of molecules on the walls of the container. We saw in Section 4.1.3 that we cannot be certain about the future behavior of an individual subatomic particle, and the same is true with human beings. We may be able to predict average behavior, but not individual behavior precisely. If we fall into a particular psychological category, such as an extravert, then our general behavior may be somewhat predictable. However, as far as an individual action is concerned, we can only say that certain courses of action will be taken with certain probabilities. We cannot be more precise! [12]

In conclusion we see that all scientific theories and laws are human attempts at explaining reality. We observe patterns in what we see and endeavor to construct laws that describe them. For example, empirical observations may suggest that certain relationships hold and then someone develops a mathematical theory to explain the relationships, or else puts forward an empirical

---

[12] The question of free will is considered in Chapter 4.

law for further investigation. However, the laws may not account for all possible situations so that "miracles" could still be part of God's plan, just as natural laws are. We elaborate on this point next.

7.3 MIRACLES AND NATURAL LAW

We cannot say that miracles violate the laws of nature when we cannot prove that the laws of nature (as we understand them) are always true. Furthermore, with the invasion of probability into science (Heisenberg's uncertainty principle), it would seem more appropriate to see events not simply as possible or impossible but as events with varying degrees of probability. You might argue that it is technically not impossible to walk on water or pass through doors as matter is largely empty space and a configuration of atoms may exist that makes those events possible. This is particularly true when quantum theory tells us that subatomic particles are actually waves and are not solid. However, from our observations of nature we would assign extremely low probabilities to such events, but *logically* not zero. We don't know that our physical laws are the same everywhere and for all time.

Some laws may operate as "if" statements. If A occurs then the law is B, but if A does not occur then the law is C. This would allow for a discontinuity in a law but still qualify as a law overall. This means that some so-called miracles would not be regarded as miracles if we had greater knowledge of our so-called laws. This is no doubt true in some cases, although some alternative explanations of biblical miracles put forward (e.g., Peter walking on water in the shallows) may require greater credulity than simply accepting the stories at face value. However, even when a natural explanation is available, the miracle is often in the timing. We see this in Exodus chapter 15 where we read of several events that could have a natural explanation concerning Moses and the people of Israel crossing the Red Sea. For example, the pillar of cloud was moved by an angel of God from in front of Israel to behind them cutting them off from the Egyptians (verses 19–20). God then drove back the waters by a strong east wind all night and in the morning the water returned to normal, swamping the Egyptians. The timing in both cases was Moses stretching out his hand.

It has been suggested that in addition to physical laws there are also spiritual laws. I believe that prayer falls into this category and that prayer is more than just trying to twist God's arm! A good example is the exercise of faith where Jesus said that with just a little faith nothing is impossible.[13] We see the power of believing prayer in the healing of mental and physical illnesses all over the world. Although we don't fully understand how it works, I believe it does happen! The question of why it does not always happen is discussed in Chapter 8 on suffering.

---

[13] Matthew 17:14–21.

7.4 BIBLICAL MIRACLES

Stories of miracles in the Bible are passed down through tradition, and only the most significant or interesting ones have been written down. An example of this selection in the New Testament is described in John 20:30 where John says that Jesus did many other miraculous signs in front of his disciples that are not recorded in his gospel. However, the ones recorded were written for the purpose of leading people to believe that Jesus is the Son of God and thus have "life." Miracles in the Bible serve several purposes such as confirming the messenger that God sends whether it be Jesus, or Moses, a prophet, an apostle, or an ordinary Christian, and revealing the nature of God, such as His justice and mercy.

We consider a number of examples. Moses needed something to convince people that God had appeared to him, so God gave him the sign of turning his rod into a snake when he threw it on the ground and it turned back into a rod when he grabbed the snake by the tail.[14] Elijah on Mount Carmel needed a sign from God that he was a true prophet when he confronted the prophets of Baal, and God sent down fire from heaven to consume the sacrifice.[15] When John the Baptist enquired of Jesus whether he was the Messiah, Jesus referred to his miracles as confirmation.[16] Nicodemus believed that Jesus was from God and that God was with him because of the miracles (signs) that he performed.[17] Jesus also said to his disciples that his works confirmed his mission.[18] Again, when Jesus forgave the sins of the paralytic, he confirmed that he was the Son of Man by healing the paralytic.[19]

Jesus regarded miracles as a sign that he was the promised Messiah.[20] Bystanders acknowledged his signs on a number of occasions,[21] and a multitude of people followed him because of his miracles.[22] After his miraculous feeding of the 5000, the people wanted to make him king.[23] When people heard that he had raised Lazarus from the dead and was coming to Jerusalem on a donkey, they greeted him with palm tree branches and as king of Israel. There were also his opponents who did not doubt his power, but claimed his power came from Satan.[24] Before continuing further, let us look at some examples of Biblical miracles.

---

[14] Exodus 4:1,5.
[15] 1 Kings: chapter 18.
[16] Luke 7:22.
[17] John 3:2.
[18] John 11:11.
[19] Mark 2:9–12.
[20] Matthew 11:2–6; John 10:37–39, 11:42 .
[21] John 9:16.
[22] John 6:2.
[23] John 6:15.
[24] Matthew 12:24; Luke 11:14.

Some Old Testament Examples

Some well known Old Testament miracles are:

- Moses and the ten plagues against Egypt.[25]
- Crossing the Red Sea.[26]
- Joshua and the fall of Jericho.[27]
- The sun standing still.[28]
- Elijah's provision of food for a widow and raising her son to life.[29]
- The three men surviving the fiery furnace.[30]
- Daniel protected in the lion's den.[31]
- The shadow going back 10 steps on a stairway.[32]

Jesus assumed that Old Testament miracles took place, for example:

- God speaking to Moses out of the burning bush.[33]
- God providing manna from heaven.[34]
- A famine in Israel resulting from Elijah's prayers.[35]
- Naaman the leper being healed when he obeyed Elisha's instructions.[36]
- Jonah being swallowed by a huge fish.[37]

It is interesting to note how very different these incidents are.

Some New Testament Examples

Some well-known New Testament miracles of Jesus are:

- Changing water into wine.[38]

---

[25] Exodus chapters 7 to 12.
[26] Exodus chapter 14.
[27] Joshua 6:1–21.
[28] Joshua 10:12–13.
[29] 1 Kings 17:7–24.
[30] Daniel 3:24–27.
[31] Daniel chapter 6.
[32] 2 Kings 20: 9–11.
[33] Luke 20:37.
[34] John 6:49.
[35] Luke 4:25.
[36] Luke 4:27.
[37] Matthew 12:40.
[38] John 2:1–11.

- Healing a paralyzed man let down from the roof in a house full of people.[39]
- Healing a blind man.[40]
- Calming a storm.[41]
- Walking on water.[42]
- Feeding the five thousand.[43]

We read that many people came to Jesus and they were all healed.[44] His healing miracles also occurred in a variety of places such as Capernaum, Galilee, Gennesaret, in the wilderness, on the mountain, in the Jordan region, in the temple, in Bethsaida, and in Jerusalem. Two important miracles are his virgin birth and his resurrection from the dead discussed in Section 6.4.3. What is particularly interesting is that many of the miracles of Jesus took place in public, for example:

- Casting out a demon in a synagogue meeting.[45]
- Healing a leper in public that brought large crowds.[46]
- Healing a lame man at a public pool.[47]
- Healing a man with a withered hand during another synagogue service.[48]
- Raising a man from the dead during a funeral procession.[49]
- Multiplying loaves and fishes in front of large crowds on two occasions.[50]
- Healing a woman bent over double for 18 years at a synagogue service.[51]

Clearly some of those who were healed had obvious infirmities that were unlikely to have been faked. There are also other miracles that created a stir and the news of these soon spread abroad.

---

[39] Mark 2:1–12; Luke 5:17-26.
[40] John 11:43–48.
[41] Matthew 8:23–27.
[42] Matthew 14:22–33.
[43] Luke 9:12–17.
[44] Matthew 4:23–24, 8:16–17, 15:30–31; Mark 1:32–34, 2:10–12; Luke 6:17-19, 7:21, 13:32.
[45] Mark 1:21–28; Luke 4:31-37.
[46] Matthew 8:1-4; Mark 1:40-45; Luke 5:12-16.
[47] John 5:1–9.
[48] Matthew 12:9-14; Mark 3:1-6; Luke 6:6-11.
[49] Luke 7:11-17.
[50] 5000 men plus women and children, Matthew 14:13–21; Mark 6:30–44; Luke 9:10–17; John 6:1–15: 4000 men, Matthew 15:32–38; Mark 8:1-9.
[51] Luke 13:10–17.

Of particular interest is the raising of Lazarus from the dead that sent some people off to report to the opposition what happened.[52] People were also very keen to meet the live Lazarus, who was readily available.[53] Some skeptics claim that the people raised from the dead weren't really dead. However, Lazarus had been sick but Jesus delayed seeing him for the purpose of bringing glory to God and His son.[54] When Jesus finally arrived, Lazarus had been dead for four days. After Lazarus was brought back to life many believed in Jesus,[55] so that the chief priests and Pharisees were worried because even they affirmed that Jesus performed many signs. They argued that if he was allowed to continue, everyone would believe in him and the Romans would destroy their nation.[56] So Jesus had to die in order to maintain their control and authority.

One thing that stands out in the raising of Lazarus is the simplicity of the narratives with no embellishment that might be expected if the stories were made up. In that lengthy story by John[57] are included unflattering incidents such as Jesus deliberately delaying his arrival, Mary and Martha apparently criticizing Jesus for his lateness, and Jesus showing strong emotions of weeping and anger that could be interpreted as human weakness. Even some of the leaders believed in Jesus, but kept quiet in case they were put out of the synagogue.[58]

The Apostles were also said to have performed miracles. For example, they performed many miraculous signs and wonders[59] and healed people who were sick or possessed by evil spirits.[60] Philip also had such a ministry[61] and Peter had a miraculous escape from prison.[62] Paul is said to have healed a crippled man,[63] cast out a demon,[64] brought someone back to life,[65] and did not die despite being bitten by a poisonous snake.[66] Jesus said that his followers would do not only miracles that he did but even greater ones in his name.[67] Of particular interest to me are stories of people resurrected from the dead. Dorcas, whom Peter raised, had been washed and laid out for burial,[68] and

---

[52] John 11.
[53] John 11:9.
[54] John 11:4.
[55] John 11:45, 12:11.
[56] John 11:47–48.
[57] John 11:1–44.
[58] John 12:42.
[59] Acts 2:43, 5:12.
[60] For example, Mark 6::7,13; Acts 5:15–16.
[61] Acts 8:6–7.
[62] Acts 12:1–19.
[63] Acts 14:8–10.
[64] Acts 16:16–18.
[65] Acts 20:9–12.
[66] Acts 28:5.
[67] John 14:12.
[68] Acts 9:36–43.

we are told many believed as a result of this event. Eutychus, whom Paul revived, was picked up dead according to Luke the physician and author of Acts who witnessed the event.[69] In the case of Jesus raising Jairus's daughter, we see from the full story that Jesus was speaking metaphorically when he said that the girl was asleep and not dead (he was laughed at for saying this), but according to Luke, her spirit returned when Jesus told her to get up.[70] The term "asleep" was an expression commonly used by Jews and Christians who believed in the resurrection of the dead.[71] Jesus used the same idea when he said that Lazarus was dead and that he was going to awaken him. However, the disciples didn't get it so Jesus had to say plainly that Lazarus was dead.[72]

It should be noted that intense miraculous activity was not evenly spread throughout biblical times but occurred in only four relatively short periods of biblical history.[73] The first period is the time of Moses when Israel escaped from Egypt and conquered the Promised Land. Hundreds of years passed through the time of the judges and the reigns of David and Solomon when there was little miraculous activity. The second period of activity was centered around the ministries of Elijah and Elisha, two of God's prophets, confirming their prophetic office. Again followed several hundred years when almost no miracles were recorded. The third period was at the time of Daniel when Israel was captive in Babylon. After Daniel's death miracles again ceased. However, during these long gaps God was still at work, but in the background. Over and above this pattern of a few key periods, God did intervene from time to time to exercise judgment and bring destruction on individuals or groups, even when other miracles were virtually absent or at least not recorded.

Five hundred years after Daniel's time led to the final period of the miraculous when Jesus began his ministry. This period extended after Jesus' earthly ministry into the early years of the church. After that time there were less miracles: Paul had a physical affliction that God did not heal;[74] Paul could not heal Epaphroditus,[75] though God eventually did it by "natural" means; Timothy suffered from frequent ailments;[76] and just before Paul's death Paul mentioned the illness of Trophimus.[77] We have to remember that early on Paul was healed of blindness by Ananias and carried out extraordinary miracles in the early stage of his ministry.[78] However, there was a later decline in miraculous activity of the early church. Since that time there have been spo-

---

[69] Acts 20:9–12.
[70] Luke 8:49–56; Mark 5:35–43; Matthew 9:18–26.
[71] 1 Corinthians 15:20; 1 Thessalonians 4:13–17.
[72] John 11:11–14.
[73] Connelly (1997: chapter 2).
[74] 2 Corinthians 12:7–9.
[75] Philippians 2:25–30.
[76] 1 Timothy 5:23.
[77] 2 Timothy 4:20.
[78] Acts 19:11.

radic revivals with people being healed and stories of God acting sovereignly in different parts of the world.

Apparently miraculous healing still occurs today and I regularly read about it from mission fields.[79] Further comments about miracles today are made in Section 7.1 where reference is also made to Keener's[80] extensive work on the subject. In particular, he compares the gospel miracles with extra-biblical miracle accounts in other Greco-Roman and Jewish literature. Although there are some very broad similarities, there are also major differences indicating that it is reasonable to conclude that the gospel stories are not due to the influence of pagan stories of magic and divination. In fact the influence is the other way round. Licona[81] refers to several reports of ancient miracle workers such as Apollonius, Onias (Honi the Circle-Drawer), Hanina ben Dosa, and Vespasian, and shows that they are either quite late (sometimes written centuries later), are far more contradictory than the gospel narratives, or have a plausible naturalistic explanation.

Signs, Wonders, and Powers

The words used to describe biblical miracles are signs, wonders, and powers. The word *sign* sometimes refers to an ordinary event, but in general it refers to a supernatural event that has a special meaning or purpose. For example, God gave Moses three signs: Moses' rod turned into a serpent, his hand turned leprous, and water from the Nile turned to blood,[82] which confirmed Moses as God's true representative.[83] God gave Gideon a personal sign to confirm His message.[84] John chose seven miracles of Jesus and called them signs to show that Jesus was the Messiah. Signs then have a purpose, namely to confirm God's message or messenger as the real thing.

The word *wonder* means a miracle that amazes those who see it. For example, Jesus set free a demon-possessed man who broadcasted throughout his locality what Jesus had done for him, and people were amazed. When Jesus calmed the storm, Matthew, Mark, and Luke each said in their gospels that the disciples were amazed and wondered what sort of man Jesus was who controlled the winds and sea.

The word *power* is used to describe a miracle that demonstrates God's power and mighty works such as the virgin birth of Jesus,[85] his miracles,[86] and his resurrection. Such a miracle would indicate the work of divine power.

---

[79] I realize that I have no way of personally verifying such stories, but they do show consistency.
[80] Keener (2011).
[81] Licona (2010:178–179).
[82] Exodus 4:1–9.
[83] Exodus 4:30–31.
[84] Judges 6:16–22.
[85] Luke 1:35.
[86] Matthew 11:20, 13:58.

When the three words occur together we have a period of intense miraculous activity. For example, Peter in his speech at Pentecost said that Jesus was attested by God to the people through mighty works, wonders, and signs that God did through him.[87] Furthermore he added the words "as you yourselves know" indicating that his audience had witnessed the miracles. The biblical definition of a miracle, then, is an amazing event (wonder) that is clearly a work of supernatural power (mighty work), and that is designed to point out something significant that God is doing (a sign).

7.4.1 Categories of Miracles

In talking about biblical miracles it is helpful to classify them by the order of clarity of the mechanism. The first category could consist of miracles whose mechanism is understood. For example the wind rolled back the Red Sea. In the same category we can include two draughts of fish (fish do swim in shoals), the calming of the storm (storms usually abate), the coin in the fish's mouth (there are fish that carry hard objects at certain seasons), and so forth. Here it is the timing and the purpose that makes these events special. A second category of miracle might be those whose mechanism we can guess at; an axehead floating (perhaps through currents and springs), manna in the desert (even this can have a "natural" explanation), and healing through psychological processes. A third category might be those miracles that may one day be explained by future science such as paranormal activity, some of the medical miracles, and the cosmic miracles in Exodus. However there is a fourth category of miracles, namely those that are beyond us and whose mechanism may remain unknown. In the New Testament these include, for example, turning water in wine, feeding the 5000, the rising of the dead, and the resurrection of Jesus.

7.4.2 God the Free Creator

As already mentioned above, there are those such as deists who argue that God is inconsistent if He works miracles contrary to His own laws of nature. There are several answers to this. First, although God is described in the Bible as creating and sustaining the world, He is not part of His creation as pantheism would maintain. As we discussed in Chapter 2, God is not bound by His creation or by time: He is a free creator and created matter, space and time out of nothing. A miracle can then be regarded as simply a sovereign act of God.

A second answer to the question of God's possible inconsistency relates to the nature of scientific law. We have seen that there is really no such thing as the laws of nature. What we see is that God has provided us with a world that

[87] Acts 2:22.

shows a certain degree of consistency, and has given us the ability to formulate what we see into laws that endeavor to explain the consistency. These laws are not final as they can be replaced or modified, and they are about repeatable events. We saw above that they also tend to be about average effects and probability rather than about individual events and certainty.

It should be noted that signs and wonders are performed not just by the "good" guys! In our definitions we included any kind of supernatural activity that could include those performed by people under any kind of spiritual influence including evil spirits. Satan can use signs and wonders to deceive people. Jesus, referring to the end-times, said that false Christs and false prophets will appear and perform great signs and miracles to deceive even the elect.[88]

7.5 CONCLUSION

We have seen that a miracle is an event out of keeping with normal activity in the world. Given our understanding of the universe and the probabilistic nature of our scientific laws, miracles are logically possible, even if extremely unlikely according to our current knowledge. If God exists and is creator and sustainer, there are occasions where He might want to step out of the shadows and intervene in some manner. As Creator He is free to do this. Such miracles reflect the generally invisible nature of God and reveal His moral character and His goodness. They may confirm a truth about God through a servant of God by confirming the servant's credentials. A miracle will also have a spiritual purpose, for example to glorify God and strengthen faith. Given two features of our world, namely quantum theory involving probability, and chaos theory where a tiny change in the initial conditions can have a huge effect later, we see that there is room for God to maneuver without a major interference with or a major contravention of our physical laws as we currently understand them. In the end the evidence comes from the historical evidence of witnesses to such events and how the witnesses are effected by the events.

We have seen that in the New Testament the disciples were involved with miraculous events not only initiated by Jesus but also by themselves.[89] This would strengthen their conviction to remain faithful to Jesus and his message, even to martyrdom. Jesus also had an impact on the general public when, for example, he was in Jerusalem at the passover feast. We read that his miraculous signs led many to believe in his name.[90] In Peter's speech on the day of Pentecost, he referred to Jesus as a man accredited by God through miracles, wonders, and signs that God did among them through Jesus, "as you

---

[88] Matthew 24: 24.
[89] Acts 5:12.
[90] John 2:23.

yourselves know."[91] There is further evidence throughout the New Testament of the impact that the miracles had on the general public.

We have also seen that there is the evidence from hostile witnesses for the miracles of Jesus. As I previously mentioned, his opponents did not doubt his power but claimed his power came from Satan.[92] High Priest Caiaphas advised the council of chief priests and Pharisees that Jesus should be killed because of his popularity as a miracle worker. If they didn't stop him everyone would believe in him, which would incur trouble from the Romans. Herod wanted to see Jesus because of his reputation, hoping he would perform a miracle.[93] Nicodemus, a pharisee and a member of the Sanhedrin, called him a Rabbi and said that they knew Jesus was a teacher come from God as no-one could perform the miraculous signs that he was doing if God was not with him.[94] When Peter and John healed a cripple, the miracle became widely known and the rulers noted that these men had been with Jesus.[95]

We see that the confirmation of just one miracle opens the door to the possibility of other miracles. For example, given the existence of the miracle of the resurrection of Jesus discussed in the previous chapter, the conditional probability of further miracles is substantially increased as it suggests, along with the evidence for God's existence given in Chapter 2, that there is a God who is at work.

---

[91] Acts 2:22.
[92] Matthew 9:33–34; John 10:19–21.
[93] Luke 23:8.
[94] John 3:2.
[95] Acts 4:13–15.

CHAPTER 8

WHY DOES GOD ALLOW SUFFERING AND EVIL?

8.1 INTRODUCTION

When I began taking an interest in this subject in my younger days it was very much from an academic point of view. Now many years later after suffering physically and psychologically myself and watching others suffer, my answers are not so slick any more! It is a very different story when we are sick ourselves or struggle with physical or emotional pain. Suffering is a challenge and this chapter is an attempt to grapple with it. Before I begin I would mention that suffering is not the same as sickness. We all get sick at some time and people vary as to how they cope with it. Sometimes pain is necessary to remove pain, such as a joint replacement or an operation such as removing a cancer.

If something unpleasant happens to you that you think is unfair, how do you react? If it is done by someone else, you probably get angry and may think of an appropriate act of retribution. If you cannot blame someone or yourself, perhaps you blame fate (whatever that is!) or luck, or if you believe in God perhaps blame God. A common reaction is to say :"If God created the world, why couldn't He have created a world without suffering and evil?" And it is not just the existence of suffering that causes problems but the amount. This raises the question of how much is too much! Some ask why does suffering happen to "good" people. Actually there are really no "good" people as we all make mistakes, even though we are made in the image of God (which unfortunately has become tarnished). However, we still feel it is not fair in the case, for example, of innocent children suffering. Also evil is not just something we observe externally, but it is also an internal reality that we all struggle with. The problem with personal evil is not, as Plato maintained, a problem of knowledge and knowing what is right, but a question of will. We still do things we know to be wrong, as Paul found.[1] In looking at the question of suffering and the existence of God we need to combine what we consider in this chapter with the evidences of God's existence given in previous chapters. In fact in this chapter we shall see that the existence of the Christian God makes it easier to understand the problem of suffering. It is in countries that have endured severe hardship where Christianity has grown the fastest, and it is in recent decades that we are seeing huge growth.[2]

There is no end of philosophical debate on the question of why God allows suffering and evil if God exists. And if He exists, how can He be a God of love, as the Bible and Christians claim. This question that poses perhaps the greatest challenge to the Christian faith and theism generally has teased philosophers for centuries. Of course if you don't believe in God, then you don't have a problem except how do you cope with suffering when it comes. Unfortunately academic arguments tend to lose their appeal when we face our own personal suffering. Nice theoretical arguments won't help when you face the slow death of a family member, the sudden death of a close friend, a painful breakdown of a relationship, or you find you have a health problem with no apparent solution. Emotions come to the surface with the right hemisphere of your brain taking over and the logic from your left hemisphere going out the back door! We may even cry out in anguish to God. Why me? Atheism provides no consolation, only extinction. Then there is the problem of other people's suffering as its distribution and intensity seems to be random and seemingly unfair, especially as it is often the innocent that suffer.

Augustine argued that if there is a God then why is there so much evil, but if there is no God then why is there so much good? The skeptic, in saying that evil exists, must also accept that good exists. In fact good and evil are not so much opposites or form a dual system, as evil is not something but is the absence or privation (lack of) of good. For example, darkness is the absence of light, cold is the absence of hot, and blindness is a privation of sight. However, to admit good exists implies that there is a moral law that tells us what is good and what is evil, and if there is a moral law we can argue as in Section 2.7.2 that there must be a moral lawgiver who created the world this way. When the argument from evil is used to challenge God's

---

[1] Romans 7:19.
[2] Johnstone and Mandryk (2001).

existence there is a problem, as it presupposes that we know the difference between good and evil, that is moral standards exist, otherwise why complain and what grounds are there for complaining? If we say that God is unjust, where do we get the idea of justice from? We need to explain the source of our notions of evil and justice.

Philosopher Alvin Plantinga[3] argues that if we really believe in wickedness then we actually have a powerful argument for God's existence. C. S. Lewis[4] raises a problem:

> If the universe is so bad, or even half so bad, how on earth did humans ever come to attribute it to the activity of a wise and good creator? ... The spectacle of the universe as revealed by experience can never have been the ground of religion: it must always have been something in spite of which religion, acquired from a different source, was held.

This raises the question of the origin of Christianity. Lewis also mentions the existence of the numinous that excites awe, and raises the question of how it became linked to morality. Nothing in the visible universe that some describe as a cold and hostile place where it is the survival of the fittest would suggest either of these two ideas.

We can also argue that evil is not a thing and therefore not something God created, but rather the creation of free will opened the possibility for evil to occur. We often distinguish between wrongful or sinful acts on the one hand and evil acts on the other depending on the degree of suffering caused to others. However, this may not be helpful as we may not know the impact of our actions on others. We all make mistakes, but it doesn't mean that we are all actually evil. We see then that if we rule out God and absolutes, we have a problem defining our terms such as evil and injustice. In fact non-belief in God does not make the problem of evil any easier to explain.

In addition to explaining the existence of goodness we also need to explain the existence of altruism, namely the concern for the welfare of others and acting to help them. We note that some altruism appears to exists in the animal kingdom, especially with species having complex social structures, for example, birds on guard duty for the rest of the flock and the experiments of Frans de Waal with chimps and other animals (see Section 2.7.1). However, without God in the equation the existence of human altruism is a major problem as, according to Hume, you cannot derive an "ought" from an "is."[5]

Suffering can turn us to God as the purpose of life for the Christian is not necessarily happiness, but rather it is to know God and continue our relationship with God into the afterlife where we experience the rewards of

---

[3] Plantinga (1993: 73).
[4] Lewis (2001b: 3).
[5] For a helpful discussion on the topic see http://www.spectacle.org/297/alt.html, accessed August, 2105.

knowing God. Admittedly suffering can also turn people away and make them bitter or cynical. We shall seek part of the answer to suffering and evil from the Bible, so that I will make biblical references from time to time. In particular, we are reminded that we have a transcendent God whose thoughts and ways are so much higher than our thoughts and ways.[6] Although this idea may provide little consolation for sufferers, it seems intellectually arrogant to rule out the idea that God may have a higher purpose that we don't understand. This is particularly true if we assume life continues after death as injustices can then be righted and God can make up for suffering in this life. After all, life here with all its woes is just a passing moment when compared to eternity. We cannot be sure that evil has no purpose just because it may seem pointless to us, as God may allow it to avoid an even greater evil or to achieve some greater good. We cannot argue against this as we are not God!

We mentioned very briefly in Section 4.1.1 when discussing chaos theory that a flutter of a butterfly's wings could led to a hurricane elsewhere, given the sensitivity of our world to changes, so that we don't know whether an evil such as the death of a child from cancer may be permitted by God because of its ripple effect in history and perhaps even in another country. Some would argue however that God allows unnecessary or excessive evil or suffering at times, and sometimes suffering seems so pointless. In dealing with this problem we note several things.

First, as we have just noted, unnecessary evil or suffering is impossible for us to know as we don't know what is unnecessary as we are not God. Part of the problem is that some physical evil can be a byproduct of a good process; rain together with hot and cold air are a necessary for food and life, but another byproduct is a tornado. Again, it is good to have food to eat but we can also get food poisoning.

Second, if God removed the worst suffering from the world then, of the suffering that was left there would be a worst instance, which could be complained about. That could then be removed and we have the next worst instance. Where do you stop? Only when there is no suffering at all and, as we shall see later, this clashes with the notion of free will.

Third, if God does not exist, then objective moral values do not exist, being simply societal or personal values. But the atheist argues that there is evil in the world so that, as already noted, objective values must exist for that person after all, indicating the existence of a moral lawgiver, as argued in Section 2.7.2. I realize that such arguments will not bring comfort to a parent whose child is dying from some terrible disease, as aspects of suffering are a mystery. In what follows in this chapter I hope to provide at least some answers.

---

[6] Isaiah 55:8–9.

8.2 POSITIVE ASPECTS OF PAIN

Pain has a positive role—it provides protection. It stops us from damaging ourselves and alerts us to a problem. If you put your hand on a hot element you remove it very quickly. The nervous system is very intricate with millions of pain sensors distributed exactly where you need them most. Nerve endings can become even more sensitive with use. Take your fingertip, which is used a lot. It is very sensitive to pressure, with an ability to detect 3 grams of pressure (you can learn to read Braille through it), but tough when it comes to pain. A force of 300 grams is needed before a sharp needle triggers pain sensors in your fingertip. However, once you hurt your fingertip it can be very painful. The back of your hand is nowhere near as tough. It starts shouting a warning with only 100 gm of needle pressure. But your vulnerable stomach area will scream at just 15 gms. That is why leprosy is such a disfiguring disease as it mainly acts as an anesthetic. Paraplegics have to continually check their bodies for possible damage.

Under our skin it is a different world again. Internal organs are seldom exposed to the danger that threatens our skin. For example a surgeon could cut my kidney with a scalpel and I feel no pain, but get a tiny kidney stone lodged and the organ will fire off an excruciating alarm rocket. Some have argued that pain as a warning system is inefficient as very serious internal problems often have little pain until it is too late. Clearly, as with any other design, designing a pain system depends on its purpose so that a compromise is usually needed. It would seem that our pain system is designed to monitor normal health, rather than the exceptional state of disease, and to help us to learn procedures of self-preservation as we go about our daily lives.

One additional feature of our pain system is that we can have referred pain, where an injury borrows the pain receptors of some other part of the body to warn us. Often pain needs to be sufficient enough to make us take some action. We need to listen to our pain as it is a gift to demand attention and to prevent us from damaging ourselves any further. There is also a link between pain and pleasure. The same nervous system can give us great pleasure. They go hand in hand. Life is full of contrasts. We cannot really appreciate joy without knowing pain and suffering. We need all colors including the black and the grey.

8.3 WHY DID GOD CREATE US?

In dealing with problem of evil and suffering there are a number of fundamental questions that can be asked. For example, Keith Ward[7] considers the question of whether God, the greatest conceivable being, can have any rea-

---

[7] Ward (1984).

son for doing anything at all, especially create the world. If He is going to do anything at all, then He will cause a change in the things that exist that won't be for the worse. If the change is for the better, then it can be argued that those things must have been imperfect to begin with. Does this mean that God is imperfect without creation? Another argument sometimes put forward is that if God is the greatest good, then it would seem that no finite state of affairs combined with God's goodness would yield a morally better state of affairs. Finite creatures cannot add anything to God at all, as we cannot do better than the best! Yet God allows evil because He has morally sufficient reasons to do so, for example it is better for humans to have free will than not have it. Why then create beings who can choose evil as this adds nothing better to God's existence or goodness? Also, God's prevention of suffering would not make the world a better place, for it would eliminate the good of moral freedom.

To endeavor to answer such questions a number of things should be noted. First, if God is infinite in some sense, adding something finite to infinity is still infinity and adding states of finite goodness cannot increase the overall goodness of reality. Here we are at our limits of comprehension and struggle with finding suitable metaphors or analogies. Second, we can at least compare finite states, namely it is better for humans to have freewill to choose between good and evil than to be just puppets that always choose good. Third, it would appear that God's motive for creation cannot be to morally improve the overall state of reality, but rather creation was for our own benefit, the benefit of the opportunity of knowing God. Fourth, God is portrayed as "our Father" in the Bible. Ward notes that parenthood changes the sorts of values that parents experience as well as adding the new values experienced by the children. He argues that God will experience similar new sorts of values through creating us much like a parent does that He would not have experienced if He hadn't created us. Loving and caring for His creatures and entering into their joys and sorrows will be a new sort of value in God and adds to the sum total of values in the universe. Fifth, God did not need to create a universe in the same way that human parents do not need to create children. However, perhaps God likes to create things. Since we are somehow made in the image of God as the Bible says, we find that we like creating things too. The universe is an amazing place with an incredible diversity of life that God somehow brought into being. God seems to be interested in variety, like a free and creative artist who has a blank canvas.

The existence of evil and suffering has been used to argue against God's existence, perhaps not so much the suffering itself but the fact that God allows it. In considering some of these arguments it should be remembered that such arguments must be considered in the context of other arguments for God's existence such as those given in Chapter 2, and in the end it is a matter of combining all the evidence both for and against.

8.4 SOME PHILOSOPHICAL ARGUMENTS

You may have heard of the following argument before.

Argument 1

(1.) God is omnipotent (all powerful) and therefore made everything perfect.

(2.) But imperfection cannot come from perfection.

(3.) So perfect creatures cannot be the origin of evil.

Or another variation:

Argument 2

(1.) If God is omnipotent He would create the best possible world.

(2.) This world is not the best possible world.

(3.) Therefore God is not omnipotent.

A very famous argument goes like this:

Argument 3

(1.) If God is all good, He would destroy evil.

(2.) If God was omnipotent, He could destroy evil.

(3.) But evil is not destroyed.

(4.) Hence there is no such God.

Before we consider these arguments we should note five things. First, being omnipotent does not mean that God can do everything. God cannot make a square circle or make an object so big He cannot move it. He cannot make the statements "A is not true" and "A is true" to be simultaneously true. God cannot do things which are inconsistent or self-contradictory. For example He cannot sin. Aquinas comments "... it is more appropriate to say such things cannot be done, than that God cannot do them."[8] Second, if God is all goodness, then He will be just, as we believe in justice and are made in the image of God. We cannot separate the two concepts of goodness and justice. Third, when atheists say it is unfair of God to allow suffering there is an element of inconsistency in their logic. If they don't believe in God they have no absolute measure of unfairness. In fact why should they believe in justice? Again it should be noted that in describing something as evil one is using some moral criteria about how life ought to be, which

---

[8]Aquinas (1920 translation: Part 1, Question 15, article 3).

suggests a reality of something that we might call Supreme Good—perhaps another name for God! This amounts to using the existence of God to try and disprove the existence of God. Suffering is more about the nature of God, not God's existence. Fourthly, what do we mean by a best possible world?[9] Best for whom? I think that a world with free will and suffering has some positives as, for example, it produces some great attributes of compassion and unconditional love. Also pain can be a great driving force as I have learnt from personal experience. These ideas we reconsider later.

8.4.1 Best Possible World

It is not easy to define the "best possible world" as we all have different views on this. Comparing two worlds would be like trying to compare two great pieces of music. How would God actually choose the best world to create? If He created any world, wouldn't He be able to improve on it? We are then back to the argument that God cannot do something that is inconsistent. If we still want to use the concept of "best," one response to that raised in Argument 2 item (3.) might be to say that this type of world is perhaps the best possible way to achieve the best possible world. We don't have the best possible world at present because, for example, it depends on us to be better, but one day it will be better. Perhaps there is no upper limit as to how good a world can be so that God created a good enough world. In a sense we live in an unfinished world as Christians are promised that there will be a new world without evil, pain, tears, or death.[10] However it is acknowledged that this kind of optimism may be of small comfort to someone who is suffering. The Christian might also argue that we live in a "fallen" world, so that the world is not as it was meant to be.[11] Furthermore, in a world where free will can cause evil and suffering, then having an incarnation (God somehow becoming flesh) and an atonement would provide a better world.

Journalist Philip Yancey,[12] in looking at the question of the best possible world, points out the difficulty of creating a superior world that has both a system of natural laws and free will, both good principles in themselves. Doing away with one problem can create another. For example, he mentions that there are 24,000 species of bacteria of which only a few score cause illness so that eliminating bacteria especially from our gut would mean we wouldn't be able to eat. When it comes to designing something we must bear in mind the purpose of the design and the fact that compromise is usually involved such as, for example, strength versus weight in design.

After 35 chapters of Job's complaints in the Old Testament book of Job, God challenges Job in chapter 38 to answer some questions. God points out

---

[9]For a technical discussion of this problem see Plantinga (1977).
[10]Revelation 21:3–4.
[11]This is discussed in more detail later.
[12]Yancey (1990: 63–64).

some of the remarkable features of nature, especially animals, and gives Job a chance to suggest improvements. Job replied that he did not understand what he was talking about and repented.[13] Another person who knew suffering was the poet John Donne, born in the 16th century, who lost his career, his wife (who died leaving him with seven children), and his health so that he became bedridden. However, in spite of it all, he came to realize that his life was not meaningless but that his suffering could be redeemed through spiritual growth and be used for the good of others. On the other hand, although God allows suffering it does not mean that suffering necessarily contributes to the world being a good place.

We shall now consider some of the above arguments. One answer to Argument 1 is as follows. If God is going to make a creature like us, what is the most perfect way that He could do it? Surely He would want to make a creature that freely loved Him. He could make us love Him but we would then just be puppets and surely that would be less than perfect. Our answer is then:

(1.) God made everything perfect.

(2.) One of the perfect things that God made was a world of free creatures.

(3.) Free will can cause evil.

(4.) Hence imperfection (evil) can arise from the perfect (indirectly).

God made evil possible, creatures make it actual. Evil is not something that God created but, as mentioned above, it is the absence or privation of good. The question of free will arose when Satan said to God that Job was only righteous and feared God because God protected him. He challenged whether Job would do the same if he was afflicted, and God allowed Satan to put Job to the test. Clearly Job's freely-given love and allegiance to God were very important to Him.

8.4.2 Further Answers

There are two aspects to the question of free will: according to Christianity and the two great commandments about loving God and our neighbors, our divine purpose is to freely enter into a loving relationship with God and freely act rightly to one another. Some philosophers in attacking the so-called free-will defense have argued that it must be possible for God to make us so that we always do what is right. After all, Jesus was described as being tempted like us but not sinning.[14] John Hick, philosopher and theologian,[15] in answer to this responded by saying that even if God implanted within us the desire

---

[13] Job is discussed further below in Section 8.11.
[14] Hebrews 4:15.
[15] Hick (1974: chapter 13).

to do what is right with respect to others (much like a hypnotist) there is still the question of being free to love God (being free in relation to the hypnotist). God wants us to freely enter into a personal relationship with Him.

Moving on to second Argument 2 above, one answer to the argument about God destroying evil is as follows:

(1.) If God is all good, He would destroy evil.

(2.) If God was omnipotent (all powerful) He could destroy evil.

(3.) But evil is not destroyed.

(4.) Hence God will one day destroy evil.

One could argue, following Augustine, that God may allow evil because He can bring good out of evil. Now evil cannot be destroyed at present without destroying free choice. However evil could be eventually destroyed if we go beyond the grave. The psalmist in Psalms 34 complained to God that evil people seemed to prosper, but when he went into the temple to meditate he realized that such people were on a slippery slope and their future uncertain. Of course if we don't believe in an afterlife, then there is no ultimate justice. It is interesting to note that Alvin Plantinga observed some time ago:

> Now, as opposed to twenty or twenty-five years ago, most atheologians have conceded that in fact there isn't any inconsistency between the existence of an omnipotent, omniscient, and wholly good God and the existence of the evil the world contains... It is heartening to see that atheologians are giving up the incompatibility hypothesis and are now prepared to concede that there is no contradiction here: that's progress.[16]

Plantinga has written extensively on this and related subjects.

8.5 FREE WILL

Since the idea of free will is part of our response to above questions, what does free will mean for us? The existence of free will has already been considered in Chapter 4 and in this section we assume that we have some kind of free will in the following discussion.

8.5.1 We Have Some Freedom

Why some? As previously mentioned, we often don't have as much freedom as we think. The type of person that we are depends not only on our physical and genetic makeup but also on our environment and past decisions. We are not only bound by the sort of person that we are but we are also bound

---

[16] Plantinga (1982: 74).

by our circumstances so that in making a decision our range of options may be very limited. Still, I believe we are ultimately responsible for our own decisions. Unfortunately when it comes to choosing between right and wrong we seem to have a bias downwards; being good seems to require an effort. There are also consequences from wrong decisions; we reap what we sow[17] and whoever sows iniquity will reap sorrow.[18] These consequences also apply to nations where competition between people also applies to competition between nations, leading to aggression and war.[19] Jesus said that those who take the sword will perish by the sword.[20]

8.5.2 We are Free to Hurt Ourselves

We do things that are bad for our minds and bodies. If we read junk material our minds can become junk! If we eat junk food our bodies can become junk! We let bad habits develop, like laziness, and we then become slaves to our habits. A lot of suffering, both mental and physical is self-induced, isn't it? Often our body gives us a warning, but we might ignore it. We also take risks and wonder why accidents happen to us. The Old Testament, especially in the Psalms and Proverbs, has many passages that warn about the painful consequences from wrong actions. If we do not exercise adequately or eat appropriately we can get sick. We also get sick because our bodies are weak or overused in some respect; as a counselor I find burnout is not uncommon. When the apostle Paul was in prison, the Phillippian congregation sent Epaphroditus to Paul to help meet his needs. Epaphroditus worked to the point of exhaustion with the result that he became ill and almost died. God in His mercy brought Epaphroditus back to health.[21]

There is a deeper aspect of our freedom that I believe has a place, and that is in the role played by epigenetics (cf. Section 2.6.3). Genes can be switched on and off, and the changes can sometimes be inherited. As we can carry harmful genes or genetic mutations through hereditary, life circumstances, and environmental effects such as pollution, these may be switched on through our own wrong actions. If we have a deleterious gene, perhaps we shouldn't blame God if its effect suddenly shows itself. Some things can be brought on by shock, tragedy, or even stress, as I have found as a counsellor. Our minds and wrong thinking can have a huge effect on our bodies, creating all kinds of illnesses and organ breakdown.

---

[17] Galatians 6:7–8.
[18] Proverbs 22:8.
[19] James 4:1–2.
[20] Matthew 26:52.
[21] Philippians 2:25–30.

8.5.3 We are Free to Hurt Others

We can hurt others deliberately. In a world of free will, physical and psychological conflict is inevitable. Because of the solidarity of the human race we can do nothing in isolation. Together we share the effects of a common good and a common evil. At a personal level we are seeing a breakdown in relationships of epidemic proportions, leaving hurt people. I meet this in the counseling room all the time. We can hurt others by careless behavior. Smoking is not only bad for us; it is also bad for the person next to us because of "side stream" effects. We can also hurt others by omission, for example failing to correct an injustice or not take proper care and leaving a situation that can be risky or even dangerous (e.g., shoddy buildings that fall down with earthquakes, or simply collapse). Through unhygienic practices people can pick up nasty bugs, even in a hospital! Also aid may not be provided in time or not at all after a disaster. Ecological disasters brought about by uncaring corporations (e.g., pollution) can hurt local people.

We can also hurt others collectively by messing up our environment. For example, the wholesale removal of forests has caused all sorts of erosion problems throughout the world. The removal of wetlands in places like the U.S. takes away natural outlets for excess water and leads to extensive flooding and much suffering. Poverty and questionable resource and environmental management can let in microbes. Our environment can physically stress us; we have noise pollution and the pollution of our air, water, and food. Young pastor Timothy apparently had a problem with the water in Ephesus, so the apostle Paul prescribed a little wine to help him ease his "stomach ailments and... frequent illnesses."[22] The misuse of antibiotics and horizontal gene transfer that can lead to antibiotic resistance being transferred from one species of bacteria to another is creating health problems now and in the future. There is also spiritual pollution where we are bombarded by advertising and the media to make us dissatisfied with how we look, with what we own, and with where we live.

8.5.4 We are Free to Ignore God

When we do this I believe we hurt ourselves the most, at the very depths of our being. We can ignore the claims of God on our lives, but we may not have any peace. I believe that our destiny is to connect with God and worship Him, and when we push Him away we are going against what we were created for. A consequence of our freedoms is the following.

---

[22] 1 Timothy 5:23.

8.6 AN IMPARTIAL WORLD

The world must be impartial if there is to be free will. In Matthew 5:45 in the New Testament Jesus says that God makes His sun rise on the evil and the good and sends rain on the just and unjust. Everyone is treated the same as far as the physical world is concerned. The world has to be a consistent place with the future relevantly resembling the past. It can be argued that this is so because of God's creativity, as there is no other good reason why it should be so.[23] Fire always burns, and not just bad people, so that we can learn to understand it and protect ourselves. A gun does not turn into a banana when the intention is to harm! The world would be an impossible place to live in if it was not consistent and God had to continually intervene. If God were to intervene, where would He stop? It should be noted that God does appear to intervene from time to time, and He alone decides when and where.[24] Such a world that avoided all suffering so that all wrong actions would have no consequences would mean that moral qualities would no longer have any point or value. Why be truthful if telling a lie had no harmful consequences? Philosopher John Hick notes:

> Perhaps the most important of all, the capacity to love would never be developed, except in a very limited sense of the word, in a world in which there is no such thing as suffering.[25]

Love grows through sharing in times of difficulty.

The world can only be consistent if it is also neutral. I can talk to you because we can both set up sound waves in the neutral common air between us. In one sense the world has to be neutral about God if we are to freely love God. Those that submit themselves to God see signs of a divine presence while those that don't see only destruction. As Hick notes,

> ... the world will be religiously ambiguous, both veiling God and revealing Him—veiling Him to ensure man's freedom of choice and revealing Him to men as they rightly exercise that freedom.[26]

Summarizing, we see that some suffering comes from the interaction of a neutral world with the existence of free-will, which can lead to sin against ourselves, against others, and against God.

8.7 NATURAL DISASTERS

The above arguments have to do with moral evil, that is evil arising because of moral decisions. We can understand a lot of this and perhaps we can accept it, as much of what happens is due to human error. For example, if we do risky things, build or work in risky places, then things can go wrong.

---

[23] This is called the argument from induction.
[24] See chapter 7 on miracles.
[25] Hick (1974: 361).
[26] Hick (1974: 318).

But what about "natural evil", namely natural disasters such as earthquakes, tornados etc. Some natural evils come from natural laws that may play an important role in our survival. For example, earthquakes are caused by plate tectonics, yet without tectonics the continents would all erode into the oceans so that there would be no life on earth. Furthermore, the existence of tectonic plates allows mineral resources to surface in the gaps and replenish the earth. If God did away with earthquakes by changing the physical laws, then this would have a huge ripple effect that could make life impossible. The same is true with other aspects of nature. For example, some diseases and viruses contribute to the total ecosystem in ways we don't yet understand. However, we struggle with the problems like handicapped children and diseases like cancer.

Why did my first wife have to struggle with cancer for 6 years before dying a painful death at the age of 47? Towards the end when she was at home I remember having to drag myself into consciousness twice every night to give her pain killers. I felt helpless in the presence of pain. There is nothing more painful than watching a person suffer. She was a courageous Christian person who set an amazing example. Why should the world be like this? Let's try and tackle this problem now.

8.7.1 Human Influence in Disasters

Natural disasters are often due to human choices or actions. For example, a fire breaks out through an accident or inappropriate behavior, and there is no proper fire exit (or the sprinklers are not working). Buildings or bridges collapse through earthquakes because of inadequate building practices. Inadequate roads can lead to motor vehicle accidents. Deforestation, global climate change, and so on, can have a disastrous effect on weather patterns causing, for example, extensive flooding or hurricanes. You only need to listen to the TV news for daily examples. Weather instability and weather extremes are now a big issue.

8.7.2 Everything Falls to Bits Sooner or Later

In Section 2.4.1 we talked about the Second Law of Thermodynamics, which implies that the amount of useful energy available in the world is steadily decreasing. Energy differences (like temperature) tend to even out. As I mentioned there, there is also a statistical interpretation of this law that says that the amount of "randomness" is increasing that I described as "everything falls to bits sooner or later." This leads to several consequences.

  (1.) Machines break down and there are accidents, on the road, in the home, at work; we have them all the time. Had any appliances pack up lately; perhaps all at once or so it seems! Some say they seem to go in threes!

There is a lot of suffering because of the failure of machines. We cannot trust man-made (or woman-made) things.

(2.) We are falling to bits too as we are part of the system. Our bodies are dying (cheerful thought isn't it!). Because our bodies are alive we can partly reverse the trend, but in case you hadn't noticed the overall trend tends to be downhill.

(3.) The world is breaking down at the macroscopic (large) level. Your back fence falls over because of erosion and your drains block because of silting or tree roots. These cannot be called acts of God! Erosion and silting are part of the leveling out process. Earthquakes are part of the process of evening out stresses in the world's crust. However, if we mess up the environment then we can expect even more problems such as those deforestation problems and strange weather patterns already mentioned.

(4.) The world is breaking down at the microscopic level. Genes and chromosomes have accidents too. These accidents are called mutations that usually occur through environmental changes. These mutations are invariably harmful and we can pass some of them on to our children. We all tend to carry some potentially lethal genes but that's okay provided we don't get too many of the wrong sort, or certain ones get switched on or off. Because of mutational accidents a good bug can become a bad bug, and there are some terrible ones around. The common cold is hard to treat because the various strains keep mutating. AIDS is difficult as the virus multiplies at over a 1000 times faster than other viruses so that its potential for mutation is much greater. You are often not fighting just one bug but a whole family of them. The aunts and uncles are just as tough as the brothers and sisters!

We see therefore that the nature of the world leads to all kinds of suffering: natural disasters, accidents, sickness, and genetic damage. However, we must keep things in perspective. Disasters upset us not just because they are painful, but because they are exceptional. People wake up most days feeling good (unless you are starving or recovering from an earthquake, flood, or drought). Most illnesses are treatable (if you have access to medical treatment). Most planes take off and land without mishap. The accident, the injury, and the tumor are life shattering, but they are generally rare. Of course that is not much comfort if you are going through such a rare event! And it does depend on what country you live in for the support you will get, if any. There are terrible inequalities in the world, but we cannot blame God for them. We should not forget that there is much good in the world. In the same way we can receive undeserved suffering we can also receive undeserved good, though some people at present may wonder where the good is!

8.8 ANIMAL SUFFERING

In Chapter 3 we considered whether animals had consciousness or not. John Hick argued that there is evidence that animals experience pain, especially throughout the invertebrate kingdom, for the following reasons:[27]

(1.) Humans have a certain genetic continuity with other forms of life.

(2.) Higher mammals behave in a similar way to humans when under intense pain.

(3.) Our sensory and nervous system is similar to other vertebrates.

(4.) The higher vertebrates can be conditioned to do things by means of their reaction to the pleasure-pain dichotomy; an electric shock for a wrong action, but a reward for a right action. We note that animals cower when there is impending pain suggesting they remember pain and consciously anticipate it, though memory may not actually require conscious awareness. Philosopher Michael Murray[28] mentions strong evidence of pain related behaviors in human beings that are not felt as pain. For example, there are well documented cases of human beings displaying both physiological and behavioral evidence that suggest they are experiencing pain, when in fact they report not actually feeling pain. We also have the instinctive reaction of removing one's hand from a hot surface before pain is felt. He concludes:

> Evidence of this sort shows us that when we see animals exhibiting types of behavior that are typically associated with feelings of pain in the case of human beings, we cannot make any confident inferences about the mental states of those animals.

We don't really know how much animals suffer.

(5.) Hick comments that animals need pain to help them survive in their environment as do humans, though we don't know about the lower vertebrates like fishes and insects. Certainly invertebrates like sea anemones that have no central nervous system are unlikely to feel pain. Whatever animals actually experience, it is needed to help them adapt to their environment.

Hick points out that, in contrast to animals, pain has a psychological component for humans in that how we experience pain depends on our mental attitude towards it. If we believe that we are entitled to palliative relief for our pain, then any pain will have a greater affect on us if we don't get that relief. By the same token, a positive attitude can assist in our healing from pain. A greater part of human misery is psychological. The physical part of

---

[27] Hick (1974: 346–347).
[28] Murray (2008: 62).

sickness may not cause as much misery as a fear of disability, anxiety about those around us coping with us, or the humiliation of helplessness and dependence on others. There is also the fear of death, which is not such a problem to animals as it is to us, though some animals (e.g., elephants) appear to be seriously affected by the death of those of their own kind. Pets are affected by the death of their master. One might ask the question of why do these subhuman forms of life exist at all, and the response might be that it is the result of evolution.

Augustine and others resolved this problem by means of the principle of plenitude that states that anything that can happen will happen eventually and that the richest and most desirable universe contains every possible kind of existence right down to bacteria. Only a small handful of known bacteria, termed pathogenic, are capable of causing disease and do so by invading the cells of a living organism. However, most bacteria will not invade another living organism, and many more bacteria are rendered harmless by our immune systems, while others, such as gut bacteria, are beneficial. Unfortunately bacteria mutate and as humans we have been irresponsible by using antibiotics too freely, and using them in animal food. Unfortunately drug companies are not prepared to spend money in research to deal with some of the more resistant bacteria that have recently appeared, especially in hospitals, such as MRSA, CRE, and *C.difficile*.

Theologically, the idea of plenitude means that God somehow expresses Himself in the creation or development of every possible level of dependent being from the highest to the lowest so that there will be imperfect and unequal beings.[29] Every level makes its own contribution to the whole, and the preying of life upon life can be regarded as a proper feature of the lower levels of animal creation. We know that population control through predation helps to maintain a balance in nature so that one species does not take over, as with plagues of insects. However we note that most animals kill only to eat and most higher animals care well for their young. Animals show not only violence and aggression but also compassion and care. We find that the so-called golden rule operates in primate colonies and in communities of other animals such as dolphins and elephants. Some might question that the evolutionary process is wasteful. Molecular biologist Denis Alexander replies:

> Wasteful compared to what? We now know that the universe with its 10^{11} galaxies each containing an average 10^{11} stars has to be this large and this old in order for us to exist. It is difficult to know what 'waste' means to the God who is the ground of all existence. Equally with those who wonder why evolution has taken 'so long' before arriving at humans, one may question the anthropocentric assumption that the only purpose of evolution is to produce humanity, when God clearly enjoys and values all his works of creation (Genesis 1; Job 38–39; Psalm 104).[30]

---

[29] For further discussion see Lovejoy (1962).
[30] Alexander (2012: 238).

Hick suggests another approach to the problem of relating sub-human life in all its varieties to the creative activity of God. Humans are embedded in this larger stream of organic life that contributes to their freedom in relation to God. Referring to mankind, Hick says:

> Seeing himself as related to the animals and, as like them, the creature of a day, made out of the dust of the earth, man is set in a situation in which awareness of God is not forced upon him but in which the possibility remains open to him of making his own free response to his unseen Maker.[31]

Man is set at a certain epistemic (knowledge) distance from God in an environment that on the surface gives the appearance of there being no God so that people may be free to choose where they centre their lives. Theologian Nicola Hoggard-Creegan[32] has the following interesting application of the so-called parable of the wheat and the weeds or "tares"[33] to evolutionary history:

> First, it makes sense of the mix of the good and evil we observe at all levels of the evolutionary progression. Second, the parable explains the close entwining of the two and the way in which goodness carries with it a dark side as often as not. Third, the parable reveals that the texture of good and evil, wheat and tares, cannot be seen without close observation.

When weeding, it is often not a good idea to pull out the weeds from our garden as it may damage other plants. Carrying over the analogy, rooting out the evil is often at the expense of the good as well. She goes on to say: "The presence of both wheat and tares obscures the glory of God and requires a moral discernment of what is going on." and reminds us how interconnected our living world is with the comment "this fragile interdependence is much closer to the image of the kingdom of God than are the previous images and pictures of competition to the death."

8.9 BIBLICAL PERSPECTIVE

We now look more closely at what the early chapters of the book of Genesis have to say about how the world came to be as it is with regard to evil and suffering. It is a controversial topic so I shall endeavor to present some of the different viewpoints. I also consider the question of how important it is to know all the answers.

8.9.1 Genesis Chapter 1

Why is the world like it is? As already mentioned, it could be argued that it is God's way of producing a theologically neutral and dynamic environment

---

[31] Hick (1974: 351).
[32] Hoggard-Creegan (2013: 84, 135).
[33] Matthew 13:24–30, 36–43).

where meaning must be sought. One question is why did God introduce the second law of thermodynamics. Could He have created a world without it? This raises the question again of whether God could have made a better world. I have already suggested above that this world is still work in progress as mentioned later. There is also a biblical explanation for the state of the world, namely that it is a "fallen" world. The ground and the animals are described as somehow becoming cursed[34] though we don't know what the world was originally like outside of the Garden of Eden. All sorts of things have crept into our world to deflect it from its original purpose. As Alister MacGrath, scientist and theologian, says:

> This is a good world gone wrong, yet which retains the memory of what it once was and the hope of what it will finally become. It is a like a country that has been invaded by an occupying force, which recalls its days of freedom in the past and eagerly awaits its liberation in the future.[35]

A key question is how literal is the story of the first few chapters of Genesis, and various views are discussed below. As there is considerable difference of opinion over this, I will confine myself to seven views; there are others with minor variations.

Literal Twenty-Four hour Days

The first view takes a literal approach and assumes that each of the seven days is a day of 24 hours, given that the evening and the morning of each day are mentioned and the days are compared to a Hebrew work-week. The days of God's creative acts are also referred to in Exodus 20:11 and 31:17, though Deuteronomy 5 doesn't mention six days of creation. Assuming some information from the genealogies, this means that the world is only about six or so thousand years old (depending on the completeness of the genealogies), therefore negating evolution and other detailed evidence relating to the age of the earth.

Longer Periods of Time

A second view first re-examines the Hebrew word "day" (*yom*). It is most commonly used for "daytime" or "daylight", and its second most common usage is as a longer period of time. For example in 2 Peter 3:8 one day is described as a thousand years with God, as in Psalm 90:4. We note that Genesis 2:4 refers to the day *yom* of the whole of creation while in Hosea 6:1–2 it mentions three days in the future that represent longer periods. The word *yom* can also mean a significant future time as in "the day of the Lord."[36] It is used to denote a period of time in 65 places in the Bible; its plural form is used as "age" six times, as "year" five times, and as "years" four times. It is also used as a month, two years, or even a lifetime, and "in that day"

---

[34] Genesis 3:14, 17, 5:29; Romans 8:19–22.
[35] McGrath (2002: 91).
[36] Joel 2:31, cf. 2 Peter 3:10.

occurs over 100 times.[37] If we take a literal stance, then an evening and a morning would describe a period from late afternoon to early morning only. However, the Hebrew word for "evening" stands for the dark part of the day while "morning" stands for the light part of the day (also called *yom*). For the Jew, the Sabbath and the Passover would begin in the evening just before sunset and go to sunset the next day so that the days for Israel began in the evening and ended in the next evening giving a picture of a 24-hour day in Genesis 1. Hebrew lexicons and dictionaries generally do not recognize the interpretation of *yom* as a period of time lasting millions of years, and when *yom* is associated in the Old Testament with a definite numeral as it is in Genesis, solar days are meant, though it doesn't rule out a metaphorical use.

Theologian Gerhard Hasel[38] points to the fact that in Genesis one "yom" is used in the singular, it has a numeral, and is preceded by "there was evening and there was morning." On the other hand, Rodney Whitefied, physicist and Hebrew scholar,[39] quoting from two authorities Gleason Archer and Norman Geisler, supports the view that since the six days each lack the definite article in the Hebrew (some English translations get it wrong), they can be viewed as a sequential pattern rather than referring to just units of time. By a detailed analysis of the use of "yom" in the Old Testament Whitefield found that the repeated pattern of the numbering of the creative times in Genesis one is unique. Each of the Hebrew numberings expressed by "yom" plus a number used in Genesis 1:8, 13, 19, 23, and 31, appear only one time in the Bible thus indicating that the creative *yom* are not ordinary (24-hour) days but could represent an extended period of time. Confused?

It is also suggested that there was a gap before the days started. Whitefied[40] points out that a pluperfect tense applies to the Hebrew verbs in verses one and two so that verse one says that in the beginning, which refers to a beginning spread over a period of time, God *had* created the heavens and the earth (a Hebrew expression for the totality of the world, namely the universe). The second verse, referring to just the earth, indicates a pluperfect ordering so that it says that the earth "had existed" or "was already existing." He believes there is a time gap between the two verses before the start of the sequence of commands modifying the environment, and a gap between verses two and three.

We read that on the third "day" the earth brought forth fruit trees that grew from seeds and bore fruit, which is not a quick process. The writer

---

[37]A good lexicon will provide a detailed listing of the various possibilities. Any extended use of the word depends on combinations with other words.
[38]Hasel (1994: Section V).
[39]http://www.godandscience.org/youngearth/yom_with_number.pdf, accessed August, 2015.
[40]Whitefield (2003) and an online abbreviated version that has a helpful description of Hebrew grammar.

did not say "And God said, 'Let there be vegetations and fruit trees' and it was so", so it seems that the writer had a time lapse in mind. In verses 20 and 24 we have a similar idea of the waters and the earth "bringing forth" living creatures. On the fifth day the waters brought forth swarms of living creatures, and the initial mention of the word "birds" could also be translated as "insects", which are needed for the pollination of plants. On the sixth "day" many thousands of animals were created along with Adam, who named all the animals; again not a quick process. In chapter two Adam said "at last" he had a companion when Eve was presented to him, suggesting duration of time. The words "at last" ("this time" or "now") are used elsewhere in Genesis to indicate a period of time or a period of waiting.[41]

The Biblical order of creation is different from the order given by fossil records, for example Genesis 1 has land plants before sea creatures, and trees before land animals, while Genesis 2 has mankind before animals. There is often considerable discussion about the meaning of the words "according to their kinds" in verses 24–25. We have to realize that the Bible is not a book of science and was addressed to people who were pre-scientific in their classification of creatures. For example the writer knew only three types of plants: grass, herbs, and trees, and similar comments apply to sea creatures, beasts, and creeping things. In Leviticus chapter 11 birds are listed with bats (as they fly), [42] weasels and mice with lizards (as they crawl),[43] and insects were credited with having four feet. Clearly not a scientific classification.

We note that the Hebrew word for create is *bara* which, when associated with God, essentially means to create out of nothing. It is used in relation to the heavens and the earth (Genesis. 1:1), sea creatures and birds (1:21) and Adam and Eve (1:27). The word *asah*, has a lot of different meanings[44] including *make* and *appoint*, and is used with all the other creative acts. On the one hand, Terry Mortenson, theologian and geological historian, has a list comparing the use of both words *bara* and *asah* with regard to creation in the Old Testament suggesting that in this case the words are interchangeable.[45] On the other hand, Whitefield[46] looks carefully at the Hebrew and concludes that the two words are different and not interchangeable in the creation account. The two words are used together in Genesis 2:3, with *asah* in the infinitive sense ("for making"), and the verse can be translated as "And God blessed the seventh day, and sanctified it because in it He had rested from all His work that He had prepared (bara) for making."

---

[41] For example, Genesis 29: 34,35; 46:30.
[42] Leviticus 11:13–19.
[43] Leviticus 11:29–30.
[44] See http://www.biblestudytools.com/lexicons/hebrew/nas/asah.html, accessed August, 2015.
[45] See http://www.answersingenesis.org/articles/aid/v2/n1/did-god-create-or-make, accessed August, 2015.
[46] Whitefield (2003).

Topical Ordering

A third view is that the "days" are to be interpreted theologically with an emphasis on the significance of creation and God's supreme role rather than as a scientific account of material origins and its details. The layout of Genesis 1 is then described as topical rather than chronological and following a literary form, a device often used in ancient literature. For example, the first three days describe how God formed the world, while the second three days describe how God populated this world with living creatures, with (loose) connections between days 1 and 4, 2 and 5, and 3 and 6, respectively. For example, we have light and darkness on day 1 with the "rulers" sun and moon on day 4, so that both days have the same purpose. (This is suggested as being a so-called Hebrew recapitulation, which is a return to the same time but with more information.) We then have sea and sky on day 2 with the rulers fish and birds on day 5, plants and land on day 3 with the rulers animals and man on day 6. Also the eight creative acts are distributed into two groups of four each.

On day 7 God rested and refreshed himself, an anthropomorphic metaphor, but there is no mention of the evening-morning pattern, which suggests that the "rest" is still continuing.[47] We also have the poetic touch with the repetition of "and God said", "and it was so", "God saw", and "there was an evening and a morning." The sun would need to be present on each day for there to be an actual morning and an evening. We also see God creating order out of chaos and having authority over the terrifying forces of nature by setting limits, separating light from dark, waters above from waters below, and seas from dry land. Clearly there is a great deal of careful literary structure in Genesis chapter one. If Genesis chapter two refers to the same creation event, which seems to be supported, then it is also topical.

Days of Assignment

A fourth view is spelt out by Old Testament scholar John Walton[48] where he takes the 24-hour days as literal, using each day to describe not the creation of something (material creation) but rather the assignment of functions to creation (functional creation) with the universe being God's cosmic temple that God resides in on day seven; for example plants were there to serve a certain function in relation to humanity. He treats Genesis 1:1 as a summary of all creation. Referring back to Aristotle's four causes (Section 2.4.2), Walton is talking about final causes, and there is some question over Walton's categories of causation, like the difference between a material cause (what something is made of) and an efficient cause (what brings something into existence) that I won't go into as I am more concerned about the general idea than philosophical overtones.[49]

---

[47] Hebrews 4:4ff.
[48] Walton (2009).
[49] See, for example, http://www.reasonablefaith.org/defenders-2-podcast/transcript/s9-07.

Days of Revelation

A fifth view is that the days refer to days of revelation to an individual. This means that they are 24 hour days. The evening and the morning describe a time of rest for the observer. Two studies originally due to P. J. Wiseman and reproduced by his son ancient historian Donald Wiseman set out the arguments.[50] However the word *asah*, which means *make*, seems to be wrongly translated as "showed", and this view has little support today.

Wiseman points out that Genesis is very ancient for a number of reasons:

(1.) There is no mythical material as in other creation stories.

(2.) The references are universal in that there is no mention of local ideas, customs, or race, but is universal, contrary to other creation accounts.

(3.) There is no mention of any event subsequent to the creation of humans.

(4.) There is no human speculation.

(5.) There is no mention of worshipping the sun, moon, or stars.

Mythology

A sixth view is that the Genesis story is purely mythological. However, this does not mean that it is not based on historical fact, the view of New Testament writers. Myth, in the technical rather than the popular sense in which the word means fabrication or false story, is a sacred narrative that serves to underlie and explain a culture and its institutions. Genesis 1 certainly has this role as it grounds everything in God and gives a reason for Israel's practice of the Sabbath. The continuity of the first two chapters with the rest of Genesis and the way the latter reads like history, suggests that we cannot just reject it. Other surrounding cultures such as the Babylonian indeed had their own creation stories, but they are full of mythological and fanciful figures. There is no comparison of these with the simple unadorned Genesis narrative of one powerful creator God, and no other deities;[51] furthermore there is no mention of polytheism or pantheism.

Darwinian Model

I refer to one more view described by philosopher Michael Ruse[52] that has roots in Eastern Christianity. Here sin is not regarded as the direct result of moral disobedience by fully mature beings, but rather is the result of the consequence of our immature, incomplete, and evolving natures. Humans are made in the image of God but only came slowly to intellectual and moral maturity. From this viewpoint, the coming of Christ and his atoning death

---

[50]Wiseman (1977).
[51]For a list of other stories see "creation myths" on the internet.
[52]Ruse (2012: 252).

on the cross was not the result of an unplanned Fall but was planned for right from the beginning as God anticipated it. Ruse uses this approach to integrate the Fall with evolution. He says that we are bound to be selfish to survive the evolutionary process, but at the same time be altruistic as it is also a good strategy in our struggle for existence. What emerges from our background is that we are a mixture of good and bad, and natural selection is a central cause of sin.

Overall Conclusion

There are other views of Genesis 1, and also further arguments supporting each of above seven views. A major problem is that there has been disagreement as to the genre of this chapter, with many suggestions such as legend, narrative, history, theology, saga, cultic liturgy, poetry, hymn, parable, story etc. By looking at the language patterns, syntax, and terminology, theologian Gerhard Hasel[53] concludes that Genesis chapters 1–11 are historical narrative-prose interspersed with some lists and poetical lines, and refers to it as prose-genealogy. This includes Genesis 1, which he maintains is not different from the rest of the book of Genesis, or the Pentateuch for that matter. He believes that compared with other ancient near Eastern literature, Genesis is unparalleled and

> is the most cohesive and profound record produced in the ancient world of 'how' and 'when' and by 'whom' and 'in what manner' the world was made. There is no parallel to it from the ancient world in any type of literature. There are bits and pieces which have been compared from various cosmogonic myths and speculations, but the biblical creation account as a unit stands unique in the ancient world in its comprehensiveness and cohesiveness.

What do we conclude from the above diverse views? First we need to focus on the big picture, namely we have a simple unadorned account of a monotheistic God creating the heavens and the earth, and that we bear something of the divine image. Second, we don't know how God did the creating but we can use scientific methods to try and find out. In our age we want to interpret everything scientifically and put our own meaning on the text, even though science continually changes. Our theology needs to be independent of our scientific worldview, though theological trends do change as well. Third, apart from the first view, Genesis one could be interpreted as accommodating evolution.

Graeme Finlay, Christian evolutionist and cell biologist[54] points out that many well-known past and present biblical teachers believe, for example, that evolution and belief in a creator are not incompatible, as God could have used evolution to bring His creation into being. The well-known Princeton theologian of the early 20th century Benjamin Warfield believed that there was no conflict between evolution and scriptural infallibility. In two places in

---

[53] Hasel (1994).
[54] Finlay (2004b: 29).

Genesis 1 we read "let the earth bring forth" vegetation or terrestrial animals and God creates man out of the dust of the earth, suggesting in both places the possibility of natural causes at play. I personally don't mind how God created the world as it is my relationship with God that is all important! What is interesting is the simplicity of the biblical account.

A question often raised is the creation of light before the sun and the earth. We could argue that according to modern cosmology, the universe was created in a burst of light with the sun and earth coming into existence billions of years later. A second explanation is that in a world where the sun and moon were generally worshipped, the latter were relegated to counter this. A third explanation is that the sun and the moon were not visible immediately and later became visible. We note that the focus is on the central place of earth because of the emphasis on man's uniqueness and responsibility under God, and not on the universe. Finlay[55] has an interesting table where he compares the language of creation with the language of science. For example, using creation language from Biblical texts, God spread out the heavens, made the stars, set the moon and stars in place, laid the earth's foundation, turned dawn to darkness, formed the mountains, created the wind, made the clouds, sent the rain, made grass grow, gave crops in their seasons, created sea creatures, created mankind, created my innermost being, and gives life and breath to everyone. Each of these has, of course, a scientific explanation.

Finlay[56] has a chapter on a theology of creation. He emphasizes that God is described as bringing creation into being by His word, a theme that runs right through the Bible, especially in the Psalms and Isaiah. In the New Testament we read in John 1:1 that God is identified with His Word and the Word became flesh, namely Jesus the Son of God, through whom all things were created and all things exist.[57] God is sustainer and preserver as well as creator,[58] again a regular theme of the Bible. We also read in the New Testament that God will create a new heaven and new earth in the future, thus restoring our fallen world.

8.9.2 Genesis Chapters 2 and 3

In Genesis Chapter two we have a second creation account that could refer to a local creation or else give a creation list in a different topical order with further details. In the latter case we can see it as the use of Jewish recapitulation mentioned above, as Adam is placed in the garden twice.[59] Also in Chapter 2, verses 4-6 we see the operation of ordinary providence with two problems mentioned in verse 5, namely no rain to grow plants and no person to cultivate

---

[55] Finlay (2004b: 43).
[56] Finlay (2004b: chapter 6).
[57] See 1 Corinthians 8:6; Colossians 1:15ff; Hebrews 1:1ff.
[58] Psalm 104.
[59] Genesis 2:8, 1).

the vegetation to secure its growth. Two solutions are provided in verses 6 and 7: God sent a rain cloud and God created man to till the garden. This suggests that God used ordinary providence like rain and ordinary growth to maintain His creation between each creative act in Genesis 1 before man arrived. Jesus himself referred to the Genesis story in an answer to a question about divorce.[60] In Romans chapter 5 and 1 Corinthians chapter 15 we have the theology given by Paul of the Genesis account, and this doesn't depend on his science. In the same way that sin entered the world through one man Adam, and death from sin for all people as all have sinned, so through one man Jesus Christ we have the free gift of salvation by grace. Jesus is compared to Adam; the first man from earth (Adam) was a man of dust while the second Adam (Jesus) is from heaven.

Spiritual Evolution

At this point it is helpful to distinguish between biological and spiritual evolution, and recognize that the question of our biological origins is a legitimate matter for science to investigate. For example, there is no reason why we cannot believe that Adam and Eve were real people and that their story is real and not mythological. However, we need to read the Genesis story in terms of the culture of the day, where God is described anthropologically as breathing into man's nostrils, planting a garden, walking in the garden, and making clothes of skin.[61] We also have symbols like the tree of life, the tree of good and evil, and a talking serpent. Theologian John Stott believes that several forms of pre-Adamic "hominid" could have existed for thousands of years before Adam and Eve did. He wrote that:

> These hominids began to advance culturally. They made cave drawings and buried their dead. ... But Adam was the first homo divinus, if I may coin a phrase, the first man to whom may be given the specific biblical designation, 'made in the image of God.'[62]

Genesis can be regarded as the story of God entering into a relationship with some of these hominids by breathing into one or more of them "the breath of life" and they received the imprint of God's image,[63] becoming spiritual beings in some sense. As mentioned in an earlier chapter, the Hebrew word "Adam" can mean either man or mankind as in Genesis 1:26–27, and perhaps is not used as a proper name until later. In a sense Adam represents all of us and our attitude to God. The identification of a name with either a person, group, or nation occurs several times in the Bible, for example, we have Israel the man (originally called Jacob) or Israel the nation. In the same way "Eve" means the mother of all living. Unfortunately, through pride and self-will, Adam and Eve disobeyed God, which led to what is described as the

---

[60] Matthew 19:4–5.
[61] Perhaps God took on human form.
[62] Stott (1984: 48ff).
[63] Genesis 1:27 and 9:6.

"fall', and the world was changed. The disobedience of Adam and Eve is our personal story too!

It would seem that Adam and Eve were the first to receive what we might regard as a special awareness of God, and perhaps there were others who received the same. In the Garden where Adam and Eve were placed there was the tree of life whose fruit would enable them to "live for ever." This suggests that in the Garden they had a special control over their bodily processes but began to die physically once outside the Garden. Initially the patriarchs lived a long time, but their length of life began to reduce.[64] There was a change after the flood showing an exponential decay in longevity down to about 70 to 80 years in King David's time.[65] Various theories have been put forward for this longevity such as the time scale was different, but these don't stand up to scrutiny.[66] In Genesis 47:9 Moses told Pharaoh he was 130 years old and he hadn't yet reached the age of those who had preceded him.

It is interesting that there are those who practice Yoga that have some aspects of physical control today (e.g., control of body temperature, circulation, and digestion), though this is not the same as control over length of life. It would also appear from Genesis 2 where it mentions the naming of the animals that Adam had some control over animals, as some have today—so-called animal whisperers. (If I whisper to animals they ask me to speak up!) We have the picture that the human spirit was in full control of the human organism, but after disobedience to God that power was somehow lost and mankind became subject to the laws of nature. This huge change was then passed on to later generations as we too have become disobedient to God. In discussing the Genesis story we can't be sure how much is metaphorical and how much is history as it is all about God creating the universe and our world, and us.

One Scenario

One suggested scenario is as follows. At some point in their process of creation (by whatever means), humans not only obtained self-consciousness but also consciousness of God. We don't know how long they existed in the Garden in this state but eventually out of pride they wanted to be independent of God and disobeyed God. The world then changed and the human species was spoilt. Life expectancy began to reduce. Initially the early patriarchs lived a long time and there are many independent traditions from archaeology that early people lived for long periods of time (e.g., the Sumerian king list and the Egyptian, Chinese, Greek, Roman and Jewish traditions). However the lifespan came down to "three score years and ten" and very much lower in the middle ages. The world got out of control as people continued to

---

[64]cf. Genesis 11.
[65]http://www.biblestudy.org/basicart/why-did-man-live-longer-before-flood-of-noah-than-after-it.html, accessed August, 2015.
[66]There was the possibility of genetic decay.

follow in the steps of Adam, and sin against God. Physical changes were also inherited, and thorns and thistles came to plague people. Also we have become hostile to nature through environmental destruction and pollution. This is one explanation for the nature of our world at present, and no doubt readers will have different views on this, depending on their view of Genesis. The key idea is that the world is currently not as God intended, but one day it will be redeemed again and restored to God's ideal.

So something happened, and there is plenty of room for speculation as to how much is fact and how much is metaphor. It could be said that either the world changed its nature, or it got out of control, or God built in the second law in His creation for some purpose when the big bang occurred. In addition to the "fall" there is a second factor, which is now described.

8.9.3 The World is a Spiritual Battleground

The Bible indicates that free will was also given to spiritual beings called angels before the world was created. Some chose to rebel against God so that a battle is now on.[67] We are caught in the crossfire. The story of Job mentioned below illustrates this struggle where Job is described as being at the centre of a spiritual "conflict" between God and Satan. We cannot always see God's purposes, and sometimes a small evil is allowed to take place in order to prevent a much greater evil, as in the story of Joseph in Section 8.11.

How does the Bible deal with the problem of evil? In the Old Testament the main focus is on monotheism and the sovereignty of God. The title Satan is used to refer to a spiritual being responsible for evil only four times there,[68] but the word is used elsewhere in a secular way with its original meaning of "adversary."[69] The problem of evil is dealt with by saying that because God is sovereign, He is the author of everything including evil.[70] Yet the message that God abhors evil and those who carry out evil comes across right through the Old Testament.[71] In the New Testament we have a clearer picture of Satan's role, where in the gospels he is held responsible for many human ills and mental afflictions. For instance, Luke 13:16 speaks about the woman bound by Satan for eighteen years, and Luke 8:28–34 describes Jesus healing the demoniac possessed by demons. Several of Jesus' parables deal with Satan's activity among men such as the parable of the sower[72] and parable of the tares.[73] Jesus seemed to regard evil as an inevitable part of

---

[67] Ephesians 6:10–13.
[68] 1 Chronicles 21:1; Psalm 109:6; Zechariah 3:11ff.; Job chapter 1.
[69] For example, 1 Samuel 29:4 and 1 Kings 11:14, 23.
[70] As with Amos 3:6; Isaiah 45:5–7;1 Samuel 18:10; and Lamentations 3:38.
[71] For example, Isaiah 13: 11.
[72] Mark 4:15.
[73] Matthew 13:24–30.

creation and did not find it necessary to offer explanations for its presence in the world. We read that Satan entered Judas[74] and desired to have Peter as well.[75] Satan also tempted Jesus in the wilderness, after which time Jesus began his ministry. The role of Satan in the future is discussed in the book of Revelation, especially chapter 20.

8.9.4 Animal Suffering and the Bible

The question sometimes raised is how can a loving God create a world where animals kill and eat each other, namely why is "nature red in tooth and claw."[76] In fact if you look at the insect world some terrible things happen there; female spiders eating their mates! The Bible is clear that non-human life is essential for our existence, as we well know with regard to the food ecosystem, the generation of carbon dioxide, seeding and pollination, and so forth. However, as mentioned above, there needs to be a balance in nature so that one species does not over-dominate; the food chain needs to be self-controlling. According to Genesis 1:29–30, God created us and animals to be herbivorous, but through the Fall it all changed. Romans 8 tells us that everything was affected by the Fall with the consequence that everything is "groaning" until the time of salvation when God will create a new heaven and a new earth.[77] This will lead to the original order being restored when animals will not kill each other and where the lion and the lamb (which could be newly created species) will live peacefully together.[78] Although there is the argument that animals are not moral creatures and therefore do not have moral rights, the Bible tells us not to abuse animals and that a moral person should take care of the needs of one's animals.[79] What about animal sacrifice? The animals were killed in a humane way and there was no waste as the meat was eaten and the skins used for clothing, much as it is with the harvesting of animals today.

Philosopher Michael Murray examines the above traditional view that animal suffering is due to the Fall when Adam and Eve disobeyed God. He argues that such explanations seem to deny pre-human animal suffering (a consequence of evolution) and do not account for why God would create such a world that would be so badly disordered by an act (or acts) of creaturely misconduct. He goes on to say that there is something deeply wrong with our world on a cosmic scale. The Fall implies that

> natural evil is ultimately rooted in moral evil. But closer inspection reveals that something more is required here ... What possible good reason could there be for creating the universe in such a way that the

---

[74]Luke 22:3.
[75]Luke 22:31.
[76]Lord Tennyson, "In Memoriam A. H. H.", 1850.
[77]2 Peter 3:13; Revelation 21:1.
[78]Isaiah 65:17, 25.
[79]Proverbs 12:10.

Fall of the first human pair could bring about a rewiring of brute nervous systems, thereby allowing for the possibility of pain and suffering?[80]

He mentions the argument that the first humans were rational and free, enjoying fellowship with God and having all that they needed, so why should they be motivated to do anything bad? In answer, he refers to philosopher Alvin Plantinga who surmises that because God made humans in His own image, and therefore with some autonomy, there was perhaps a high probability that they would want to be like God in being the center of the universe, and this could lead to the Fall. Plantinga says: "Perhaps a substantial probability of falling into this condition is built into the very nature of free creatures."[81]

Murray's answer to the disruption following the Fall is that perhaps the origin of suffering (and animal pain) is ultimately explained not just by the Fall of Adam, but by the Fall and interference of Satan in creation.[82] Satan was an exalted angelic being who chose to rebel against God and was demoted! As the tempting serpent in the Garden of Eden, he had a hand in Adam and Eve's fall. We don't know how symbolic is the story nor what process led to the world being affected by the Fall.

Several other arguments are considered by Murray to explain the suffering of animals. The first that he mentions, considered by Swinburne, argues that

> animals have a capacity for pain and suffering which confers a type of moral significance, or perhaps nobility, on animals and on certain types of their behavior. Moreover, the goodness of this moral significance outweighs the suffering animals in fact endure, and further would be unattainable without the possibility of that pain and suffering.

The second refers to the beneficial effects of pain, mentioned above, to protect an animal's survival as it does for humans. The third assumes that as pain and suffering have a soul-making effect for humans, the same could be true for animals who, like humans, are perhaps eventually resurrected to immortality. Keith Ward[83] argues that animal immortality, like human immortality, provides a justification for the existence of this world so that any suffering or troubles in this life are minor compared with the prospect of eternal happiness. He adds:

> Immortality, for animals as well as humans, is a necessary condition of any acceptable theodicy; that necessity, together with all the other arguments for God, is one of the main reasons for believing in immortality.

Relevant to this comment is Romans 8:18–25, which refers to the eventual liberation of the whole of creation.

Some have argued that since animals were part of the original creation, they must have a role to play in the future. Murray[84] suggests that perhaps

---

[80] Murray (2008: 84).
[81] Plantinga (2000: 212–213).
[82] Isaiah 14:12–15; Ezekial 28:12–19.
[83] Ward (1982: 201–202).
[84] Murray (2008 :129).

animals were created with the dormant potential to experience pain, but it was the Fall that led to the experience of pain through the transformation of the created order. He argues that it is good that the universe is naturally regular, providing freedom of choice and a diverse creative order, and that it has gone from chaos to order with evolution providing human precursors that will experience some pain in their evolutionary development. He contends[85] that animal pain and suffering might be explained in part by each of the above arguments. It should be noted that there is nothing inherent in the Genesis account that excludes animal death before the fall. Death from the curse concerns man's (eventual) spiritual death.[86]

In the next section we again refer to the future restoration of our world, which will also extend to the animal kingdom. Isaiah 11:6–9 says that at that time the wolf and lamb, the leopard and goat, the calf and the lion, and the cow and the bear will coexist peacefully together, and children will be present. Also poisonous snakes will no longer be harmful and the lion will eat straw. Such a reference either emphasizes metaphorically the extensive nature of the peace or else suggests that some animals may be immortal. The question is which ones and how many, and whether they will be different from what they are now. C. S. Lewis[87] suggests that the selfhood of some animals is due to their contact with humans so that an animal's personality is due to the animal's master and therefore may contribute to the master's immortality by being immortal itself.

8.9.5 The World is to Be Restored

In talking about decay Paul says in Romans 8:18-22 that one day creation will be free from the bondage of decay and will share our freedom as children of God. The current pain of creation is like that of the pain of childbirth, but God will remove sorrow and make all things new.[88] The New Testament teaches that God's objective is to establish His kingdom on earth with us.[89] There will be a New Heaven and a New Earth without suffering, death, and predation in nature.

8.9.6 Suffering and Other Religions

In contrast to many world views, Christianity accepts the reality of evil and suffering while giving some reasons for it and offering God-given strength to endure it. Hinduism sees evil as an illusion so that the problem cannot really be dealt with since the problem is denied. Yet Hinduism is full of purification

---

[85] Murray (2008: 196).
[86] Genesis 3:19; Romans 5:12.
[87] Lewis (2001b: 126–127).
[88] Revelation 21:1–4.
[89] Revelation 21:3.

rites to appease God and win God's favor. It also blames suffering on a person's own actions, usually from a previous life. Evil and good are simply ignorance or enlightened awareness, respectively, so that the cruel actions of others are attributed to their misunderstanding. Both Hinduism and Buddhism are dominated by reincarnation where one pays for past sins. In Buddhism there are the four noble truths and the eightfold path, as described in Section 6.6.7. New-Age followers tend see anything negative in their lives as an illusion, and therefore nothing about their lives is wrong; morality is relative for them. A materialist will have difficulty defining evil as he or she would not have any way of measuring evil as there are no absolutes. Some would even deny the existence of evil or call it something else (e.g., going against a society's norms).

We see that other religions tend to leave a person's failures on their shoulders with not a great deal of hope of becoming a better person. But in Christianity we have a God who reaches toward us by grace, and provides a way for salvation[90] and a way for us to know Him.[91]

8.10 SOME TRADITIONAL ANSWERS

For the materialist there is no explanation for suffering, only bad luck. For a religious person there are the following suggestions that are often put forward as to why there is suffering.

8.10.1 Suffering is God's Judgement

The Jews believed that suffering was due to God's judgment on personal sin or parents' sin. If you suffered, you or your parents must have sinned. My response to this is as follows.

> (1.) I believe that God judges sin, either now or eventually.
> We see this with both individuals and nations. In Amos we read that Israel was spiritually bankrupt. The poor were oppressed and Israel's religion was corrupt. Amos a shepherd was called by God to warn them of judgement. God often used pagan nations to bring judgement on Israel because of their sin and following after other gods. In Isaiah 10:5 Assyria is described as the rod of God's anger and the staff of God's fury. However these nations themselves did not escape the judgement of God either. God had a special covenant relationship with Israel whereby they were rewarded if they did what God told them and were punished when they were disobedient, after much explicit warning. I don't believe such rules apply in the same way today, though our actions do have

---
[90] John 3:16.
[91] See Chapter 11.

consequences. Also God is described in the Bible as being very long suffering.

(2.) The sins of our parents can affect us physically and psychologically.
Today we see the same cause and effect. A father is an alcoholic and a wife beater. The last thing the son wants is to grow up like his father so his attention is focussed on the behavior of his father rather than on positive behavior. As you tend to become what you focus on, the son then turns out to follow the same path with the same repercussions in further generations still. We can therefore hurt our children by giving them our hang-ups. This idea was mentioned under the topic of epigenetics (Section 2.6.3).

In the Old Testament the effect of turning away from God is described as profoundly affecting children through several generations.[92] However the clear teaching of the Old Testament[93] is that if we are righteous then we are not to blame for our children's actions or our parents' decisions.

(3.) Jesus did not agree with the Jewish approach in several places in the New Testament.
He inferred that suffering just happens without giving specific reasons. Luke 13:1–5 refers to Jesus being informed about a brutal massacre of some Galileans in the Temple while they were engaged in offering sacrifices. This took place on the orders of Pilate, the Roman governor, who must have suspected an insurrection and was determined to nip it in the bud. The implied question to Jesus was: "What crime had these victims committed that they suffered this tragic fate?" Jesus linked this event with an accident which involved the deaths of eighteen people, presumably workmen engaged in repairs to the tower of Siloam that collapsed on top of them. In both cases Jesus dismisses the suggestion that the men who died were more sinful than those who lived. Again he refuses to accept the popular Jewish idea at the time that disasters and calamities are God's punishment for sins committed by the victims. However, he assumes that no one is completely guiltless. All are sinful, and should take tragedies as a warning. We should repent and be prepared to meet God at any time. Life is precarious and we need to make our peace with God.

Another example is given in John 9:1–3. Jesus explained to his disciples that the man they saw blind from birth was not blind because he or his parents had sinned, as believed by Judaism, but that God's power might come to be seen at work in him (when he was healed).

---

[92] Exodus 20:5.
[93] Ezekial 18; Jeremiah 31:29–30.

8.10.2 Satan Can Afflict People

We read that some supernatural beings that God created rebelled against God and have since interfered with us and our world. God does allow Satan to afflict people as we see in the story of Job. In Luke 16:13 Jesus speaks about a daughter of Abraham being bound by Satan for eighteen years. Satan can have an indirect effect by tempting people to sin. For example, in John 5:14 Jesus told the paralyzed man he had healed at the pool of Bethesda to sin no more so that nothing worse would befall him. He also told the women caught in adultery much the same thing. For the man let down through the roof of a house by four friends Jesus forgave him his sins first before healing him.

8.10.3 Reincarnation: Suffering from Sin in Past Lives

Suffering is sometimes explained by the doctrine of reincarnation where rewards and punishments are determined by the way we live in previous existences as well as how we live today. If we live a bad life we may come back as something unpleasant! It is an attractive belief for some because it reconciles the suffering in the world with the justice of God; suffering is the outworking of so-called bad Karma (actions). Reincarnation is supposed to open up the possibility to have a second (or third etc.) chance and allows for ultimate redemption. In fact we would expect society to get better given the multitude of chances to improve, but this hasn't happened as there is no evidence of a significant moral improvement, though this could be debated as it depends on how we define "improvement." Some psychologists have assisted their patients to regress into the past using for example hypnotism to recall so-called events from past lives to help overcome some presenting problems. Even if the psychologist doesn't believe that the past events are real, the key is that it can work.

There are a number of other problems with reincarnation. Its fundamental weakness lies in its failure to explain how suffering does actually benefit us. If I was a murderer in my last life and I am born deformed in this life, it may be a judgment I deserve. But since I cannot recall the circumstances under which I was prompted to commit murder, how can such a judgment teach me repentance? How can I repent of sins whose origin and nature I have forgotten? Furthermore, reincarnation, while professing to explain the inequalities of life, succeeds only in making them disappear into the mists of the past. We are forced to ask what caused the first inequality and the initial suffering, and why were we all different to start with given the earth had a beginning? Also in pantheistic systems there is no moral standard for right and wrong. Bad Karma is not a moral law, but a system of retribution with no fundamental guidelines to tell us what to do. In Buddhism, for example, we have rules of expediency that are voluntarily assumed. Sins must be punished and cannot be forgiven, which is contrary to Christian grace (God's unmerited favor) that we can receive from God. As we noted above, Jesus made it clear that a person's unfortunate experiences in life are not necessarily because of sin.[94]

What about the sufferer? Unfortunately it is not much comfort to be told that if you try hard, things will be better the next time you are born. We want to make the best of this life here and now! The god of reincarnation is like a big impersonal computer working out the exact retribution for our sins in different lives. One of the worst features of the doctrine of reincarnation is that it can lead to fatalism and paralyze the desire to improve the social environment. Why bother to help people? It's their own fault that they are where they are! Reincarnation is anti-humanitarian in that if you help someone they will have to suffer even more to work off their debt to bad Karma. According to traditional Hindu belief, helping others does not help increase their good Karma, only your own. "The social compassion that exists in India is the result of non-Hindu, largely Christian, influence. Hinduism did not produce Mother Teresa."[95]

8.10.4 Suffering is All in the Mind

One approach to suffering is to deny its existence; if we are positive enough and have enough faith we can be healed. There are several religious groups that teach this, for example, Christian Science founded by Mary Baker Eddy taught that evil, sin, and sickness were an illusion, and are therefore not real. Now there is some truth in what they say as we know that what we think can determine our health; positive thinking is important.[96] A related idea is to deny the physical and concentrate on the spiritual, the foundation of much of Eastern philosophy and religion. It teaches that the body is not important and is the source of sin. However, in the Bible the body is not downgraded, but rather we are told to glorify God in our bodies.[97]

8.11 SOME OLD TESTAMENT RESPONSES

The Old Testament has a good number of complaints and laments about suffering, and I believe it is helpful to see how some people in the Old Testament coped with suffering. I have chosen Joseph, Jeremiah, Elijah, Job, and David the psalmist; five very different men from very different backgrounds. They all suffered even though they were servants of God. In each case suffering had a long-term purpose. I have also included the writer of Ecclesiastes (the Preacher) as he provides a very different perspective.

Joseph

---

[94] Luke 13:4–5.
[95] Geisler (1999: 642).
[96] For example Proverbs 3:7–8, 14:30, 15:30, and 16:24.
[97] 1 Corinthians 3:16.

We read that the Lord was with Joseph when he was sold by his brothers to the Egyptian Potiphar, captain of Pharaoh's guard. Joseph rose up to being in charge of the household. Because of Joseph, God blessed the house. However Potiphar's wife had designs on him and when he refused her attentions she falsely accused him. He was put in the prison where king's prisoners were kept. However we read that the Lord was with Joseph and showed him steadfast love, and gave him favor with the keeper of the prison. In spite of this Joseph spent two years in prison. What an unjust situation! Joseph must have thought that God had forgotten about him. God however was working out His purposes and Joseph's time came. Because he was able to interpret the king's dream and warn of an impending famine he rose to be second in charge at the age of 30. The punch line comes in the last chapter of Genesis, chapter 50 verse 20, when his brothers find out what Joseph had become. Joseph said to his brothers that although they meant evil against him, God had other plans and meant it for good. He was able to provide for his brothers and their families during the famine as well as for thousands of Egyptians.

These brothers were the ancestors of the 12 tribes of Israel and, through Joseph, were preserved by God's providence. Unfortunately Joseph had to suffer for it initially. The lesson that Joseph learnt that still applies today is that we cannot see the whole picture so we need to trust God during difficult times. Sometimes God allows a small evil to take place to prevent a much greater evil. In addition, although Joseph initially suffered, his character was strengthened and refined by the painful process that helped him to become the just and powerful man that he was.

Jeremiah

From a young age Jeremiah was a set apart by God to be a prophet. Unfortunately he had the unenviable task of bringing a very unpopular message of doom and gloom to Israel. Consequently he suffered for it by being beaten several times, being slung in the stocks by the chief officer of the temple, being imprisoned several times, and being cast into the mire at the bottom of a water cistern. He had to continually stand up against false prophets. Right through the 52 chapters of the book of Jeremiah we read that no one heeded the words he brought from the Lord. How frustrating for him! In chapter 20 verses 11, 13, 14, and 18 we have many contrasts. His enemies won't overcome him so he can praise the Lord, yet in the very next verse (verse 14) he curses the day he was born and then complains about spending his days in shame. It was not a cry of remorse but the cry of a servant of God who was finding his mission too much for him.

Elijah

Elijah went through a major crisis at Mount Carmel, described in 1 Kings 18:16–46 and 19:1–18, where he saw the power of God demonstrated, but then succumbed to "post adrenaline depression." He had experienced a mountain top experience when God responded to his prayer and sent fire down from heaven to consume the sacrifice, the wood, the stones, the dust, and the water. Now he was scared off by a threat from the queen of Israel, Jezebel, and was out in the wilderness. He was depressed and exhausted, weak with hunger, and needed sleep. He had come down from a great spiritual high with a bump and he asked God that he might die, as he felt he was no better than his fathers. Yet God dealt kindly with Elijah. For example God did not criticize him but looked after his physical needs at Beersheba and then counseled him at Mount Horeb giving him an object lesson about His presence and providing encouraging words about other prophets. God did four things when He counseled Elijah, as follows: (a) God waited until Elijah had enough food and sleep, (b) God waited until Elijah had enough rest before asking him what he was doing there, (c) God encouraged him to talk about his problem and ventilate his feelings, and (d) God gave him a task to do.

Job

Job was apparently a real person as he is mentioned elsewhere.[98] Even if he wasn't, the story has a profound message. As we saw above, Job lost everything because he was caught in the middle of a spiritual debate between God and an Adversary. His suffering was very severe, losing his children, possessions, his wife's support, and finally his health (Job chapters 1 and 2). Three friends of his came and sat with Job for seven days and nights, not saying a word because they saw how much he was suffering. In his anguish Job cursed the day he was born (chapter 3) and wanted to die. His three friends followed the traditional Jewish view of suffering and tried to tell him that he must have sinned and should repent of his sin. However, Job complained that they were wrong (26:2–6 and chapter 31) and that his suffering was unmerited (see chapter 1). In chapter 29 he bemoaned the loss of the good old days when God watched over him. He wanted to know why God allowed all this suffering to happen to him. When he complained that God was unjust to him he was taken to task by a young friend called Elihu who said that Job was wrong to say that God was unjust as He could do no wrong (34:10–35: 15).

God finally speaks up in chapter 38 and it is clear that God has been present all along listening to the three friends. He criticizes Job for his ignorance and reminds Job of His greatness and power and of Job's limitations, and questions his accusation against almighty God. Job responded by saying he was unworthy and out of line, that he had no answers, and would say no more (40:1–5, 42:3–5). Although he had asked God for answers, God did not give His reasons for Job's suffering. God simply told him that He is sovereign and inferred that Job would not have the ability to grasp the reasons as he was not God. Job had the faith that God was still on his side, was ultimately just, and did have reasons even if Job did not know them. God blessed him abundantly for his response and completely restored him so that he ended

---

[98] Ezekial 14:14, 20; James 5:11.

up better off than before. We learn from this that although we may not see God's reason for suffering, there is no reason why there isn't a reason!

What is interesting is that God had more sympathy for Job's outbursts than for his friends' pious comments and arguments, and asked Job to pray for his friends and would accept Job's prayer. However his friends did at least sit in silence beside Job for seven days and seven nights before speaking; perhaps their best time with him! One message from Job seems to be that when we suffer it is not a question of whether God is responsible or not, but rather how will we respond to the suffering. In the end it wasn't just about Job's resignation and passivity, but rather about an absolute and unshakeable trust in the incomprehensible God. Even Jesus did not explain suffering but endured it to the bitter end. Another message stems from Job 42:5 where Job says that he had heard of God but now sees God; suffering had brought him into a greater knowledge of God, and God used the suffering to draw Job closer to Himself.

David

In Psalm 73, David the psalmist struggled with the question of why do the wicked prosper, as did Job,[99] Jeremiah,[100] and Habbakkuk.[101] The wicked seemed to be strong and healthy and did not suffer as much as others, nor had the same troubles. He got his answer when he went into God's temple. Here he realized that the wicked are on a slippery slope ending in a fall to destruction. We therefore shouldn't be jealous of the temporary success of the wicked as there is a final judgement. We are to focus on God as He is continually with us,[102] and our relationship with God is permanent and will survive even the dissolution of our bodies.[103] Suffering is transformed into hope because our relationship with God does not end at death. God will not let us go.

One lesson from the above examples is that God is not always interested in our immediate personal happiness. He is more concerned that His will is done, and we can get hurt in the line of duty. Sometimes a personal cross and suffering are involved,[104] but there is a good conclusion.

The Preacher

The Book of Ecclesiastes (the preacher) is an example of someone struggling with the futility of life. He describes everything as vanity including wisdom, riches, and pleasure, as they all pass and you cannot take anything with you when you die. He complained that everyone will be snared into a state of adversity and despair sooner or later (Ecclesiastes 9:11), and it makes no

---

[99] Job 21:7–15.
[100] Jeremiah 12:1.
[101] Habbakkuk 1:1–4.
[102] Psalm 73:23.
[103] Psalm 73:26.
[104] Luke 9:23; 2 Timothy 3:12.

difference whether you are wise or foolish, and righteous or wicked (9:2ff.). He says there is no such person as a righteous one who does good and never sins (7:20). The Preacher sees the burden of toil that God has laid on the human race (3:10), though the burden is not necessarily evil, and he concludes that pleasures are fleeting, evil and oppression are prevalent, and suffering is inevitable, irrespective of one's moral worth. Sound familiar? Readers may be surprised to find such a book as this in the Bible, yet for many it describes life as they see it. However, there are some positive things that I want to take from this book.

First, God through His providence, has made everything beautiful in its own time (3:11), and has put eternity (hidden time) into people's hearts, but with a limitation. We cannot find out what work God has done from the beginning to the end. This means that we cannot fathom God's actions, but we need to accept that everything is done well, even though we might not see it right now. We see only the middle of God's work, and therefore need to suspend judgement. Second, we should be content with and make use of what we have, enjoying the pleasures that God has given us (3:12–14). Third, God will judge both the righteous and the wicked, according to their deeds, so there will ultimately be justice (3:17, 12:14). Although sinful people may do evil and prolong their lives, it will not go well with them, while it will be well for those who fear God (8:12-13). Finally we are told it is our duty to fear God and keep His commandments (12:14).

8.12 SOME POSITIVE ASPECTS OF SUFFERING

In his Section we shall consider some of the positive things that can come out of suffering. The reader may not agree that they are all positive!

(1.) *Suffering can wake us up.* The wake-up call could be physical as we realize that we need to change our life style if our health is involved. It can also be a spiritual wake-up to remind us of our mortality. C. S. Lewis said: "God whispers to us in our pleasures, speaks in our conscience, but shouts in our pains: it is His megaphone to rouse a deaf world."[105] God reminds us through the megaphone of pain that we need Him, and that a real sense of satisfaction and hope can only come from Him and not from things or people. Although it can be harder to think about God when all is going well, suffering can lead to a greater intimacy with God.[106] God does not do this as a narcissistic parent that needs attention, but rather because we are created to be in relationship with Him. We can only be truly happy when we are in that relationship; we need God rather than God needs us.

---

[105] Lewis (2001b: 83).
[106] Psalm 121.

At the age of 82, two years after he retired, my father had a serious accident and lost a leg. He was unconscious and on a life support system for four weeks. It was an uphill struggle for him. Yet he told me that the experience was a blessing for him because while he was in hospital he had a chance to think. His Christian faith was rejuvenated by the experience and he made the most of his later years dying at 97 years old. I guess we can all look back and see some positives from our suffering, but there will be some events that seem purposeless as we are not God!

(2.) *Suffering can be a refining process to make us fit people for God's kingdom.* One philosopher John Hick[107] calls it "soul making." Philosophers James Moreland and Bill Craig (2003) point out that "Many evils occur in life that may be utterly pointless with respect to the goal of human happiness; but they may not be pointless with respect to producing a deeper knowledge of God." The Bible mentions four aspects of this idea.

(a) *Suffering can be a disciplinary process.* A good parent disciplines the child that he or she loves. For example, a child has to learn that you don't play with electrical plugs so that perhaps a slap on the hand or other form of punishment is needed. The writer to the Hebrews[108] says that we should pay attention when God corrects us and not be discouraged. Like a parent, God corrects everyone He loves, and disciplines everyone He accepts as a son or daughter. The writer continues with the comment that if we accepted the discipline of our parents we should also accept the discipline of God.[109] It is true that sickness can be a sign of God's reproof, and repentance can lead to healing.[110] However it is important to realize that not all suffering comes from God to teach us a lesson, otherwise we might as well resign ourselves as Muslims do to fatalism. If it is the will of Allah, why fight it or help others to fight it?

(b) *Suffering can develop character.* It makes us strive. It can teach us perseverance and self-sacrifice. We can allow children to stumble and fall when they are learning to walk and the same thing can happen to us with God. In Malachi 3:2 God is described as a strong soap or a refiners fire that burns up the dross and purifies the silver and the gold. The New Testament speaks about trials teaching us endurance, steadfastness, and character, leading to hope.[111] Suffering and evil can contain the potential for good, but we may not avail ourselves of it. Our suffering is wasted if we don't learn from

---

[107] Hick (1963).
[108] Hebrews 12:5–6, 10.
[109] See also Job 5:17.
[110] 2 Kings 20:4–6; 2 Chronicles 26:19–20; 1 Corinthians 11:30; James 5:15–16.
[111] James 1:2–4; Romans 5:3–5.

it! God is more interested in character-building than our happiness. Jesus learned obedience through what he suffered[112] and set us an example.[113]

(c) *Suffering removes any sense of self-reliance and pride.* With suffering we are forced to rely on God's presence and grace as Paul learnt,[114] and on others who are healthy. It helps us to distinguish between needs and wants, necessities and luxuries, to rest our security on people rather than on things that can be taken away, and to appreciate any little pleasures in life. (When I was writing this I was recovering steadily from a knee replacement that didn't go according to plan because of bleeding that led to an extra spell in hospital and severe extended pain.) Paul knew what it was to battle with an undisclosed ailment that he called a "thorn in the flesh." It interfered with his busy ministry, but he came to realize that it was there for a purpose that he might learn that God's grace was sufficient for him and that through his weakness God could reveal His power. God delivered Paul from many trials, which are listed in 2 Corinthians 11:16–28, but allowed this one.

Philosopher Peter Kreeft[115] comments that it is often the people who are comfortable, outside observers that use suffering to argue against the existence of God, while sufferers themselves are often made into stronger believers. Theologian Hans Küng[116] makes the profound comment:

> It is only if there is a God that we can look at all this immense suffering of the world. It is only in trusting faith in the incomprehensible, always greater God that man can stride in justifiable hope through that broad, deep rivera; conscious of the fact that a hand is stretched out to him across the dark gulf of suffering and evil.

(d) *Suffering provides one with an eternal perspective.* A Christian's real citizenship is in heaven.[117] Paul says that our present sufferings won't compare with our future glory.[118] Our resurrection bodies won't experience pain and suffering (and probably won't be carbon based!).

(3.) *Suffering can be for God's glory.* Through healing (sometimes miraculous), glory can be brought to God so that others may be turned to

---

[112] Hebrews 5:8.
[113] 1 Peter 2:21–24.
[114] 2 Corinthians 12:7–10.
[115] Strobel (2000: 67), Kreeft (1986).
[116] Küng (1985: 431).
[117] Philippians 3:20.
[118] Romans 8:18; 1 Corinthians 2:9; 2 Corinthians 4:17–18.

God. I have read many stories of this happening overseas and even locally, e.g., sight or hearing restored, or simply just recovering.

(3.) *Suffering develops an empathy within us for others who suffer.* The story of the Good Samaritan in Luke chapter 10 exemplifies this. It is because suffering often seems so indiscriminate that it evokes such a deep response. We do not respond the same way when suffering is merited, but unmerited suffering brings out the best in other people. Also, God is concerned about creating a community of believers who love and care for one another.[119] This is character building in a Christian setting. However we can put this idea in a more general setting. The following quote from a New Zealand magazine called the Grapevine in 1983 throws some light on the apparent lack of action from God when things go wrong.

> Maybe God shows his opposition to cancer by not eliminating it or making it happen only to bad people, but by moving friends and neighbors to share the loneliness, ease the burden.... Maybe God is doing something when those who suffer inspire others with their amazing hope and courage. The fire which ravages a township is hardly an "act of God." But the effort people make to save lives even at the risk of their own, the determination to rebuild, the generosity shown - maybe these qualify as acts of God.... When people with prolonged illness demonstrate a double-dose of strength and good-humor.... When widows find the courage, after the funeral, to pick up the pieces of their lives and face the world alone (and we can add broken marriages as well).... When the bed-ridden express enormous gratitude for the tiniest act of kindness, for a day of sunshine, for an hour free of pain.... When the parents of a retarded child are enabled to wake up each morning and face their responsibilities bravely.... Maybe these too are "miracles"...evidence that we are not alone when we suffer. It is because suffering is so indiscriminate that it evokes such a response. We don't respond the same way when suffering is merited. Unmerited suffering brings out the best.

It seems then that God does not waste suffering. We shouldn't waste suffering either. I believe as a counselor that there is something positive we can learn from every bit of suffering that comes our way. Instead of asking "Why has this happened to me?", we can ask "What have I learnt?" Instead of asking "What is the meaning of all this?" we can ask "What meaning can I give to this?" Viktor Frankl, a psychiatrist who was a prisoner of war, made the following insightful comment: "Man is ready and willing to shoulder any suffering as soon and as long as he can see a meaning in it."[120] Sometimes however it is very hard to find the meaning!

---

[119] 2 Corinthians 1:3–7.
[120] Frankl (1992).

Author Philip Yancey[121] makes an important point when he talks about the so-called "health and wealth" theology where health and success are regarded as God's rewards for us for living the right kind of life. Of course God does miraculously heal people and He does bring financial wealth to some. However the second part of Hebrews chapter 11 describes people of faith who have suffered terribly; there is no mention of health and wealth there! Paul, who suffered a great deal in God's service, said that present afflictions are nothing compared with the eternal future glory with God.[122]

8.13 AN APPROACH TO SUFFERING

In this Section I want to give few ideas about facing suffering. Although perhaps not so relevant in a book of apologetics it is important to present a Christian approach to suffering. The most important idea is the unjustified suffering and death of Jesus.

8.13.1 The Death of Christ

The first thing for a person to realize is that God understands when we suffer. The Bible teaches that Jesus Christ the Son and God the Father were united in some way before the foundation of the world. However the Son emptied and humbled himself and became a human for the purpose of dying on a cross to identify with our suffering.[123] In fact his whole life and not just his death was a sacrifice. Jesus was the human face of God, the mediator between God and man, and came to show us what God is like from a human perspective. When Jesus became flesh he was subject to the forces of nature, for example to hunger, thirst and suffering, and fulfilled the role of the suffering servant described in Isaiah 53 when he was crucified. He was driven to the greatest possible distance from the Father when he took on the sin of mankind in his crucifixion, as holiness and sin are infinitely distant from one another. Jesus entered the very depths when he cried out "My God, my God, why have you forsaken me", quoting Psalm 22. Yet God's love extended across the barrier, and the death and suffering of Jesus were transformed into victory over sin and death. In the very depths of Jesus' despair God was showing His greatest love.

In a mysterious way God experienced suffering through Jesus, and the death of Jesus is the greatest example of unjustified suffering; the suffering of a totally innocent man. It is not a question of Jesus dying to appease an angry God but rather Jesus representing us sinful people in some way and enlarging the concept of the Old Testament sacrifices for sin. He transformed suffering

---

[121] Yancey (1990: 98–99).
[122] 2 Corinthians 4:16–18.
[123] Philippians 2:5–11.

into something wonderful for the benefit of all, as his death became part of our redemption in reconnecting us with God, and his resurrection points to our future resurrection.

In the same way, our suffering can be transformed and used for the benefit of all whom we come in contact with. God does not necessarily remove suffering, but gets alongside us and comforts us so that we can comfort others.[124] God is not oblivious to the suffering in His world and is even aware of the sparrow falling to the ground.[125] He also suffers from the sins of His people as described in the Old Testament book of Hosea.[126] The account of Paul's conversion on the road to Damascus included the words of Jesus, "Why are you persecuting me?" indicating that Jesus still suffered when his people suffered through Paul's persecution of them. Paul mentioned several times that we may share in Christ's sufferings.[127] As he says in 1 Corinthians 1:22–23, the crucifixion was a stumbling block to the Jews and folly to the Gentiles.

I now wish to look at a number of practical aspects of suffering.

Fear

As Yancey[128] points out, fear is the greatest enemy to recovery from illness and suffering. In fact fear can augment any pain as the accompanying anxiety can stop a person from relaxing and managing his or her suffering. Several years ago when I was dying from a staph. aureus (MRSA) infection that had begun in my left lung leading to pneumonia I had to face serious life threatening open-heart surgery. At the time the odds were not very good, but surprisingly I was not afraid. Apart from extremely high doses of potent antibiotics, I was not on any major painkillers as fever was my main problem. However I knew that many people were praying for me and my wife was confident of my recovery. I could trust God for whatever the outcome. Although it was a very difficult time for me and difficult to pray, I knew could rely on others to pray for me, resting in the fact that God's perfect love casts out fear.[129] As a trained counsellor I know how important it is just to be with people who are suffering especially those going through a period of grief through the loss of a loved done.[130] As mentioned above, Job's comforters spent some time just being silent with Job. We may not always know what to say, but our presence speaks a lot and helps to calm any fear. From my own experience I know that talking can be a great distractor from pain. Suffering is a lonely experience and only God knows how we are really feeling. We can hand over our fear to Him knowing that we are in God's hands.

---

[124] 2 Corinthians 1:3–5.
[125] Matthew 10:29
[126] Hosea 11:1–4, 8.
[127] 1 Colossians 1:24; 2 Corinthians 1:5; Philippians 3:10; Romans 8:16–17.
[128] Yancey (1990: 175).
[129] 1 John 3:18.
[130] See my chapter on grief in Seber (2013).

Faith

Paul was convinced that absolutely nothing could separate him from the love of God.[131] We need to have the faith that we can trust God with our future even if it means going to be with Him. A big problem is our ignorance and our spiritual immaturity. Consider a mother who permits her two year old son to undergo painful heart surgery to save his life. The child is convinced of his mother's love, not because he understands what is is going on, but because of his mother's intimate care and presence through his painful experience. The story of Job indicates that he had this kind of faith in God that the child had in his mother.

Helplessness

When we become very ill we end up relying on other people, including hospital staff, which can engender a feeling of helplessness. I believe it is very important for people to try and do as much as possible for themselves even though kind people will offer to do things for them. Even carrying out the smallest task helps a person to regain some power and control over his or her life. However in our helplessness we can turn to God for comfort.

Thankfulness

When we are very ill we may not feel particularly thankful! Sufferers can succumb to despair and depression. However, by focusing on even the smallest blessings, and thanking God for them, can lift a person's spirit. Paul in Philippians chapter 4 spoke about rejoicing in the Lord even though he was in prison. He said not to be anxious about anything, but we should pray to God with thanksgiving and we would have the peace of God that would garrison our hearts. We are told to give thanks *in* all circumstances, but that does not necessarily mean *for* all circumstances (e.g., those caused through sinful acts by others).

Hope

Yancey[132] notes that when it comes to suffering people sometimes confuse hope, a belief that something good lies ahead, with optimism or wishful thinking as the latter can imply a denial of reality. He said,

> True hope is honest. It allows a person to believe that even when she falls down and the worst has happened, still she has not reached the end of the road. She can stand up and continue. Realistic hope permits a dying person to confront reality, but at the same time gives strength to go on living.

He believes that realistic "success stories" about people who succeed in spite of extreme difficulties help to engender hope. We know scientifically that hope, faith, and a purpose in life are medicinally good for us. Christians know that no matter the outcome of their lives, they have the final hope that one day

---

[131] Romans 8:38–39.
[132] Yancey (1990: 214).

they will be with God and then be totally whole. Philippians 3:20–21 speaks about the Christian's citizenship being in heaven where their bodies will be transformed.

8.13.2 Conclusion

There is much more that can be said about the role of the Christian faith in suffering, but this is not my aim in writing this book. However, for a good discussion of this topic the reader is referred to Part 5 of Yancey,[133] which speaks about how faith helps and how the Holy Spirit is our comforter and counsellor who groans on our behalf.[134] If underserved suffering comes the way of Christians they can either be bitter or they can recognize that they have a particular role to play in dealing with the suffering. This leads to an acceptance and asking God to use the suffering to transform their lives so that others may recognize the source of their strength and be challenged by it.

---

[133] Yancey (1990).
[134] John 16:7; Romans 8:26–27.

CHAPTER 9

IS CHRISTIANITY A BLESSING OR A CURSE TO SOCIETY?

9.1 SHAMEFUL EVENTS

There are a number of atheists such as Dawkins, Harris, and others today who maintain that religion is harmful to society. Certainly the events of 9/11 have highlighted some of the negative aspects of religion, and religion has been regarded as being responsible for a lot of violence and bloodshed, which is true. Sometimes political divisions are reinforced by religious divisions as in the case of Northern Ireland. Critics of the Christian church often point an accusing finger at events such as the Crusades, the Inquisition, and the Salem witch trials that were all carried out supposedly in the name of Jesus as examples of Christianity violently imposing its will on other people. The Crusades were a response to Muslim aggression, and violence was directed against Jews as well as Christians. There were of course other factors in addition to the church's input such as politics. In the witch trials there was hysteria and belief in astral appearances when a person could be in two places at once, making it hard to prove one's innocence!

Unfortunately it is true that these and similar atrocities are an indictment on how people can depart so far from the teachings of Jesus, and are a terrible blot on church history. The perpetrators of such events, however, were not true followers of Jesus, and Jesus had some harsh words to say about those who did various things in his name that were not in God's will. Unfortunately today we read about counterfeit believers who fall from grace very publicly and, using the analogy of fruit trees, Jesus said we will know them by their "fruits."[1] However God can forgive such a person who comes back to God in repentance, as God did with King David in the Old Testament.

Dinesh D'Souza[2] points out that much of the non-Christian slant on the above events is not true and gives evidence that numbers were exaggerated. In fact such events are not typical of the Christian church, which has done a great deal for society. Christianity originally spread very rapidly in spite of persecution because of Christians and their kindness, their care for the poor and needy, their charity, their love for one another, their sharing of possessions, and their spiritual empowerment by God's holy spirit; people were won over. The Catholic church has come in for a lot of criticism recently over sexual abuse, but people are often unaware of the selfless and sacrificial work carried out by many nuns and priests in helping the poor and needy in many parts of the world. They, along with other Christians, have gone to people in parts of the world that the world has neglected, like those living on a rubbish dump, in slums, or in a leper colony.

Christians have also established a wide range of aid organizations like the Red Cross, hospitals, homeless shelters, orphanages, rehabilitation groups, and many others. Atheists who argue that many of the current conflicts are religious wars are in fact not supported by the media. The conflicts have more to do with ethnic rivalry, politics, or autonomy and self-determination. What atheists do not fully acknowledge are the huge mass murders carried out by atheist regimes such as for example communism, nazism, and Maoism or, as D'Souza shows, they come up with alternative explanations that don't stand up to scrutiny. He concludes that "Atheism, not religion, is responsible for the worst mass murders of history."[3] Some have compared secular tolerance and religious intolerance. This is clearly not a true comparison as we had severe secular intolerance in the case of the French and Russian Revolutions and China's cultural revolution when repression was carried out on a huge scale. In the twentieth century there were huge efforts carried out in Soviet Russia, communist China, and in Cambodia by the Khmer Rouge to control religion, which led to more oppression, not less. Alister McGrath in his history of atheism stated:

> "The 20th century gave rise to one of the greatest and most distressing paradoxes of human history: that the greatest intolerance and violence of that century were practiced by those who believed that religion caused intolerance and violence."[4]

[1] Matthew 7:18–23.
[2] D'Souza (2007: chapter 18).
[3] D'Souza (2007: chapter 19).
[4] McGrath (2004: 230).

Writer Nick Spencer refers to various kinds of atheists, and in his introduction he says

> ... attempts to build atheist societies ... in Russia, China, Albania, North Korea and elsewhere ended humiliating, enslaving and killing on a scale that made previous religious wars look like playground scuffles.[5]

9.2 MISSIONARIES

Missionaries have had some bad press at times. They have been accused of being insensitive to local cultures and traditional spiritual practices, and although it was true in some cases it is not generally the case now. As missionaries were sometimes linked to the pecuniary interests of colonial powers that led to exploitation, they got blamed for some of the inappropriate things that happened. However, missionaries did and are still doing an enormous amount of good for people who are struggling in terms of such things as education, health, and the promotion of local industries, not to mention spiritual benefits that critics sometimes ignore. Often Christian organizations are the first on the scene when there are natural disasters. In New Zealand, the Maoris bought guns off the traders and could have killed each other off, but they learnt peace from the missionaries. Sociologist Bob Woodberry[6] makes a strong case that Protestant missionaries "heavily influenced the rise and spread of stable democracy around the world." He also argues and provides statistical evidence that they were

> a crucial catalyst initiating the development and spread of religious liberty, mass education, mass printing, newspapers, voluntary organizations, and colonial reforms, thereby creating the conditions that made stable democracy more likely.

In Africa for example where Christian missions struggled, they led to immense enlightenment and liberation from the fear of demons, and from irrationality, cruelty, and impersonal, unsocial, and unhistorical attitudes.[7]

9.3 HEALTH BENEFITS OF CHRISTIANITY

In the past some people (e.g., Freud, Dawkins) have claimed that Christianity is not good for a person's psychological well-being. Although there will be some Christians who have unhealthy attitudes, research supports the idea that committed Christians tend to be healthier, live longer, suffer fewer psychological problems like depression, and are more community-minded than non-believers or nominal believers. Also, as a general rule, they are caring members of society, and will give a lot more time and money to help the needy than their neighbors. There are a large number of studies supporting these

---

[5]Spencer (2014).
[6]Woodberry (2012).
[7]Küng (1984: 101).

findings. For example, educationalist Brian Udermann[8] surveyed MEDLINE from 1976 to 1999 and concluded: "A strong body of evidence appears to reveal a positive relationship between spiritual commitment and health status", and "Study after study has suggested that a relationship does exist between the strength of one's spirituality and one's overall health." The same was found to be true in recovery from various illnesses.

An extremely comprehensive survey of over 1200 studies and over 40 reviews of studies of religion and health is given by Harold Koenig, Dana King, and Verna Carson (2012). It covers an extensive range of physical and mental health issues and endeavors to give psychological, social, behavioral, and physiological mechanisms as to how religion might effect physical and psychological health. Again we find that a majority of studies show that religious beliefs and practices are associated with greater hope, optimism, self-esteem, and self-reported happiness as well as less loneliness, depression, anxiety, drug abuse, and so forth. I know as a statistician that a strong positive correlation does not necessarily imply causality, but a persistent positive correlation over a variety of studies provides very strong evidence for causality. The smoking lobby used this causality argument for a long time with regard to smoking and health until the evidence became overwhelming.

Some have argued that religious beliefs are the result of natural selection because of the health benefits and not because the beliefs are true. However, how the beliefs have arisen tells us nothing about the rationality and validity of the beliefs, and the same arguments can be leveled at atheism. According to evolution, our cognitive mechanisms were selected for so-called enhanced fitness, and there is no reason for them to produce true beliefs. For the Christian, a supernatural explanation of the beliefs is conversion, whereby God gives the power to completely transform a person. The word "conversion" in the Greek means to "turn around." Other metaphors used in the Bible are being "born again" and being "raised from the dead," all indicating a transformation from within.

9.4 SCIENCE FLOURISHED UNDER A CHRISTIAN UMBRELLA

For the first fourteen centuries AD there was generally harmony between science and theology. This was mainly because the church had a monopoly of learning in education, not only in the classics but also in mathematics and astronomy. Aristotle had produced a comprehensive scientific theory of the world and Thomas Aquinas, the famous theologian and philosopher, had reconciled this with dogmatic theology. As far as the church was concerned, the ultimate had been reached in understanding the world. Any questions of fact were not settled by experiment but by authority, the authority being the

---

[8] Udermann (2000).

church. It has been argued by some skeptics that Christianity held up scientific progress. This is not true. Even at the time of Galileo when the telescope he invented in 1609 showed Aristotle's theories of the solar system to be false, the majority of scientists and philosophers in Europe still stood solidly with the church. They believed that science had to be ultimately compatible with a Christian view of the world. Also, if God created the world, then studying the world was a means of learning about God.

Galileo, who was a devout catholic and had support from the church for a number of years, not only raised questions with his telescope, but also aroused dissension because he dogmatically went beyond the available evidence. It is interesting that historian of science John Heilbron[9] studied Roman Catholic cathedrals as solar observatories and concluded that the Church gave more support than any other institution for the study of astronomy for over six centuries. Catholic missionaries, and especially the Jesuits, set up global scientific networks and took European science to Asia and the Americas, bringing back new knowledge of the New World (e.g., places, plants, and peoples) back to Europe.[10] Historian John Stenhouse comments:

> During the eighteenth century Protestant missionaries took the lead from the Catholics. They built a global network of schools, hospitals, universities, lecture halls, museums, tracts, books, and so on to disseminate the latest Western knowledge. Millions of Africans, Asians, Muslims, and Pacific Islanders first encountered Western science in a missionary school, dispensary, hospital, or book.[11]

If you were asked the question "How many teeth has a horse?" you would not just sit down and meditate for many hours until you came up with the answer. You would go and get the answer straight from the horse's mouth! Such is the spirit of enquiry today; we no longer just sit down like the ancient Greeks and reason out the way the world ticks—we go and experiment and take measurements, though we need to know what measurements to take. Theories are usually invented in people's minds and then they devise experiments to test the theory. Sometimes experiments throw up puzzling phenomena that lead to a search for a new theory. Science owes a considerable debt to Christian scientists in the seventeenth century who believed that nature was not an aspect of God but was rather God's intelligible creation that they could explore and manage. Because God was rational they believed that the world was a rational and ordered place and an appropriate focus of scientific enquiry.

In England there was certainly no conflict between Science and Christianity, and there was no persecution of the new experimenters. In particular, the Royal Society was founded "to the glory of God" in 1662 when a Royal charter was signed. Many of the dedicated pioneers of the new scientific movement were themselves devoted students of the Bible. For example Robert Boyle

---

[9] Heilbron (1999).
[10] O'Malley et al. (1999).
[11] Stenhouse (2004).

(1627–1691), one of the first founders and first fellows of the Royal Society as well as being the "father of modern chemistry", was a deeply religious man, spending much time over the translation of the Bible into various languages and writing many theological works. John Ray (1627–1705), who laid the foundations of botany and zoology, was a man of deep religious conviction.

One of the greatest figures of the time was Isaac Newton (1642–1727) who, in his Principia Mathematica (1687), invented calculus (or at least a more modern form of it), developed his laws of mechanics, and in particular derived his famous inverse square law of gravitation. Everything in the universe, including the heavens, moved according to fixed universal laws. For Newton this did not conflict with his idea of God; God had not been ousted out of His universe but rather it was an argument for His existence as He was responsible for such a marvelous system. John Ray saw this unity in the structures he observed in plants and animals, and expressed this emphatically in his famous book "The Wisdom of God Manifested in the Works of Creation" (1691). Before 1860 most scientists were Christians and included famous names like Michael Faraday, Charles Babbage, James Simpson, Gregor Mendel, Louis Pasteur, William Kelvin, Joseph Lister, and James Clerk Maxwell.

There was a conflict at an early stage stage in history, though not basically between Science and Christianity but rather between experimental science and the ideas of Aristotle and Greek pseudo-science. D'Souza[12] describes the debate as between two kinds of religious dogma about discovering God's hand in creation: deductive reasoning where you arrive at a conclusion through logic alone, and the use of experiments where you go out and actually measure things. He points out that " historians are virtually unanimous in holding that the whole science versus religion story is a nineteenth century fabrication." He mentions two well-known fabricators, John William Draper who wrote the *History of the conflict between religion and science* in 1874, which is "full of whoppers and lies", and Andrew Dickson White, who wrote a two-volume study *History of the warfare of science and theology in Christendom* in 1896 that is equally misleading.[13]

Sociologist Rodney Stark[14] in his second chapter speaks about the religious origins of science and argues that there was no inherent conflict between religion and science; instead Christian theology was essential for the rise of science. He insisted that Christian monotheism was a powerful and positive force that mobilized millions of Europeans to achieve admirable goals. He deplored the efforts of historians

> to dismiss the role of religion in producing 'good' things such as the rise of science or the end of slavery, and the corresponding efforts to blame religion for practically everything 'bad'.[15]

---

[12] D'Souza (2007: 95–96, 102–103).
[13] For a critique see Lindberg and Numbers (1986: 1–18).
[14] Stark (2003).
[15] Stark (2003: 12).

D'Souza also mentions that there are some myths about Galileo that atheists like to perpetuate to support the idea that scientists were persecuted by the church.

Unfortunately the seeds of conflict were sown and for the next three centuries after the founding of the Royal Society we find the pressure on the church slowly increasing and the gap between the church and the new science becoming wider. Both parties were to blame for this. The theologians on the one hand clung to views that were difficult to defend, and they were slow to recognize the right of science to offer its own explanations. The scientists on the other hand were slow to realize that they were not concerned with ultimate problems and that there were certain questions outside the realm of scientific investigation. In fact scientific knowledge is not the only kind of knowledge.

In contrast to Dawkins, some atheists believe that science cannot explain everything so that scientific belief can be compatible with religious belief. If you were to ask me why is the water in the jug boiling, I could explain that it has to do with the laws of chemistry about molecules and heat. However, another explanation is that the water is boiling so I can make a cup of coffee. Both explanations are valid. One is a scientific explanation and the second is a personal explanation. The philosopher Keith Ward[16] claims that a personal explanation is a perfectly satisfactory form of explanation that does not seem to be reducible to a scientific explanation. Philosopher Richard Swinburne[17] discusses these two explanations in detail. In the end, science, and physics in particular, cannot answer such questions about purpose, intentions, and awareness, or even about the nature of reality, given today's quantum theory.

Some philosophers endeavored to extend Newton's laws even further; from the physical universe to man himself. Reality, after all, was just compact billiard balls, whether little ones like atoms, or big ones like planets, all obeying Newton's laws of mechanics. Surely it would be reasonable to carry this over to atoms in living organisms, and even in our brains. Of course this extension of mechanics to such atoms at the time had no scientific backing, but it did lead to a very materialistic philosophy in which free will was considered an illusion and God was regarded as unnecessary. This particular view, usually referred to as reductionism, was strongly represented by a number of French philosophers in the 18h century such as Laplace, Voltaire, and Rousseau. Such views, discussed in Chapter 3, are held by many psychologists today.

It is interesting to note that many of American's earliest colleges and universities began as Christian institutions. For example, Harvard university, the oldest institution of higher learning in the U.S., was founded in 1636 and named after its first benefactor, the Reverend John Harvard. Yale university,

---

[16]Ward (2008a: 23).
[17]Swinburne (2004: chapter 2).

founded in 1701, traces its roots to 17th-century clergymen who sought to establish a college to train clergy and political leaders for the colony. About 90% of American colleges were church-founded later at the time of the civil war in the 1860s. Also it was under the auspices of the church that the first medical research institutions and the first observatories were established and supported. D'Souza notes that: "From the Middle Ages to the Enlightenment, a period of several centuries, the church did more for Western science than any other institution."[18]

9.5 CHRISTIANITY AND SOCIETY

In looking at the ancient societies of Greece and Rome, we see that Christianity spear-headed a number of revolutionary ideas. In the form of a Jewish morning prayer there was a sentence in which a Jewish man gave thanks that God had not made him "a gentile, a slave, or a woman." Paul turned this around and said that there is is no difference between Jew or Greek, male or female, bond or free, as all are one in Christ.[19]

9.5.1 Jews and Gentiles

The Jews were the chosen people of God because of God's covenants through Abraham, Moses, and David, and consequently the gentiles or non-Jews were looked down upon. The Talmud that contains the written traditions of the Pharisees says some very negative things about the gentiles. However Paul, himself a Jew of high standing, explained that both Jew and Gentile had access to God through Jesus Christ. In this sense there is no difference between the Jew and the non-Jew in the Christian church because of the establishment of a new covenant. In the words of Ephesians 2:14, the wall of hostility between the two groups has been broken down as they are reconciled through the cross of Jesus.

9.5.2 Husband and wife

As regards to the Jewish law, the wife was not a person but a thing so that she had no legal rights and was absolutely in her husband's possession. At the time when Paul wrote his letter the husband could divorce his wife for just about anything. All he had to do was to have a bill of divorcement correctly written out by a Rabbi and to hand it to his wife in the presence of two witnesses. The only other condition was that the woman's dowry had to be returned. In contrast, the wife had no rights of divorce except in one of two very special cases, for example her husband was a leper or an apostate. We

---

[18] D'Souza (2007: 95).
[19] Galatians 3:23.

find this ready divorce in Islam's Sharia law with regard to women, where women are not treated the same as men and their rights do not have the same legal status as men. Sharia treats Muslims and non-Muslims differently in court.

The Greek situation with regard to women was worse. Prostitution was a major part of Greek life. The Greek expected his wife to run his home and care for his legitimate children, but found his pleasure and companionship elsewhere. To make matters worse, there was no legal procedure of divorce—it was simply a matter of whim. The one security that the wife had was that her dowry must be returned. Consequently family life was almost extinct and fidelity was non-existent.

In the Roman world the situation had become even worse than with the Greeks. The mighty Romans had conquered the known world and originally divorce was unheard of in the Roman republic. Unfortunately there was a rapid decline in marital relationships so that by the time of Paul writing, family life was wrecked and the marriage bond on the way out.

The Christian approach to marriage that Paul describes in Ephesians 5:21–33 is very different. He had to communicate his message to a male-dominated society and work within the framework already established. Rather than advocating getting rid of marriage to create more freedom or getting a better deal for women, Paul's revolutionary idea was to transform marriage from both sides and to set a new standard where love rather than legality was the bond. As the passage in Ephesians is sometimes misunderstood and misused, a word of explanation is appropriate at this point. Verses 22 and 23 that say that wives should be subject to their husbands and the husband is head of the wife are sometimes used by husbands to justify their controlling behavior. However, one has to consider the context, namely verse 21. This verse talks about being subject to one another out of reverence for Christ, which according to the original Greek means *mutual* submission. As already noted above, society was such that women at the time were already seen as being in a supposedly submissive role so that verse 22 describes the manner of submission, namely "as to the Lord." In verse 23 Paul used the daring analogy of the husband-wife relationship being like the relationship between the church and Jesus Christ, where the church is the bride and Jesus the bridegroom. It tells men how to love their wives and it explains what it means for a husband to be first among equals; he must be prepared to die for his wife. Authority brings responsibility. We then have the injunction for a husband to love his wife as his own body (verses 28–29).

The equality in marriage is endorsed by Paul when he says that each partner has authority over the other partner's body;[20] and they are mutually dependent.[21] Other texts that speak about fellow-heirs can be brought into

---

[20] 1 Corinthians 7:4.
[21] 1 Corinthians 11:11–12

the discussion.[22] This equality extended to adultery, which was prohibited for Christians, and cut across the double standard in a society where men did as they pleased.

Today we see other problems relating to women in societies. For example, the practice of female genital mutilation of women can cause serious health problems.[23] Wearing the burqa and in fact any total body covering for women can lead to vitamin D deficiency and resulting physical problems, especially as osteoporosis can be a problem for women generally as they age. It is important that they receive sunlight on their skin, even if done in private. Not allowing women to be educated and condoning the physical abuse against a recalcitrant wife in certain circumstances is a violation of human rights. We still see gender inequalities in the workplace, and female abuse is very prevalent in some cultures including Western ones.

9.5.3 Parents and children

If Christianity did much for raising the status of women it did even more for children. In any civilization children are generally loved, but in pre-Christian and heathen civilizations there could exist cruelty that would be totally unacceptable in a country like New Zealand, for example, where Christian principles form the fabric of society. In contrast we see in the Roman civilization certain features that made life perilous for children. The first was the father's power or "patria potestas", where the child became the property of the father so that he had absolute power over his children for their entire lives until he died. As theologian William Barclay[24] notes, the father could sell his children as slaves or even inflict the death penalty on them. Although the father's power was seldom carried to its limits because of public opinion, there were some instances when a father executed his son.

A second feature that made life perilous for a child was the custom of child exposure. A new-born child was placed at the father's feet. If he picked up the child the child was accepted, but if he walked away the child was thrown out. Furthermore, if a child was sickly or deformed it had little chance of being allowed to live. It is into this situation that Paul wrote his advice to Christian parents and children. He didn't advocate taking away the father's power but asks for a transformation of the relationship. First of all he reiterates one of the ten commandments that children should honor their parents by obedience.[25] He then turns to the fathers and tells them not to provoke their children to anger, and in Colossians 3:21 he adds the words

---

[22] 1 Peter 3:1–7.
[23] http://www.answering-islam.org/authors/roark/women_suffering.html, accessed August, 2015.
[24] Barclay (1958: 208).
[25] Ephesians 6:1–3.

"lest they should be discouraged." The children are to be brought up in the discipline and instruction of the Lord.

9.5.4 Masters and slaves

There are five passages that refer to the relationship between a master and slave.[26] When Paul wrote to Christian slaves he must have been writing to a very large number in the Christian church as there were apparently something like 60 million slaves in the Roman Empire at the time. It has been estimated that about 85-90 percent of the inhabitants of Rome and the Italian peninsula were slaves or of slave origin. A Roman citizen didn't work as it was undignified, so that slaves did virtually all the work, even to the extent of being doctors, teachers, or personal secretaries. Generally the life of a slave could be pretty grim, as a slave in law was just a thing rather than a person, and the master had the power of life and death over the slave. A slave who was too old or sick to work could be thrown out like a broken tool. However, slaves could do quite well financially and could eventually buy their way out of slavery. They were not too different from the average free person, nor were they distinguishable in terms of race, speech, or clothing.

Paul does not tell the slaves to rise up and rebel, but rather tells them to be good and obedient slaves, even if the master is a believer, and serve their masters as they would serve Christ, even when the master is not around. Paul has a word for Christian masters also. He points out that although a man may be a master of men he is still a servant of God and one day both master and slave will stand on an equal footing before God, as God is no respecter of persons. It is important remember that our ideas of slavery tend to come from modern day slavery where slaves were treated badly, and which was far worse than in New testament times. For example, African slavery was based on race and resourced through kidnapping.

Paul does not condone slavery, which existed in many centuries before Christianity and was widely practiced throughout the ancient world as an indispensable institution. But because it was such an integral part of life at the time there was little he could do to change society except from within. However, the enslavement of fellow Christians was discouraged right from the start. It is interesting that Paul interceded with a master called Philemon (see Paul's letter by that name) on behalf of his runaway slave. Paul suggests that the separation might have been so that Philemon could have his slave back as a brother and not as a slave. Christians were the first group in history to start an anti-slavery movement, and it was through the influence of Christianity and Christian men like Wilberforce in Great Britain, John Woolman in America, and many others who devoted their lives to the cause that slavery was eventually abolished. There was a lot of opposition, even

---

[26]Ephesians 6:5–9; Colossians 3:22–4; 1 Timothy 6:1–2; Titus 2:9–10; 1 Peter 2:18.

from some church leaders as the slave trade was very lucrative; abolition cost Britain a huge amount of money in compensation to slave owners in the colonies.

A fundamental idea behind abolition is that you cannot maintain the unequal status of both master and slave when both owe their allegiance to one master, God. The slave would eventually be freed. As far as the Old testament is concerned, although slavery was countenanced then, the law demanded that slaves eventually be set free.[27] Pastor Timothy Keller sums up abolition well: "Christianity's self-correcting apparatus, its critique of religiously supported act of injustice, had asserted itself."[28] In dealing with injustice, sometimes supported by the church, the emphasis was not on liberalizing Christianity but on taking Christianity more seriously, and getting back to the fundamentals of the Christian faith.

Keller goes on to give further examples of Christian activism such as the Civil Rights movement in the United States in the mid-twentieth century, the abolition of apartheid in South Africa in the 1990's, and the bringing down of totalitarian regimes such as communism in Eastern Europe in the latter part of the twentieth century with the Berlin wall coming down in 1990. We can thank people like Martin Luther King, Jr, Desmond Tutu, and the polish priest Jerzy Popieluszko for their leadership and role in bringing in the changes. The German theologian Hans Küng writes:

> It cannot be denied that active Christians took the lead in the struggle for social justice in South America, for peace in Vietnam, for the rights of blacks in the U.S.A. and South Africa, and also—it should not be forgotten—for the reconciliation and unification of Europe after two world wars.[29]

9.5.5 Other social contributions

D'Souza[30] mentioned several contributions that Christianity made to society, namely:

- developing a new importance of family life brought about by the changes mentioned above

- introducing the idea of romantic love and using it as a basis for getting married and preserving a happy marriage

- introducing consent on the part of both a man and the woman for marriage. In some countries marriages are still arranged and people are often pressured into marriage against their wills.

---

[27] Exodus 21:2; Leviticus 25:40.
[28] Keller (2008: 64).
[29] Küng (1984: 30).
[30] D'Souza (2007: 58–78).

- extending respect to ordinary people, as reflected in the new political institutions in the West

- introducing the notion of political and social accountability through a new model of servant leadership that Jesus introduced.[31] Leaders are judged by how well they respond to the concerns and needs of the people they lead.

- Christian civilization created the basic rules of modern economics, which led to capitalism. D'Souza argues that

  > capitalism satisfied the Christian demand for an institution that channels selfish human desire toward the betterment of society. Some critics accuse capitalism of being a selfish system, but the selfishness is not in capitalism—it is in human nature.

  The Bible does not condemn money, but rather the love of money.

- Capitalism brought prosperity and the idea of progress. Most of us today will want our children to live better lives than us. There is also a belief in moral progress. History is seen as going somewhere in a single direction rather than going in cycles where past mistakes reoccur, a view held by many cultures such as the Chinese and Indian cultures.

- Christianity made a very important contribution to society with its emphasis on compassion. Large relief programs are set in motion when there is a major disaster in some part of the world whether it be famine, floods, earthquakes, or tsunamis.

- Many great figures who have dedicated their lives to the service of others such as Mother Teresa and Albert Schweitzer have come out of Christianity. We don't generally find this in other religions or secular society.

- The modern concept of freedom whereby we have the right to express our opinions, choose where we work and live etc. is basically inherited from Christianity.

As we noted above, Christianity introduced the idea of the equality of all people, as every person is important to God and God is "no respecter of persons"; something missing in many cultures today. "Another Christian concept, no less crazy: the concept of equality of souls before God. This concept furnishes the prototype of all theories of equal rights." says philosopher Friedrich Nietzsche.[32] D'Souza adds: "This Christian idea was the propelling force behind the campaign to end slavery, the movement for democracy and

---

[31] Mark 10:43; Luke 22:27.
[32] Nietzsche (1968: 401).

popular self-government, and also the successful attempt to articulate an international doctrine for human rights."[33] We have already seen how Christianity endeavored to transform marriage from within and elevate the status of women rather than contest patriarchy. Jesus broke all the social conventions and permitted women to travel with him, even of low social status, and be part of his circle of friends. In the early church women predominated because of their new status, which didn't go down too well with the Romans, given their view of women. It is interesting that Jesus' twelve disciples were ordinary men from many different backgrounds, for example fishermen and a tax collector, and today Jesus calls people from all walks of life to follow him. He is no respecter of personal status.

It can be argued that those who would remove Christianity are in danger that, in time, all the rights we so freely enjoy will be downgraded and lost. We are seeing this beginning to happen in Western civilization, especially with the secularization of Europe and the U.S., and the rise in skepticism. However, according to Keller,[34] there has also been a rise in those following traditional religious faiths as well, as seen for example in Africa, Latin America, and Asia. He continues, "the world is polarizing over religion. It is getting both more religious and less religious at the same time" and "We have come to a cultural moment in which both skeptics and believers feel their existence is threatened because both secular skepticism and religious faith are on the rise in significant, powerful ways." Technological advancement is not bringing an inevitable secularization after all!

9.6 CONCLUSION

Contrary to some views about Christianity being harmful to society we have seen the contrary to be true, namely in many aspects of society Christianity has been good for people. Of course there are always individual exceptions in any group of people that would otherwise display some general characteristics, but here we are talking about society as a whole. The probability that Christians makes a positive contribution to society is greater than the probability that they make a negative contribution. The Christian church has made enormous beneficial contributions to society, and has also inspired the greatest art, music, and literature.

---

[33] D'Souza (2007: 67).
[34] Keller (2008: introduction).

CHAPTER 10

WHAT ABOUT EVOLUTION?

10.1 THE NATURE OF EVOLUTION

It was with some hesitation that I started writing this chapter because of the controversy it stirs up not only between non-theists and theists, but also among Christians and even among evolutionists themselves, whether Christian or otherwise, because of some problems associated with Darwinism. There are those who take the Genesis story literally and believe that the earth was created less than 10,000 years ago, which implies that evolution is not true. Some would divorce the Bible from science and fully endorse evolution, while some Christian evolutionists maintain that Adam and Eve were real people and that the world underwent a discontinuity called the "fall" in the Bible, as explained in Genesis.[1] Theistic evolution takes several forms depending on the number of times a theist believes that God intervened, ranging from zero to many times. For example, one view is that God performed at least three supernatural acts of creation, namely the creation of matter, first life (involving the first cell), and the human soul. On the other hand the whole process from

---

[1] For a sample of four views see Carlson (2000).

beginning to end can be regarded as a continuous process somehow guided by God leading to a final cause; a kind of progressive evolution. In this chapter I am not going to take up a particular position, but simply look at evidence for evolution along with some problems with it. What I do find is that there is evidence both for and against the current understanding of evolution, which I shall endeavor to describe.

Before proceeding further I need to first of all define what is popularly meant by (biological) evolution, usually referred to as Darwinism or more recently neo-Darwinism. It is a theory of evolution that is a synthesis of Darwin's theory in terms of natural selection (the "survival of the fittest") and modern population genetics that explains the surfacing of new species in terms of genetic mutations. Here mutations due to "random" copying errors in DNA cause variation within a population of individual organisms and that natural selection acts upon these variations so that those that adapt best dominate the next generation gene pool. A prerequisite is that such changes are faithfully inherited. One might also add to the topic such things as genetic drift and sexual selection. In its essence, evolution is simply about change. Given my discussion on randomness in Section 4.1.1, mutations are random simply in the sense that they occur irrespective of their usefulness to the host organism and not necessarily merely by chance. As the evolutionary biologist Ernst Mayr commented:

> When it is said that a mutation or variation is random, the statement simply means that there is no correlation between the production of new genotypes and the adaptational needs of an organism in a given environment.[2]

It is important to distinguish between prebiotic and biological evolution as the former involves chemical evolution and is more contentious; it is discussed in Section 10.4.1.

The theory of evolution, although couched in biological terms, has tended to become a philosophical system that means different things to different people. For some it is simply a theory about change whereby all organisms from bacteria to animals and plants can be observed to change over time so that new species are obtained. For others it is a more comprehensive theory that involves life from non-living matter with a particular life-form being the originator of all life, the so-called "initial Darwinian ancestor." Sometimes there is confusion between observation and theory, a common problem with science, and there is considerable difference of opinion about what drives evolution. Although we might end up creating something in the laboratory, it doesn't mean that is how it all began; some theoretical ideas may be incapable of proof. It is important to realize however how much the subject is still changing so that theories will come and go. This is seen, for example, in the three topics paleontology (the study of fossils), epigenetics (discussed in Section

---

[2] Mayr (1988: 98).

2.6.3), and evolutionary developmental biology (evolution of development) mentioned in Section 2.6.4 and discussed further below.

Evolutionary biologist Frank Ryan raises another issue, namely that natural selection is not sufficient for our evolution. He says:

> ... natural selection alone could not have given rise to the evolution of life, and its subsequent diversity, ... Great evolutionary forces such as symbiosis and hybridization, are of vital importance to the variation ....[3]

Although evolution is often thought of as just a fight for survival and is sometimes unfortunately couched in the form of a "selfish gene" metaphor, there is a form of cooperation underlying evolution such as stronger members looking after the weaker members, mothers sacrificing for their young, and we read of interesting stories where one species adopts the orphans of another species. There is even greater cooperation in the case of symbiosis where two species help one another to survive rather than compete, for example insects and flowering plants working together in pollination, bats and fruit trees collaborating in seed dispersal, and especially viruses interacting with their hosts. A well known one is where sharks allow small fish to clean parasites and debris from their mouths and skin. Theologian Nicola Hoggard-Creegan[4] reminds us of how interconnected our living world is and how "this fragile interdependence is much closer to the image of the kingdom of God than are the previous images and pictures of competition to the death."

10.1.1 Hox Genes

Evolutionary developmental biology, a combination of the two fields evolutionary biology and developmental biology, or evo-devo for short, is a field of biology that investigates how the embryos of different organisms develop in order to try and determine the ancestral relationship between them and to discover how developmental processes evolved. This topic looks particularly at the evolution of embryonic development from a single cell to birth to see how novel features arose like feathers, for example. Such development has always puzzled scientists until a set of developmental genes called Hox genes discovered in 1983 paved the way to a new understanding: for example fruit flies have 104 Hox genes whereas humans have 255.[5] These genes guide overall body architecture and the development of a body plan. They bind to the DNA of other genes producing a whole range of processes leading to eyes, limbs, hearts, and other complex structures. A single mutation in one of these genes can dramatically change an organism, thus suggesting how radical changes in body design can take place.

---

[3] Ryan (2009: 125–126).
[4] Hoggard-Creegan (2013: 114).
[5] Holland (2013).

Hox is short for homeobox, and refers to a region of DNA within a particular gene that allows that gene to turn on or off other genes once it has been translated into a protein. Different *Hox* genes are expressed at different locations along the body and can signal other genes to activate some anatomical characteristic specific to a particular region of the body. Often only a few genes are involved, for example just two (*BMP4* and *calmodulin*) can lead to the wide variation in beak shapes and sizes of different Darwin's finch species; it depends on their degree of expression. The gene *BMP4* is also involved in evolution of the African fish species cichlids with regard to jaw development. What happens is that very different species may use the same genes in embryo development, but they are regulated differently (e.g, fruit flies and human beings with the so-called *PAX-6* family switching on eye development).

Often very little changes are needed in the *Hox* genes and surprisingly few genes are involved; we don't need new genes arising every time we get a new species. Also the genetic tools for building organs like fingers, toes, and limbs were in place for a long time before the environment led to their actual development, as in the transition of fish to land, for example. It was not necessarily a question of mutation leading to many new genes but rather of gene switches. Jerry Coyne, an evolutionary biologist at the University of Chicago, has said, "The evolutionary conservatism of these genes across long-diverged species is staggering."[6] For example the mouse *Hox* gene *PAX-6* previously mentioned that triggers eye formation can induce in the fruit fly *Drosophila* eyes all over the fly's body including the wings. Such conservatism (existing across species) suggests that the genes have a very ancient history, but according to Coyne they cannot account for diversity. He says we need to exercise some caution in applying various theories as those special genes and their expression may not be the only answer to what drives evolution. Some switches can lead to loss of traits through devolution! Clearly neo-Darwinism does not provide all the answers as evolution is not just about what genes there are. We still have lot to learn about *Hox* genes.

10.1.2 Evolution of Everything

One thing that I do find inappropriate is that some people take the theory of evolution out of its biological context and apply it to almost all of life—the evolution of everything. For example Georg Hegel, professor of philosophy in Berlin (1813), expounded an evolutionary view of the whole world in which everything was evolving, including the scientific, aesthetic, legal, political, and religious through thesis, antithesis, and synthesis. Karl Marx used this idea of evolution in his "Das Kapital" (1867) to show that the goal of history was the evolution of a classless society. If humans were truly evolving then they

---

[6] Coyne (2005).

would pass through the religious stage and finally have no need of religion; it being a man-made natural process. We find similar inappropriate theoretical extensions today to religious and cultural evolution. However, the gap between the "haves" and the "have nots" gets bigger throughout the world, crime rates increase, mental health is a growing problem, and worldwide strife is on the increase. There is not much sign of any improvement, which suggests that evolutionary theory should stick to biology and genetics.

It should be clear that the theory of evolution as such does not necessarily entail any particular atheistic, agnostic or religious understanding of the world—it is simply a question of what science tells us. Writer Dinesh D'Souza[7] points out that "As a theory, evolution is not hostile to religion. Far from disproving design, evolution actually reveals the mode by which design has been executed." However, if materialism is put forward as the underlying basis for evolution, we are faced with the fact that inanimate matter managed to organize itself so that it can contemplate itself! Such questions are considered below.

10.1.3 Can We Trust Evolution to Get it Right?

Biologist Richard Dawkins[8] comments that; "Our brains themselves are evolved organs... evolved to help us survive" and admits that we cannot completely trust our senses. Evolution's role is to preserve adaptive behavior, not true belief. This idea is supported by other atheists such as the prominent philosopher Thomas Nagel,[9] and not trusting our beliefs and powers of reason can also apply to a belief in evolution. Here naturalism clashes with science and scientific thinking as evolution only shapes our cognitive faculties for survival and not for truth or reality. Evolution can only work on behavior and eliminate non-adaptive behavior; mental states are perhaps irrelevant to survival. Hence our mental states are not necessarily reliable if they have arisen through undirected evolution. The same problem applies to Dawkins's theory of memes, as atheism can also be a meme. Philosopher Edward Feser[10] argues that if the competition between memes for survival is actually what determines all our thinking, then we can have no confidence in our beliefs or arguments from such beliefs as being true or compelling. He goes on to say:

> For if the meme theory is correct, then our beliefs seem true, and our favored arguments seem correct, simply because they were the ones that happened for whatever reason to prevail in the struggle for "memetic" survival, not because they reflect objective reality to us.

He notes that

---

[7] D'Souza (2007: 132).
[8] Dawkins (2006: 367).
[9] Nagel (1997: 134–135).
[10] Feser (2008: 245, 240).

no brain process could count as a "meme" except when interpreted as such by some mind. Hence since "memes" presuppose the existence of mind they cannot explain the mind.[11]

In this chapter we shall consider some evidences for evolution, but also highlight some of its serious problems.

10.2 SOME EVIDENCES FOR EVOLUTION

In a book of this size it is difficult to do full justice to the evidence for evolution, especially as I am not a biologist, although I have been involved at a research level in statistical ecology and human blood genetics. Instead I will give a brief summary of some evidences supporting evolution.

(1.) There is a huge amount of variation in virtually every population so that an environmental change can lead to some parts of the population adapting more than others (e.g., introduced plants growing in the Antarctica). This response can be rapid and can lead to all kinds of novel organisms. All that is needed is a small selective advantage, say less than 1%.

(2.) There is a unity to life as all living things are formed using the same genetic code. This code is almost universal, apart from a few minor and rare exceptions, whereby triplets of nucleotides (called codons) determine which amino acid is added next to a protein chain. Further comments concerning human molecular genetics are made in the next section.

(3.) The fossil records in the rocks show only single-celled organisms in the oldest rocks with a progression relating to the age of the rocks to invertebrates, fish, amphibians, reptiles, mammals, and finally man. The order is as predicted by evolution.

(4.) Natural selection explains why strains of bacteria and pests develop resistance to human attempts at control.

(5.) Evolution helps to explain why some features of living things seem to be either poorly designed or serve no apparent functional purpose as with vestigial organs. For example, there are flightless beetles with wings, snakes with tiny legs under their skin, virgin whiptail lizards that engage in fake sex, blind fish (*Astyanax mexicanus*), hind leg bones in whales (*Archaeocetus* has two vestigial hind legs that protrude from its body but which seem to serve little or no function), wings on flightless birds (ostriches, cassowaries, kiwi, and kakapo), non-functioning eyes as

---
[11] Feser (2008: 240).

with some cave dwellers and burrowers, and dandelions that have the proper organs (stamen and pistil) necessary for sexual reproduction like all flowers, but do not use them. Sometimes an organ is adapted in an animal for a different use (e.g., wings for balance).

(6.) The existence of certain pseudogenes in humans shows links with other primates. This topic is discussed in more detail below.

(7.) The existence of poor design in some living creatures, discussed further below.

(8.) Embryos in many species often appear similar to one another in early developmental stages suggesting evolutionary development.

(9.) Nearly every part of the world has its own unique and distinctive set of animals and plants. For example, the mammals of Australia are mostly marsupials or "pouched mammals" and are generally unrelated to those found anywhere else except for some distantly related marsupials in South America. However, similarity can occur through so-called "convergent" evolution. For example flying insects, birds, and bats have all independently evolved the capacity of flight, having "converged" on this useful trait; such features are called "analogous" and their common function arises in spite of their evolutionary ancestors being very dissimilar or unrelated. Analogous features are in contrast to so-called "homologous" features, which have a common origin, but not necessarily having a similar function. The bat wing, for example is homologous to human and other mammal forearms. (If you wave your arms a lot you still won't fly!)

(10.) Some have argued against evolution as it hasn't been observed before our eyes. However, one example of real-time evolutionary changes is the Long-Term Evolution Experiment where changes in the bacterium E. Coli (a prokaryote, that is without a nucleus) have been monitored since 1988 through thousands of generations by Richard Lenski's group at Michigan State University (cf. http://myxo.css.msu.edu/ecoli/, accessed August, 2015). The LTEE began with a single cell of *E. coli* that was used to generate twelve genetically identical lines of cells. The researchers have noted that the lab populations have evolved to increase cell size, grow more efficiently on glucose, and then proliferate more rapidly when transferred to fresh media. A citrate-using (Cit+) variant finally evolved in one population after 31,500 generations, causing an increase in population size and diversity. However, there was no evidence of new complex systems evolving, which might suggest a limitation of mutational changes as seen also in the malarial parasite (an organism, or eukaryote, having a nucleus), and the HIV virus.[12]

---

[12] Behe (2007).

(11.) The process of adaption is very common in nature as we find different species have developed different methods for coping with the environment. We see amazing adaptions with animals, but it perhaps not so well known about plants, for example mechanisms for climbing. Consider tendrils on vines that can be used to support the plant or access nutrients and absorb and dissipate energy when there are strong winds. If the tendril finds no anchor, the vine "decides" to abort its investment and integrated sensors come into play. The vine can alter its body plan by a chromatin modification relating to the DNA of the vine (an example of epigenetics).

Evolutionary changes can take place in four different ways. First there is microbial evolution through natural selection, where for example bacteria mutate to become antibiotic resistant, viruses are able to host-hop (e.g., HIV and SARS), and the rise of well-known drug-resistant strains of malaria parasites. Second, there is microevolution in which there is a change in allele frequencies that occur over time within a population. This change can be due to mutation, selection, or genetic drift. For example, we can get variations within a species due to selection pressures and genetic drift as with the peppered moth where the wing color shifted from predominately white to predominantly black in the North of England with environmental changes there (e.g., blackened trees). Third, there is speciation when one species evolves to a closely related sister species and some of the best examples arise on archipelagos. For example, we have the finches of the Galapagos Islands where there is variation of beak size and shape on different islands, as observed by Charles Darwin (though the birds can interbreed, which raises questions about their speciation to begin with).

Variation at these three levels is well supported, but there is some disagreement with the fourth level usually referred to as macroevolution whereby there is change above the species level such as for example the development of the family of horses, and the formation of vertebrates from invertebrates and birds from reptiles. The question is whether macroevolution is a continuation of microevolution or a different process depending on one's views on gradualism (changes take place gradually) and the role of natural selection. Biochemist Douglas Theobald,[13] for example, presents a case for common descent, the phylogenetic tree, and macroevolution. However, developments in this topic are ongoing. For example, Journalist Laura Spinney says

> If you want to know how all living things are related, don't bother looking in any textbook thats more than a few years old. Chances are that the tree of life you find there will be wrong.[14]

It is clear that evolution is a useful theory for specific variation but it has some potential challenges in explaining the overall picture. It is not a question

---

[13]Theobald (2012).
[14]Spinney (2008).

of order out of chaos, but rather it seems like the unfolding of an order that was already there, but hidden.

10.2.1 Vestigial Organs in Humans

Some vestigial organs in living things have already been mentioned. We now consider some in humans that are regarded as vestigial. For example the vomeronasal organ that is responsible for detecting pheromone signals no longer functions, as essentially all the genes related to pheromone detection in our DNA are defunct. It is significant that the list of vestigial organs has shrunk from around 100, when the notion of a vestigial organ was originally promoted, to a very short list. We keep finding uses for such organs. For example, the endocrine glands that were once thought to be vestigial are now known for their importance in producing hormones. The thymus is involved in protecting the body against disease. The appendix does have a number of useful roles like the production of certain endocrine cells in the human fetus, having some immune functions with young adults, and being the home for some good bacteria.[15] Interestingly enough, the only mammals other than humans known to have appendices are rabbits, opossums and wombats, and their appendices are markedly different from the human appendix.

Tonsils may be more important to help children fight off diseases. There are some barely used nerves and muscles that seem to serve no useful purpose such as the plantaris muscle of the foot that is missing for some people. There are also muscles that some people can use to move their ears (I can only do eyebrows). However the muscle of the outer ear helps protect people against freezing in colder climates. The coccyx is necessary for sitting with comfort, and provides a means of attachment of certain muscles. We have a "third eyelid" or *nictitating membrane*, but this is used for collecting foreign material that gets in the eye. Some organs may of course actually be leftovers from an earlier period of human development but not necessarily from a prehuman species. We may be devolving not evolving!

10.2.2 Evidence of Poor Design

The question of poor design in living things has lead to some controversy. Some of the so-called animal design defects are:

(1.) The panda's "thumb" (radial sesamoid bones). It is not that the thumb does not work, as it works well when the panda uses it for its main purpose of stripping leaves, but rather it is described as poorly designed. A proper thumb is regarded as being "better." Pandas dine almost solely on bamboo, but because of their digestive system they only absorb about

---

[15]See, for example, http://www.sciencedaily.com/releases/2007/10/071008102334.htm, accessed August, 2015.

17% of what they eat so they spend most of their life eating and sleeping. (Sounds like some young people I know!) Their digestive system suggests carnivorous ancestry.

(2.) In many animals there is the inefficient route of the current laryngeal nerve that travels from the brain to the larynx by looping around the aortic arch. Not so great for the giraffe!

(3.) Mammals have lost tetrachromatic vision as compared to other four-legged creatures. (Some fish, birds, and insects have these four kinds of cone cells, whereas humans and closely related primates generally have three. It appears that some women may have four. Almost all other mammals have two.)

(4.) There seems an unnecessary use of materials in the African locust as nerve cells start in the abdomen but connect in the wing.

(5.) A number of enzymes (e.g., RuBisCO and nitrogenase) seem to function very inefficiently.

(6.) There are apparent anomalies with regard to bones and bird flight. For example, strong but heavy bones not suited for flight occur in animals like bats, while unstable, light, and even hollow bones that are more suited for flight occur in birds that cannot fly, like penguins and ostriches.

(7.) The intricate reproductive devices in orchids seem to have been put together by a committee (perhaps a university one called argumentum) from components commonly having different functions in other flowers.

(8.) Non-coding or "junk" DNA is present in most organisms. However, epigenetics suggests something different in the form of switches (see Section 2.6.3). Epigenetics has, for example, been used to explain why Homo sapiens survived but not Neanderthals nor the Denisovans, archaic forms of Homo sapiens, discovered recently in Siberia.[16] On the negative side, the researchers found that many of the genes whose activity is unique to modern humans are linked to diseases like Alzheimers and schizophrenia. They comment that perhaps these recent changes in our brain may underlie some of the psychiatric disorders that are so common in humans today.

(9.) Some species have genes for non-existent features. For example birds do not have teeth but have several degenerated genes for them.

(10.) Fetal teeth exist in toothless mammals, for example anteaters and baleen whales, but the teeth are reabsorbed.

---

[16]Gokhman et al. (2014).

There a number of so-called human design defects and some are given as follows.

(1.) There are several related to the human reproductive systems. For example, in females there is a gap between the ovary and fallopian tube that can lead to ectopic pregnancies. As the birth canal passes through the pelvis there can be problems if the baby's head is too large for the opening; a common cause of miscarriages. Females of other primates do not experience this difficulty. The clitoris in females is not in the best place for stimulation in sexual intercourse. In males the testes develop initially in the abdomen and come down later into the scrotum leaving weaknesses in the abdominal wall that can eventually lead to hernias later.

(2.) Some argue that the human spinal column is inefficient as it can commonly be malformed. The human coccyx or "tail" has some anomalies.

(4.) There are a great many human congenital diseases and genetic disorders. For example we have some 120 cancer predisposition genes alone.[17]

(5.) Human faces are much flatter than those of other primates, but they all have the same set of teeth. Consequently humans have a number of problems such as overcrowding of teeth, wisdom teeth, and poor sinus drainage.

(6.) The visual nerve fibers in the human eye converge to form the optic nerve, which crosses the retina (in order to reach the brain) and thus creates a blind spot; squids and octopuses do not have this.

In looking at the above depressing selection, a number of questions arise. Discussing a systems design requires looking carefully at the purpose of the design and what it is trying to achieve; tradeoffs are inevitable such as balancing strength versus lightness in a vehicle. Also any design has to be put in the context of the whole ecosystem rather than just the individual, as a "super" creature may imbalance or wipe out a part of the system. One popular example of so-called sub-optimality, referred to briefly above, is the vertebrate eye in which the neural wiring of the retina is on the side facing the incoming light, and the the retina is upside down. The latter is a tradeoff that allows the eye to process the vast amount of oxygen needed for vertebrates. It creates a slight blind spot but we have two eyes and the two blind spots don't overlap. What is interesting is the newly-found optimizing role of radial glial cells called Müller cells that act like optical fibers cells and seem to mediate the image transfer through the vertebrate retina with minimal distortion and low loss, thus supporting the feature of an inverted retina as an optical system. An authority in retina physiology, George Ayoub, concludes:

> The vertebrate retina provides an excellent example of a functional—though non-intuitive—design. The design of the retina is responsible

[17]Finlay, personal communication.

for its high acuity and sensitivity. It is simply untrue that the retina is demonstrably suboptimal, nor is it easy to conceive how it might be modified without significantly decreasing its function.[18]

Biochemist Michael Denton[19] also discussed the functioning of the eye in some detail and concluded:

> The more deeply the design of the vertebrate retina is considered the more it appears that virtually every feature is necessary and that in redesigning from first principles an eye capable of the highest possible resolution (within the constraints imposed by the wavelength of light) and of the highest possible sensitivity (capable of detecting an individual photon of light) we would end up recreating the vertebrate eye–complete with an inverted retina and a choriocapillaris separated from the photoreceptor layer by a supportive epithelium layer and so forth.

10.3 EVIDENCE FOR HUMAN EVOLUTION

Human genetics is a highly technical subject that is rapidly advancing and changing so I can only give a brief overview. It appears that genetics alone cannot explain the sudden appearance of modern "cultural man" as humans have to be born into a culture in order to learn to talk. Apparently children brought up by wild animals can never learn our basic human skills and learn language. Although many animals have evolved special ways of producing sound for communications, no monkey or ape uses its tongue, lips and other moveable parts as in human speech. Also, the human supralaryngeal pathway is entirely different from all other mammals. [20] Cell biologist Graeme Finlay in a lecture on "Human genetics and the image of God" given in 2006 in Cambridge makes the comment: "Genetics are unable to describe fully what make us human" and "We're in the age of the genome, but we can still recognize that it takes much more than genes to make the human."[21] In addition to establishing genetic links with the primates he said that:

> Indeed it is our relationships with other people (and, I believe, with God) that constitute us as human. The basis of christian faith is that a personal God has revealed himself in personal terms as a person, Jesus of Nazareth.

Finlay[22] brings together the case for the evolution of humans and we shall consider some of the evidence below.

---

[18] Ayoub (1996).
[19] Denton (1999).
[20] Berry and Jeeves (2008: 11).
[21] http://www.faraday.st-edmunds.cam.ac.uk/CIS/Finlay/lecture.htm, accessed August, 2105.
[22] Finlay (2013).

10.3.1 Pseudogenes

The human genome project has shown that humans have about 25,000 to 30,000 protein-coding genes in their DNA (enough to build a person?), that is genes for building proteins, as well as about 10,000 to 20,000 so called *Pseudogenes*. The latter are derelict (fossilized) genes that are recognizable either as degraded remains of genes that are still active in other species, or as scrambled copies of authentic genes that are genes that can still code for producing proteins in our DNA. The fact that humans have a lot of pseudogenes could suggest that we have a long genetic history indicating the many changes that have taken place. For example, Finlay[23] mentions the pseudogene (*MYH16*) that he describes as originally being a functional gene in pre-human primates with segment

Pre-human primates...*CAT AGC ACC GCA CCC CAT TTT GTC*....

However humans contain a highly damaging mutation causing a deletion (starred), namely

Humans with deletion...*CAT AGC **C GCA CCC CAT TTT GTC*....

Grouping what's left in threes we get a new sequence of triples

Humans (new triples)...*CAT AGC CGC ACC CCA TTT TGT C*....

The information is now garbled and the coding ability of the gene is lost. A similar example where a gene is inactivated in only humans arises in the olfactory receptor (OR) gene family of 1000 members where, for humans, a very high proportion of OR genes are pseudogenes, although many of the OR pseudogenes possess shared mutations with chimps, gorillas etc, suggesting that those mutations occurred in common ancestors. All sampled dolphin olfactory receptor genes are pseudogenes, which relate to the fact that both humans and dolphins no longer require an acute sense of smell. Humans have also experienced hair loss and have lost the ability to make vitamin C. Finlay[24] gives further examples of pseudogenes and mutations.

10.3.2 Establishing Connections

To begin the detective work of tracing what some have suggested to be our ancestry we start with what has been the usual approach of comparing human and chimpanzee DNA sequences that can be directly aligned; these have been quoted as about 98.4% identical. However, if the insertions and duplications mentioned below are considered, the value drops by about 5%. In commenting

---

[23] Finlay (2004a, 2013: chapter 3).
[24] Finlay (2013).

on these numbers, writer Jeremy Taylor[25] says: "The infamous 1.6% counts for little", and he shows up other differences in his book using genomics, neuroscience, and cognitive psychology. He believes that we need to shift our focus away from apes, and says that chimps are not useful for making comparisons as we are not descended from the chimpanzee. Nevertheless, our common ancestor with chimps would seem to be our closest common ancestor with any other living species.

When sequences are examined for inversion, Taylor notes that

> the chromosomes are all but blotted out by a blizzard of red lines denoting inverted sequence....the extent to which you estimate the difference between chimp and human genomes depends entirely on where you look and how deeply.[26]

We see that inversions can be used to demonstrate evolutionary relationships just as well as any other complex mutations as for example a recent paper and its interesting graphs on the gibbon genome.[27] There are also differences in the number of repetitions of gene sequences where a greater number of copies create a greater genetic effect. Most of the difference from chimps is made up of noncoding DNA that can have little or no effect on brain, body, or behavior (if the non-coding RNA is not functional), but proteins can have an effect and 80% of proteins are different with humans and chimps. As with epigenetics, it is the gene switches (regulatory sequences) that play a fundamental role in gene expression. Deletions also occur and include about 200 loss-of-function mutations leading to humans losing some gene functions compared to chimps. Between 6-8% of shared genes show pronounced splicing-level differences between the two species, and if small inserted elements of so-called "junk DNA" are taken into account the percentage in common goes down to 87%.[28] Summing up, Taylor says that differences include

> deletions, inversions, copy number differences, and splice variants. Differences in timing and rates of expression of identical or similar genes, and the role of 'master-controller' transcription factor genes further act to amplify the genetic distance between humans and chimps.[29]

What is interesting is that small changes in some genes can make a big difference. For example, versions of FOXP2 gene involved in the regulation of gene expression relate to bird song, bat echolocation, and human speech. Two amino-acid substitutions distinguish the human FOXP2 protein from that found in chimpanzees, but only one of these two changes is unique to humans.[30] What we find is that although diverse species can have a similar gene, its expression can be very different, as mentioned above with the FOXP2 gene.

---

[25]Taylor (2009: 103).
[26]Taylor (2009: x).
[27]Carbone et al. (2014).
[28]Taylor (2009: 128).
[29]Taylor: (2009: 299).
[30]For a further discussion of this gene and regulatory genes see Carroll (2005b).

Bearing in mind the above comments, we now look at comparisons made among primates. At most genetic loci we are closely related to chimps, but at some loci chimps and gorillas are the closest. Using such genetic mapping techniques we shall see below that the chromosomes of humans and the great apes (chimps, gorillas, orangutans) may be rearranged to form what appears to be an ancestral set where the structures of most of the chromosomes belonging to the common ancestor have been unambiguously derived. Chromosomes of other primates (e.g., gibbons, macaques) can be obtained by cutting and pasting blocks of human chromosomal material. Graeme Finlay notes three lines of evidence for ancestral connections: (1) Our chromosomes have evolved through well-known processes of cutting and pasting as just mentioned; (2) We share with other species uniquely rearranged genes; and (3) We share with other species unique random additions to our DNA. We now consider some examples of these processes.[31]

The first example relates to the *COX8H* pseudogene. This is active in New World Monkeys (NWMs) such as spider monkeys, but is inactivated in humans, apes, and Old World Monkeys (OWMs) such as baboons where they all contain the same 14 base deletion. Detecting such changes can perhaps help us determine links between species. For example, if two or more species have a gene altered in the same place, then the probability of it occurring independently and at random in the same place in all those species is extremely small, thus suggesting that they all come from the same ancestor where the mutation originally occurred. With detective work, tree diagrams can then be drawn showing connections between species.

Our second example relates to chromosome number 2 that in humans is an end-to-end (telomere-to-telomere) fusion of two chromosomes that are separate in the great apes, as mentioned above.[32] On either side of the central join of chromosome 2 in humans the DNA sequence corresponds to those of the separate chromosomes possessed by the other great apes.

In our third example we note that humans and chimpanzees also differ because of rearrangements with chromosomes in which inversion occurs where a chromosome breaks in two places, the intervening segment is reversed, and the bits rejoined. Such a process distinguishes, for example, the human chromosome 17 from the chimp equivalent that has been determined to be a derivative. Work on other breakpoints has been underway.

The fourth example relates to the male Y chromosome, which lacking the typical partner to pair with (the pairing suppressing rearrangement) has gone through extensive remodeling. For instance, in humans but not in the great apes, a particular segment of the X chromosome has been copied (*transposed*) into the Y chromosome where it has been subsequently inverted into a dis-

---

[31]Finlay (2003, 2004a, 2013).
[32]Telomeres are pieces of DNA, related to longevity, at the ends of the chromosomes that prevent them from sticking together end-to-end and causing cell death.

tant site. Also, a section of chromosome 1 has been copied into the the Y chromosome of humans and the two chimp species, suggesting that humans and chimps descended from a common ancestor that gave rise to the shared trait. Such copied segments of DNA constitute approximately 5% of the human genome. Humans and chimps (but not other primates) share duplicated chromosomal segments that have generated two copies of a novel gene and produced the unstable "CMT" genetic region involved in neurological diseases. Another duplication connecting humans, chimps, and gorillas is associated with the development of a certain type of leukemia, CML.

Of particular interest is the development of colored vision, previously mentioned. New World Monkeys have two-color vision, whereas humans, apes, and Old World Monkeys have three-color vision. Our three-color vision is due a single gene encoding a color pigment protein (opsin) on the X-chromosome being duplicated to give two opsin genes.

In the process of producing proteins, DNA is copied ("transcribed") into RNA, which on very rare occasions is copied back into DNA ("reverse transcribed") by enzymes called *reverse transcriptase*. These enzymes are produced by virus-like genetic entities that exist within cells. It is then possible that some such genetic changes become inherited and provide future markers of genetic history. About 40% of human DNA consist of blocks of genetic material that have been reversed transcribed and spliced into chromosomal DNA giving us an accumulation of such DNA.

10.3.3   Retroviruses

We see from just the few examples given above that there are a number of genetic markers providing apparent connections between the primates. One set of markers is the so-called class of retroviruses, which infect the host DNA by inserting their microscopic genomes (called proviruses) pretty much at random into the host DNA.[33] Some of these can become permanent so that they are passed on to future generations and are called endogenous retroviruses or ERVs, discovered in the early 1980's in the human genome. There are thousands of these in human DNA, making up nearly 8% of the genome. Retroviruses can cause diseases such as cancer in humans and other species, and retroviral insertion into DNA can be regarded as a special type of genetic mutation. A well-known example of a retrovirus is the HIV virus causing AIDS. Work published in 1999 that investigated the insertion sites of six ERVs showed connections among humans and the other primates that has led to a suggested ancestral line, namely

$$NWM \to OWM \to gibbon \to orangutan \to gorilla \to chimp/bonobo \to human$$

---

[33] Finlay (2004b).

This genetic evidence provides support for the idea of a common descent among the primates, a major feature of evolutionary theory. A second feature of the theory is the concept of natural selection where a genetic variation can give an advantage to an organism enabling it to reproduce more efficiently than one without the variation and therefore increasing the frequency of the variant in the population gene pool. As a general rule, most mutational-type changes are either damaging to the host or don't offer any advantages so that natural selection does not function to promulgate the changes when they first happen. Instead ERVs start to accumulate mutations that mostly end up as fossil genes, though a small proportion can end up as genes that can produce proteins. In fact a few now appear to be essential for human development (e.g., placental development), and the ERVs can no longer be considered as junk DNA.

10.3.4 Genetic Parasites

Our DNA can also end up containing segments of DNA from genetic parasites, some of which we have already referred to as ERVs, or HERVs for human ones. Usually such insertions do not have much of an effect as genomes seem to be able to tolerate some additional bits of foreign DNA. There are three main categories of parasitic DNA: (1) "Long terminal repeat" or LRT elements (which include the HERVs), (2) LINE elements ("long interspersed elements") including "L1" sequences, and (3) SINE elements ("short interspersed elements") including "Alu" sequences. It is argued[34] that if two species possess the same block of DNA consisting of one of the previous three types inserted in the same site in their DNA, then they must have inherited it from a common ancestor; the chance of it occurring independently would be very small. LINEs in particular are excellent markers of evolutionary relationships. These are all *transposable elements* or TEs ("jumping genes") that multiply haphazardly in the genome, and there are others.[35] Our DNA is not as simple as we thought!

There is also a wide range of mutations that are generally harmful (though not all) that lead to pseudogenes or new functions. Genetic connections like those mentioned above have been found to exist with and between other mammals whose gene sequences have been studied. Establishing genomes of species is therefore an ongoing project.

10.4 IN THE BEGINNING...

Beginning with chemical evolution we now work through some of the stages that were required for biological evolution to have occurred. We start by

---

[34]Finlay (2004b: chapter 3).
[35]See Finlay (2008, 2013: chapter 2).

looking at what the earth's original atmosphere might have been to start off evolution at a chemical level. In order for appropriate organic molecules to form, it was originally thought that the earth's primitive atmosphere provided a particular organic ocean "soup" that enabled appropriate molecules such as proteins to form. For this to happen the atmosphere was assumed to be strongly "reducing", with various views on the substances present such as water and some of the following: carbon dioxide, methane, ammonia, hydrogen, and nitrogen. What is important is that the primitive atmosphere could contain only the smallest amount of free oxygen, otherwise organic molecules could not be formed. It appears that there was more oxygen than originally considered so that the early atmosphere was less reducing than first thought and possibly even oxidizing. For example, geochemical analysis indicates that the composition of Precambrian deposits is short of nitrogen.

10.4.1 A Prebiotic Soup?

Physical Chemist Charles Thaxton and colleagues[36] looked at evidence for a prebiotic soup and came to the following conclusions:

(1.) There seems to be no geological evidence indicating that an organic soup, or even a small organic pond, ever existed on earth.

(2.) Concentrations of organic substances in the ocean would be insufficient for the substances to combine appropriately, as dilution processes would have taken place. For example, destructive processes would tend to dominate synthesis. Without going into the authors' technical details, consider just the action of solar ultraviolet light. Experiments, which I will comment on below, used discharge methods to create substances such as amino acids in the laboratory. Unfortunately solar ultraviolet light breaks down the same organic molecules, and they are also open to "attack" by other substances; they therefore would have had a short lifetime. The authors concluded that the idea that life emerged from an oceanic soup of organic chemicals was a "most implausible hypothesis" and referred to the notion as "the myth of the prebiotic soup."

(3.) The time frame for chemical evolution to take place seems far too short under random chemical conditions. The finding of molecular and microfossils showed that life existed very early on, thus indicating that it appeared in a comparatively short time of less that 10^7 to 10^8 years. A mechanism is needed to explain this relatively short evolutionary time frame.

(4.) Early pre-experiments are flawed for at least two reasons. First, more realistic atmospheric conditions are needed. Second, potentially interfering cross-reactions have been largely ignored. For example, in going

---

[36]Thaxton et al. (1984: chapters 4 and 5).

from free amino-acids to polypeptides the presence of sugars or aldehydes will cause interfering cross-reactions. Unfortunately, in most cases, the experimental conditions have been so artificially simplified that they are nothing like what they might have been on the primitive earth.

According to engineer Walter Bradley,[37] from 1980 on NASA scientists showed that the primitive earth had only minimal amounts of the reducing gases methane, ammonia, or hydrogen, but instead consisted of the inert gases of carbon dioxide and nitrogen, and of course water. In 1952 Miller, a graduate student of the nobel Laureate Harold Urey, carried out his famous experiment in which he applied a gas discharge, simulating lightning, to what was was regarded as a mock-up primitive atmosphere and obtained amino acids, the building blocks of life. He aimed to test the Oparin-Haldane hypothesis that complex molecules could have arisen from simpler ingredients through natural processes such as, for example, lightning, volcanic activity, and ultraviolet. The experiment was hailed as a breakthrough to show that life could have evolved from nonliving chemicals, and that life was inevitable. Unfortunately, the atmosphere he simulated consisted of a reducing environment of methane, ammonia, hydrogen, and water vapor, and not carbon dioxide and nitrogen, for which the experiment does not work. Miller's experiment, although ground-breaking, is now more of historical interest.

It seems then that the pre-biotic soup did not exist for if it did it would have been rich in amino acids that would have led to considerable nitrogen content in early organic matter. Instead the content was relatively low—just 0.015 percent. This indicates that there was not a substantial soup present when the pre-cambrian sediments were formed (unless it existed for only a brief period of time). Currently there is an interest in volcanic vents possibly providing the necessary chains of amino acid needed for life. At this stage it is only speculation.

10.4.2 Evolution of Proteins

Given the early conditions of the earth's atmosphere we have the initial problem of obtaining amino acids of which there are 20 used for life. Then we have the problem of forming basic protein molecules and more complex ones like insulin that consist of long and irregular non-repetitive sequences of amino acids. The order is as critical for the protein to carry out its function as the order of letters in a word is for giving the word its meaning. Many people have endeavored to calculate various probabilities for proteins and DNA to form through random processes. For a protein, the first complication is that every amino acid has two types, a left-handed version or L-form, and a right-handed version, or D-form. Functioning proteins use only L-forms, although both forms have roughly the same frequency in nature.

---

[37]Strobel (2000: 135).

A second complication is that the amino acids must form a chemical bond call a peptide bond so that the amino acids can join together. Peptide and non-peptide bonds occur at roughly the same frequency. The third complication is that we have to get the right ordering of the amino acids, of which there are 20 to choose from. Assuming statistical independence, we can calculate the probability of getting at random a prescribed short protein of 100 L-form amino acids with 99 correct bonds as:

$$\left(\frac{1}{20}\right)^{100} \times \left(\frac{1}{2}\right)^{99} \times \left(\frac{1}{2}\right)^{100} \approx 10^{-190}.$$

Now this is an over-simplification as some positions can tolerate alternatives (though not too many)[38] which, with other possible factors (like some dependence), increase this probability. However, even allowing for these adjustments, we still seem end up with a number like 1 chance in N, where N is a number greater than the estimated number of molecules in the universe. And this is for a comparatively short protein! In contrast take an extraordinarily complex protein and enzyme complex like nitrogenase that is a catalyst that splits the bonds in a nitrogen gas molecule to make soluble nitrates and, as it were, "fix" nitrogen. There is only one known family of enzymes that accomplishes this process and it has about 25,000 atoms with about 2000 amino acid residues. With regard to the formation of further proteins we note that the human body has about 100,000 proteins. This raises the question of having enough time for the development of proteins to take place, though natural selection will change the probabilities.

The probabilities will also be small for the random formation of other biological building blocks such as DNA and RNA, which work hand in hand to produce proteins with the correct sequencing of the amino acids. It is not appropriate to discuss all the intricate interactions between DNA and RNA here except to make a few general comments. DNA carries the genetic information for the synthesis of the proteins, while there are several types of RNA including its messenger form (mRNA), transfer form (tRNA), and its ribosomal form (rRNA), plus a number of others. Here mRNA carries information from DNA by enzymes to ribosomes, which are molecular machines consisting of proteins and rRNA. These machines "read" the mRNA and use tRNA to translate the four-letter genetic code into the twenty-letter code of amino acids and catalyze the formation of the protein. Biologists Gordon Mills and Dean Kenyon[39] make the following comment:

> Primordial DNA synthesis would have required the presence of specific enzymes, but how could these enzymes be synthesized without the genetic information in DNA and without RNA for translating that information into the amino acid sequence of the protein enzymes? In other words, proteins are required for DNA synthesis and DNA is required for protein synthesis.

---

[38] See the discussion of Axe's work on folded proteins in Meyer (2013: chapter 10).
[39] Mills and Kenyon (1996).

This gives us the "chicken and egg problem." Which came first the chicken or the egg? (A simpler egg and a simpler chicken beginning with what?)

The authors discussed and rejected various hypotheses based on RNA developing first at random and becoming self-replicating, and concluded:

> We consider that historical biology should be open to all empirical possibilities, including design—and see the molecular biological system of organisms, of which RNA is so stunning a part, as exemplars of design.

As Meyer[40] notes,

> it doesn't matter what pre-biotic conditions are assumed along with the most favorable reaction rates, the probability of obtaining information-rich bio-macromolecules at random is in the words of physicist Ilya Prigogine and his colleagues, " vanishingly small . . . even on the scale of . . . billions of years."[41]

As a mathematician specializing in statistics I remember reading with interest papers coming from the Wistar conference in 1966 entitled "Mathematical challenges to the Neo-Darwinian interpretation of evolution" involving mathematicians, engineers, and concerned scientists. The topic of combinatorics that studies the number of ways a group of objects can be combined or arranged (and is connected with probability) played a prominent role in the discussions. Although more sophisticated models for calculating probabilities have since arisen, like the discussion above on protein formation, the problems raised by that conference still hold today and haven't been given the prominence they deserve. The probabilities are still extremely small.

Thaxton et al. sum up the position well with the comment "that 'chance' should be abandoned as an acceptable model for coding of the macromolecules essential in living systems."[42] This now seems to be an accepted view and it appears to me that evolution has a direction that suggests a final cause (and not just natural selection), as well as some randomness. Philosopher Alvin Plantinga[43] notes that you cannot rule out scientifically the possibility that evolution is directed or purposeful, or even guided by God.

The regulation of a gene is much more complicated than originally thought. It is influenced by multiple pieces of regulatory code that are located both close to and far from the gene itself, and by noncoding RNA, that is not coding for proteins. It appears that RNA decoded from DNA as a basic unit of heredity is perhaps more fundamental than the traditional concept of the gene itself. Not all genetic information is stored using the standard genetic code, and protein synthesis in human mitochondria relies on a genetic code that differs from the standard code. A major question was how the genetic code originated in the first place since proteins are required for DNA

---

[40] Meyer (2012: 272).
[41] Prigogine et al. (1972: 23).
[42] Thaxton et al. (1984: 146).
[43] Plantinga (2011: xii).

synthesis and DNA is required for protein synthesis; the "chicken and egg" problem again.

It seems that RNA molecules with their diverse roles in biological systems could have been first on the scene (the RNA World hypothesis). Although some RNA molecules code for proteins and act to translate codons to amino acids, others (called ribozymes) are long enough to fold into specific three-dimensional shapes that endow them with the ability to catalyze specific chemical transformations. They are responsible for several processes including protein synthesis, transfer RNA (tRNA) processing, and self-splicing of certain introns (intervening non-coding sequences). However, as geophysicist and philosopher of science, Stephen Meyer says that

> any RNA molecule capable of even limited function must have possessed considerable (specified) information content. Yet explaining how the building blocks of RNA arranged themselves into functionally specified sequences has proven no easier than explaining how the constituent parts of DNA might have done so, especially given the high probability of destructive cross - reactions between desirable and undesirable molecules in any realistic pre - biotic soup.[44]

10.4.3 Evolution of the Cell

We have moved from the initial atmospheric conditions to proteins, and now we consider the evolution of a cell. A living cell differs from a nonliving system in at least three ways: it must process information, store information, and replicate. Bruce Alberts, who was president of the National Academy of Sciences, in an editorial described the cell as a factory containing an elaborate network of interlocking assembly lines each consisting of a set of large protein machines.[45] He used the word machines because these protein assemblies contain highly coordinated moving parts—like machines designed by humans. We see these machines, which we now look at, at work in an amazing way in the construction of proteins.[46]

First, a molecular machine unwinds part of the DNA helix shaped molecule in a process called transcription to expose the genetic instructions for assembling the particular protein. These instructions are then copied by another machine to form a molecule known as messenger RNA that moves out of the cell nucleus to the ribosome, a molecular factory. Here there is an assembly line where a sequence of amino acids is formed from amino acids transported from other parts of the cell and then linked into a long properly sequenced chain. This is then moved from the ribosome to another molecular machine that helps fold it into the precise shape necessary for its function. Finally the protein is released and moved by another machine to the exact place where it is needed. The cell has features that seem to indicate that it was designed. Physicist Jeff Tallon in a local newspaper article commented:

---

[44]Meyer (2012: 277).
[45]Alberts (1998).
[46]See also Rana (2008).

> The cell ... is a marvel of complexity — a miniature city, with information systems, transportation systems, refuse collection, factories, ambulances, police and gatekeepers. Each human cell has half a million ribosomes, and to pick just one of the 100,000 proteins, haemoglobin is produced by ribosomes in bone marrow at a rate of 100 million million protein molecules every second.

Under just a "random" model the extremely small probabilities just keep on multiplying.

Combining the short time frame for chemical evolution to occur together with the infinitesimal probabilities involved, I believe the scientists Cyril Ponnamperuma and Carl Woese summed it up well with the comment:

> Not only was the time too short, but the mathematical odds of assembling a living organism are so astronomical that nobody still believes that random chance accounts for the origin of life.[47]

Paleontologist Simon Conway-Morris takes the view that evolution was inevitable and seemingly guided by invisible constraints that lead to convergence in which two or more lineages have independently evolved similar structures and functions. He makes the comment:

> The heart of the problem I believe is to explain how it might be that we, a product of evolution, possess an overwhelming sense of purpose and moral identity yet arose by processes that were seemingly without meaning.[48]

Woese[49] argues that we need a new biology for the twenty-first century, and the reductionism of the twentieth century (which in fact underlies materialism) and molecular biology, are no longer adequate. This is echoed by some leading evolutionary biologists, especially those associated with a group of scientists known as the "Altenberg 16" who doubt whether mutation and natural selection can drive evolution. Paleontologist Stephen Gould said that neoDarwinism "is effectively dead, despite its persistence as textbook orthodoxy";[50] support for this view has been steadily growing. As we shall see below, one of the problems is the origin of new biological information.

10.4.4 Animal Fossils

We first consider the fossil records of animals. Multicellular life can be traced back to the pre-Cambrian, and one-celled bacteria to more than 3.5 billion years ago. The Cambrian was a geological period that probably started a little more than 540 million years ago. It began with an explosion of life sometimes referred to as the "biological big bang" because it gave rise to the sudden appearance of most of the major phyla that are alive today, as

---

[47]Ponnamperuma and Woese, Newsweek, August 6, 1979.
[48]Conway-Morris (2003: 2).
[49]Woese (2004).
[50]Gould (1980: 120).

well as some that are now extinct. Paleobiolgists Douglas Erwin and James Valentine[51] have written an extensive and detailed book on the subject and describe an even earlier explosion in the Ediacaran "period" that they dubbed the "Avalon explosion."

The Cambrian explosion was of a comparatively short duration of about twenty-five million years or so, which is a small fraction of the earth's history, and asks the question of whether there was sufficient time for the necessary evolutionary changes to take place. This explosion of life occurred at the same geological time all round the world. For example, we have the famous Burgess shale site, due to volcanic activity, in Yoho National Park connecting with Marble Canyon in Kootenay National Park Canada, and in Southern China the Cambrian Explosion has been documented in fine detail in the Chengyiang bed. Before the explosion there were some jellyfish, sponges, and worms, and no evidence of a long history of gradual divergence.

After the "explosion" there were, for example, representatives of the arthropods (leading to insects, crabs and so forth), echinoderms (including modern starfish and sea urchins), and chordates, including modern vertebrates (leading to mammals later). These animals, although fundamentally different in body structure, suddenly appeared fully developed. There were between twenty and thirty-five completely novel body plans that appeared. One example is the trilobite that suddenly appeared on the stage with an articulated body, a complicated nervous system, and compound eyes. Among the Chengyiang animals they found 136 different kinds of animals representing diversity in the level of phyla and classes. Further extensive deposits from volcanic activity have been found in North America such as the Florissant beds in Central Colorado.

We see then that we have a sudden explosion of genetic information, for example Cambrian animals would have needed complex proteins like lysyl oxidase that requires four hundred amino acids in today's animals. But after the explosion there was stasis, that is basic body plans remained distinct for a very long time. This maintenance of stability within species raises some questions about neo-Darwinism. As Erwin and Valentine acknowledge:

> The patterns of disparity observed during the Cambrian pose two unresolved questions. First, what evolutionary processes produced the gaps between the morphologies of major clades?[52] Second, why have the morphological boundaries of these body plans remained relatively stable over the past half a billion years? After all, there is no a priori reason clades could not display a pattern of rapid exploration of morphologic space, coupled by a subsequent expansion of that space during the Phanerozoic, but it is not a pattern commonly observed among the bilaterian metazoans.[53]

---

[51] Erwin and Valentine (2013).
[52] A clade is a group of organisms believed to comprise all the evolutionary descendants of a common ancestor.
[53] Erwin and Valentine (2013: 330).

According to neo-Darwinists, small differences in structure occur before large differences in structure and body plans, but with the Cambrian explosion the reverse took place with major differences occurring first, followed by minor variations later; for example phyla before classes, classes before orders, and orders before families. Jun-Yuan Chen, a Chinese paleontologist and acknowledged expert on the Cambrian explosion, described what happened as "top-down" evolution, to contrast it with the "bottom-up" evolution required by Darwin's theory. Apparently the number of animal phyla become fewer with time, not greater, and new phyla have not continued to appear in all the ages since the early Cambrian period.

Various theories have been put forward to try and account for the lack of fossilized precursors in the Precambian period such as the smallness of the animals or the lack of hard body parts, which is not the case as there is counter evidence for these theories. Even though conditions for the preservation of ancestral forms, whether soft-bodied or microscopic, were ideal (even sponge embryos are found in similar strata), the precursors are absent. Other arguments like incomplete sampling to explain the lack of precursors do not stand up statistically. Paleontologist Michael Foote, for example, believes that we actually have a representative sample of morphological diversity.[54]

More recently, with the developments in genetics, there are indications that genetic divergence may have began very much earlier than what the fossils suggest so that there would appear to be an extremely ancient Precambrian ancestor of the Cambrian animals. However we still have the problem of the missing ancestral fossils as well as an extremely wide divergence of suggested dates (say by a billion years) for the appearance of an ancestor. The dates depend on which molecules and genes are studied as well as which estimation methods are used. Clearly the Cambrian explosion provides a major problem for neo-Darwinism and for the so-called "tree of life" that shows divergence from a common ancestor (the deep-divergence hypothesis). Attempts to construct trees have led to some contradictions. So-called convergent evolution whereby animals with very different body plans end up with similar features also creates problems for finding a common ancestor and for constructing an appropriate tree.

Transitional Forms

Although DNA builds proteins as well as providing switches in epigenetics, other structures have to be formed so that with the Cambrian explosion there is the question of how the hierarchical arrangement of cells, tissues, organs, and body plans developed? Some in the past have argued that we will eventually find transitional forms before this period, yet discoveries over the past 150 years have shown that the explosion was even more abrupt and extensive than originally thought. It is suggested that we would expect missing links to be rare as there would be fewer than both their ancestral and descendent

---

[54]Foote (1997: 181).

forms. Speciation can in fact be a very rapid process after a gene shift, and the change can be very marked. Paleontologist Roger Cuffey[55] lists a large number of transitional forms with references including many "crossing from one higher taxon into another."[56] A lot of fossils have been found recently so that we need to maintain an open mind on this question of transitional forms. A good example is the Tiktaalik fish that has a flat head with eyes on top and gills and lungs. On the other hand, postulating that more transitional forms will eventually be found is a bit like replacing the God-of-the gaps argument by the evolution-of-the gaps argument.

At present there are still missing transitional forms between major species, but some forms have been put forward as transitional such as those for frogs and salamanders, turtles, snakes, bats, flatfish, horses, whales (the recent find of *Himalayacetus subathuensis*), and even humans.[57] In the past the archaeopteryx was described as the missing link between reptiles and birds. Others believed it to be a different species of extinct bird with modern feathers and teeth, and not part bird part reptile. It has now been suggested that it was a dinosaur as a new small dinosaur from China shares traits with Archaeopteryx indicating that the new find may have been an ancestor.

The horse is another case in point as there is not a continuous series. In fact it can be argued that there is devolution in some cases. For example, the earlier Eohippus has 18 ribs, the later Orohippus 15, the earlier Pliohippus 19 and the later Equus Scotti 18. The smallest (dog-sized) animal in the series (Eohippus) is not a horse but a rock badger. In dealing with transitions there is some debate as to what constitutes an ancestor. What is also interesting is that some creatures, for example the Coelacanth, a sturdy fin fish from the Devonian Period that has been put forward as half fish and half reptile and is really a fish, looks identical to fossils specimens 60 million years ago. Other attempts at explaining the gaps have led to theories like "punctuated equilibrium", for which there appears to be little evidence.

The underlying similarity—or "homology"—between species can be explained by either design or "descent with modification." One explanation is that the similarities come from similar genes. This, however, is problematic as there are some cases where similar features come from different genes, and there are plenty of cases where similar genes have produced very different features (e.g., eyes in mice, octopuses, and fruit flies). The genes are so similar that if you put the mouse eye gene into a fruit fly with its own eye gene missing, the fruit fly will develop its eyes as it normally would. A serious problem relating to descent with modification is the existence of "orphan" genes or more technically "ORFans" (for "open reading frames of unknown

---

[55] Moore and Cuffey (1972).
[56] For a more recent list see Theobald (2012).
[57] See Prothero (2007) and the internet, such as http://www.talkorigins.org/faqs/comdesc/section1.html, accessed August, 2015.

origin"). These are simply "new" genes that lack sequence similarity to any other known genes, and they turn up all over the place. Although the number varies between species, roughly 10-20% of each genome's protein-coding sequence contains such genes, and more are turning up as investigations continue.

When it comes to using mutations of various kinds such as exon shuffling, gene duplication, gene transfer and so on, along with natural selection, we have to have genes in the first place for change to take place. With regard to mutations, in order for them to change body plans the mutations have to occur very early on, and it is these ones that tend to be harmful. The genes involving the major changes needed for macroevolution tend to either not vary, or vary to the detriment of the organism. In the words of the geneticist John McDonald,

> Those [genetic] *loci* that are obviously variable within natural populations do not seem to lie at the basis of any major adaptive changes, while those *loci* that seemingly do constitute the foundation of many if not most major adaptive changes are not variable within natural populations,[58]

and he refers to the problem as a "great Darwinian paradox."

A question often raised is what about the *Hox* genes mentioned in Section 10.1.1 as small changes in these can have a major effect on the organism as they affect body plans.[59] There are three problems with regard to mutations of *Hox* genes.

First, *Hox* genes orchestrate many other different genes and so it is perhaps not surprising that experimentally generated macro-mutations such as used in fruit fliy and mice experiments for example tend to be either harmful or inhibiting. We are also beginning to find mutations in the human *Hox* genes that have been shown to cause congenital malformations. Second, as emphasized by Meyer,[60] the genes are expressed after the beginning of animal development and certainly after the body plan is being set in place, so that they do not determine overall body-plan formation. The argument is that the genes determine local differentiation of body parts within a body plan, but global body-plan specification is regulated by entirely different genes. Third, the *Hox* genes function as switches for turning other genes on and off and in themselves do not contain the necessary information for building structural parts; they only alter genes that are there. It would seem then that mutations in the *Hox* genes could not build body parts by themselves and they cannot regulate genes if the genes are not there! Certainly research in this area will be interesting to follow as the *Hox* genes no doubt play a fundamental role in body-plan development.

---

[58]McDonald (1983: 93).
[59]e.g., Gilbert (2000: chapter 22) and http://www.ncbi.nlm.nih.gov/books/NBK9978/, accessed August, 2105.
[60]Meyer (2013: 319).

10.4.5 Human Evolution

As we have climbed up the evolutionary ladder we have now arrived at humans. According to human evolution, the human line first emerged in East Africa about 6 or 7 million years ago, when we split off from the apes from a common ancestor. We are told that we were up on our hind legs, namely bipedal, before 4 million years ago; our brains conveniently exploded up in size about 2 million years ago; and modern humans emerged a mere hundred thousand or so years ago. Java man, constructed from a skull cap, femur, and three teeth was not an apeman but a "a true member of the human family."[61]

Humanoid Fossils

There are quite a number of humanoid fossils, many of which are not considered direct ancestors to *Homo sapiens*, but are loosely related to direct ancestors.[62] However, there are apparently a number of human species, and a type of bar graph showing the different ones and their timespans is given by the Smithsonian Institute.[63] There is some debate about the number of species, and the pictures given there are a testimony to vivid imagination! As noted by geneticist Richard Berry,[64] the earliest remains of *Homo erectus* were found widely distributed in the upper Pleistocene about a million years ago. Its skeleton was somewhat like ours, but with a smaller brain, bony ridges over the eyes, a ridge on the back of the skull to attach powerful neck muscles (could do with those now!), flat face with no chin, and large upper incisors. Like some of us today he was an erect walker, meat eater, and toolmaker.

It has been suggested that Adam was a Neolithic farmer in Turkey, and anthropologist Victor Pearce[65] listed some evidences supporting his theory that link in well with cultural aspects of Genesis chapters two to four. For example, the cultural surrounding of Genesis places Adam and his successors after the Old Stone Age and at the threshold of the Bronze Age, a time frame called the Neolithic Age (New Stone Age). We find references to tents, farming, livestock, musical instruments, implements of bronze, and even iron that give us a rough timeframe for Adam and his immediate generations. A major question is who or what produced the dramatic cultural changes, especially those about 4,000 BC, that led to the sudden appearance of complex language, organized religion, writing, mining, mineral smelting, and complex city societies? Some scientists suggested that the first "protolanguage" evolved slowly but at the same time they trace all Indo-European languages back to only one protolanguage that originated in the Middle East.

---

[61] Lubenow (1992: 87).
[62] See, for example the summary by Berry and Jeeves (2008: 7–8).
[63] http://humanorigins.si.edu/evidence/human-fossils/species, accessed August, 2015.
[64] Berry (1988: 68).
[65] Pearce (1969).

When it comes to human evolution we find there are very few fossils that are believed to be of creatures ancestral to humans, and often all we have are skull fragments or teeth. Imagination runs riot when artists try to reconstruct the whole person! Also the times that separate the fossils are so great that we cannot say anything definite about their supposed ancestral connections. It would be like constructing a plot from a large book by looking at a random selection of a small number of pages. Henry Gee, a chief science writer for Nature and an evolutionist, comments on this when he said that all the evidence for human evolution between about 10 and 5 million years ago "can be fitted into a small box." Further, the conventional picture of human evolution as lines of ancestry and descent is "a completely human invention created after the fact, shaped to accord with human prejudices," and:

> To take a line of fossils and claim that they represent a lineage is not a scientific hypothesis that can be tested, but an assertion that carries the same validity as a bedtime story—amusing, perhaps even instructive, but not scientific.[66]

Types of men earlier than Homo sapiens of the Upper Paleolithic (i.e. before 30,000 BC) have left no progeny; for example, Austral opithieinae, Homo erectus, and Homo neanderthalensis have all died out, but there is evidence of interbreeding of Neanderthals with Homo sapiens.[67] No doubt the picture will change as more fossils are found.[68]

Thought Processes

We have looked at our physical origins, but what about our mental origins? In Chapter 2 we saw that we appear to have a mind that is more than just a brain so that what then is the role of evolution in forming our mental abilities and our beliefs? As noted there, the truth or falsity of a belief can only be affected by natural selection if the belief can influence behavior, which will either be selected for or selected against. As philosopher Edward Feser argues, natural selection tends to maximize the capacity of an organism to survive and reproduce, and

> there is no reason to assume that having a true system of beliefs really is what is most conducive to survival: maybe our environment is such that we have been able to survive and reproduce as well as we have only because we have developed a mostly false system of beliefs.[69]

If all we have are purely neurophysiological properties underlying thought, then there is no way that the truth or falsity of a belief can have any effect on behavior. Feser goes on to argue that the validity of a belief will depend on its content so that evolution cannot account for the reliability of our thought processes. This resonates with a comment made by Hawking quoted in Section 2.2 under "string theory", which essentially raises the same question about

---

[66] Gee (1999: 23, 32, 113-117, 202).
[67] See "neanderthals and modern man" on the internet.
[68] For a recent discovery of a fossil jaw and comments see http://www.bbc.com/news/science-environment-31718336, accessed August, 2015.
[69] Feser (2006: 153).

validity. Philosopher Alvin Plantinga follows a similar line in providing an evolutionary argument against naturalism.[70]

Some have likened our brain to a digital computer. However, a computer simply runs programs that produce meaningless results unless interpreted as meaningful by us. It is a mechanical device we can use and it is something we observe; we are the ones who give it significance. Referring to comments by Searle, Feser says:

> If computation is observer-relative, then that means that its existence presupposes the existence of observers, and thus the existence of minds; so obviously, it cannot be appealed to in order to explain observers or minds themselves. That would be to put the cart before the horse.

He concludes that:

> the mind's ability to think in accordance with the laws of logic cannot be explained in terms of the brain's running a certain kind of program.

Computers run on algorithms or "rules", but these can have a variety of interpretations unless we appeal to a mind that interprets the rules. We cannot therefore explain the mind in terms of algorithms!

10.4.6 Second Law of Thermodynamics

We discussed this law in Section 2.4.1, and we mention it again as it has some bearing on chemical evolution as it is a process of going from the simple to the more complex. It is true that the earth is not an isolated system but rather a closed system that receives energy from the sun. However, as important macromolecules of living systems like proteins, DNA etc. are more energy rich than their precursors (amino acids, heterocyclic bases, phosphates, and sugars), the second law would predict that such macromolecules will not spontaneously form. As Thaxton et al. (1984: 121) comment:

> It seems safe to conclude that systems near equilibrium (whether isolated or closed) can *never* produce the degree of complexity intrinsic in living systems. Instead they will move spontaneously toward maximizing entropy, or randomness.

Unfortunately long time periods don't help as the system will move in the wrong direction, namely toward equilibrium. Nobel Laureate Ilya Prigogine and colleagues[71] make an equally strong point:

> The probability that at ordinary temperatures a macroscopic number of molecules is assembled to give rise to the highly ordered structures and to the coordinated functions characterizing living organism is vanishingly small. The idea of spontaneous genesis of life in its present form is therefore highly improbable even on the scale of the billions of years during which prebiotic evolution occurred.

---
[70]Plantinga (2012).
[71]Prigogine et al. (1972: 23).

Prigogine developed a more general formulation of the laws of thermodynamics that included nonlinear, irreversible processes such as autocatalytic activity.[72] Here there are some systems, for example in fluid dynamics, that obey nonlinear laws that produce two kinds of behavior. Near equilibrium, order in the system breaks down, but if the system is driven far from equilibrium through energy passing through it at a high rate, ordering may suddenly appear. We therefore need a system that is constrained in some way so that the energy flow maintains the system in a state far from equilibrium. A complex state such as an ordered state that is highly unlikely under equilibrium can then be maintained indefinitely. There are two problems with this. First, as we have previously mentioned, ordering is not the same as complexity, as the macromolecules of biological interest are highly irregular. Second, as the ordering depends on boundary constraints, a system cannot be more complex and contain more information than that of its environment. Biochemist Manfred Eigen in 1971 showed that selection would not lead to evolutionary development in an open system unless the system was maintained far from equilibrium, and that the reaction would have to be autocatalytic (the reaction product is itself the catalyst for that reaction) and be capable of self-replication.

With regard to autocatalysis, Stuart Kaufmann[73] put forward his so-called autocatalytic set theory to explain the origin of life. Although not rejecting natural selection as part of the evolutionary process, he maintains that a combination of natural selection and mutation as in Neo-Darwinism is not the sole source of order. His theory in its simplest form postulates that life is a collection of molecules that catalyze each other by chance, thus producing a stable whole; life started without DNA. However, if life did not start with DNA or RNA or proto-RNA, Kauffman has to describe the transition from autocatalytic sets to DNA-based life. Another difficulty with the theory is that autocatalytic sets depend on reversible reactions so that if there are enough different molecules to allow a randomly chosen chemical to catalyze a randomly chosen reaction, then we can expect a randomly chosen chemical to inhibit a randomly chosen reaction. So processes that build complexity also tend to destroy it. Like other computer models, Kaufmann's model is not complete with regard to contributing factors, and doesn't include inhibition, for example. Many metabolisms are possible, but how did we get ours? Other suggestions for the creation of life have been put forward such as vents in the ocean, life from clay, and life from outer space. They all have serious problems and can be laid to rest!

[72] See Prigogine and Nicolis (1977).
[73] Kaufmann (1995); see also Hordijk, Hein, and Steel (2010).

10.5 IRREDUCIBLY COMPLEX SYSTEM

A topic that has some considerable debate is the so-called irreducibly complex (IC) system. The term was introduced by biochemist Michael Behe who defined such a system as one "composed of several well-matched, interacting parts that contribute to the basic function, wherein the removal of any one of the parts causes the system to effectively cease functioning."[74] His example of such a system is the standard mouse trap consisting of a platform, a hammer, a spring, a catch, and a holding bar. Each part is essential for the mouse trap to do its job. In contrast, a system is cumulatively complex if the components of the system can be arranged sequentially so that the successive removal of components never leads to the complete loss of function (e.g., as in a house or a city). Natural selection and mutation could account for a cumulatively complex system, but according to Behe:

> An irreducibly complex system cannot be produced ... by slight, successive modification of a precursor system, because any precursor to an irreducibly complex system that is a missing part is by definition nonfunctional ... Since natural selection can only choose systems that are already working, then if a biological system cannot be produced gradually it would have to arise as an integrated unit, in one fell swoop, for natural selection to have anything to act on.

In other words, natural selection only preserves things that perform a function, namely survival to the next generation. Irreducibly complex systems do not function until all the parts are present simultaneously and working together, though this is an over-simplification. The argument is that if natural selection is going to produce such a system, it has to produce it all at once or not at all. This would not be a problem if the systems in question were simple. But they are not.

The irreducibly complex biochemical systems Behe considers are protein machines consisting of numerous distinct proteins, each indispensable for function; together they are beyond what natural selection can muster in a single generation. However, nature is full of amazing adaptions and it is unlikely that God directly intervened on every occasion, as some would argue. Clearly another explanation is needed and it need not rule God right out of the picture.

Examples

One such biochemical system that Behe refers to as an irreducibly complex system is the bacterial flagellum. The flagellum is a whip-like rotary motor that enables a bacterium to navigate through its environment using sensory systems that tell it when to turn on and when to turn off so it guides it to whatever it is seeking. The flagellum includes an acid-powered rotary engine, a stator, various rings, bushings, and a drive shaft. The propellor

---

[74]Behe (1996: 39).

can stop spinning within a quarter of a turn and yet instantly start spinning the opposite way at 10,000 rpms. Behe[75] says that a flagellum is of the order of a couple of microns in length where a micron is 1/20,000 of an inch, with most of its length a propellor. The motor is only about 1/100,000 of an inch. The intricate machinery of this molecular motor requires the coordinated interaction of some thirty to thirty-five complex protein parts. Yet the absence of any one of these parts results in the complete loss of motor function. However, a partial system might still have some biological function such as the type III secretory system; it has some proteins homologous to the flagellum, though there are problems in arguing from homologies.[76] For a suggested evolutionary development of the Flagellum see http://www.talkdesign.org/faqs/flagellum.html, and for a response to this see www.evolutionnews.org/2011/03/michael_behe_hasnt_been_refute044801.html, both accessed August, 2015.

A second example is the cilium, a whiplike hair on the surface of eukaryotic cells; a cell contains about 200 of them. In the respiratory tract they beat together in harmony to sweep mucous toward one's throat for elimination. A single cilium consists of about 200 protein parts and is made up of rods, linkers, and motors forming a complex system that appears to be irreducible with regard to these three components.

A third example is the so-called Krebs cycle, an intricate biochemical pathway consisting of nine enzymes and a number of cofactors playing a central role in cellular metabolism. Melendez-Hevia et al. say that:

> In the Krebs cycle problem the intermediary stages were also useful, but for different purposes, and, therefore, its complete design was a very clear case of opportunism.[77]

and "In this case, a chemical engineer who was looking for the best design of the process could not have found a better design than the cycle which works in living cells." The words "opportunism" and "best design" suggest something more than chance, namely direction. It does not seem surprising that a protein can play different roles, depending on what other proteins it teems up with. Building materials in a house can be combined in a variety of ways to produce various types of houses or structures.

A number of other processes have been regarded as irreducibly complex such as the human immune system, electron transport, telomere synthesis, photosynthesis, blood clotting, material transportation within cells, and the synthesis of nucleotides. Almost any biological structure of interest such as gills, lungs, feathers, etc. possess multiple necessary components. In particular, some animals and insects have amazing features. For example, a certain species of snakes called bolyerines are boa-like but have the unique ability

---

[75] Strobel 2004: 254).
[76] See Section 10.4.4.
[77] Melendez-Hevia, Waddell, and Cascante (1996).

among vertebrates to fold the front half of their upper jaw backwards when they attack prey. Such a mechanism requires a great number of special features involving bones, joints, tissues, and ligaments along with specialized nerves all working together.

Another example is the tiny bombardier beetle that can shoot an explosive "bomb" of gases out of its tail end four or five times in succession. It has "twin exhausts" at its tail that produce two chemicals providing an explosive mixture along with two enzymes that make the chemical reaction go millions of times faster. The beetle uses a special inhibitor chemical to keep the mixture from reacting inside and preventing a premature explosion. A great many processes have to work together or else no beetle! Explosive chemicals, inhibitor, enzymes, glands, combustion tubes, sensory communication, muscles to direct the combustion tubes, and reflex nervous systems, all have to work perfectly.

Irreducible complexity, as defined above, has not been generally accepted by the scientific community. An argument put forward against the irreducible complexity of say the flagellum of the *E. coli* bacterium is that of its approximately 40 different kinds of proteins only 23 are common to all other bacterial flagella studied so far suggesting it is possible to make considerable changes to the system without spoiling it. Of the 23 proteins only two are unique to flagella and the others all closely resemble proteins that carry out other functions in the cell. However, I don't find this argument particularly convincing. It's like comparing different models of an electric appliance. They might have similar functioning parts (e.g., a button, a spring, a heating element) but each one has its own design and structure.

A more powerful argument uses the fact that evolution also takes away parts. For example, evolution might add a part, part 1 say, that is initially just advantageous and improves things slightly, but is not essential. However, later evolution leads to part 2 being changed or removed that now makes part 1 essential. This can happen several times so that all that is left are parts that were just favorable but are now all essential. A question here is why would part 2 be changed or removed. Perhaps because part 1 with part 2 changed or removed is more adaptive than part 1 plus part 2. Since this scenario is possible, it cannot be ruled out, though the word "complex" still applies and still raises statistical problems. We do, however, need to avoid using a a God-of-the gaps argument to account for such complexity, although complexity is still a problem.

Some have suggested how evolution might get round this complexity problem by using a kind of Darwinian optimization principle and natural selection, and have demonstrated how it might happen by generating some target word sequence on a computer using simulation.[78] This has been achieved in a relatively short number of generations for some examples, but suffers from

---

[78] For example Dawkins (1996: 46–47).

several flaws. To begin with, the target is pre-chosen and the algorithm is programmed to reach the target through changing the probability. Molecules in real life don't have a target sequence "in mind" and, in between, sequences may not confer any evolutionary advantage. In fact they may be a disadvantage. What is missing from the experiment is the idea of one word having a selective advantage of over another, both being short of the target. This applies to two proteins that are both ineffectual in developing a hypothetical protocell, and having to choose one.

Another approach to get round chance has been to use the idea of self-organization discussed above where energy is applied to a system under non-equilibrium conditions. In biological systems the idea of autocatalysis, already mentioned, has been used where a particular molecule acts as its own catalysis to rapidly reproduce itself. It is found that particular amino acids do tend to form linkages more readily with some amino acids than with others that could produce some sort of self-organization, but this does not tie in with what we see in protein sequences of amino acids. Chemical affinities do not generate complex specific sequences; they tend to produce redundancy and repetition (e.g., as in crystals) instead of unpredictable and aperiodic sequencing, and thus reduce the ability to convey new information.[79] Order is not the same as information, and the problem is the generation of information as self-organization cannot provide new information.

10.6 IS EVOLUTION COMPATIBLE WITH THE NATURE OF GOD?

This is the question asked by Graeme Finlay,[80] who raises the problem of evolutionary dead ends and why they happened. There were five major catastrophic extinctions occurring around 443 (Ordovician-Silurian), 359 (Late Devonian), 251 (Permian-Triassic), 201 (Triassic-Jurassic), and 66 (Cretaceous-Tertiary) million years ago, respectively.[81] During the largest extinction event, the Permian-Triassic, popularly known as the Great Dying, some 57% of all families and 83% of all genera became extinct including a mass extinction of insects; marine invertebrates suffered the greatest losses. Apparently all life on Earth today is descended from about the 4% of species that survived. The second largest extinction, the Triassic-Jurassic for which there is a very accurate date, led to the destruction of more than 95 percent of terrestrial megaflora species, plus all the animal species dependent upon these plants. At least half of the species now known to have been living on Earth at that time went extinct. However, the mass speciation that followed was surprisingly quick and robust. Within 10,000 years or less, large theropod

---

[79] For further comments on this problem see Meyer (1999: 82–91).
[80] Finlay (2004b: chapter 7).
[81] There were other mass extinctions as well; cf. Benton (1995).

dinosaurs appeared, and in less than 100,000 years dinosaur species diversity attained a stable maximum. Not only were the new species complex creatures but they appeared amid hostile environmental conditions. The short time between the extinction and Jurassic speciation is surprisingly short and requires an explanation.

The last extinction event wiped out the dinosaurs, and it is generally put down to a huge asteroid colliding with the Earth at the end of the Cretaceous period. A huge amount of sediment was blown into the atmosphere blocking the light and causing the atmosphere to cool dramatically, which led to the extinction of the (largely cold blooded) dinosaurs. Finlay notes the parallel disasters in God's dealing with Israel when His plans seem to come to nothing, for example the destruction of the kingdom of Israel (Samaria destroyed in 722 BC) and of Judah (Jerusalem destroyed in 587 BC). He argues that God has given to His creation, human or otherwise, a certain degree of of freedom. He also mentions that out of disasters God can achieve new beginnings and new directions. For example, the demise of the dinosaurs perhaps made room for the large-sized mammals to emerge. Also Finlay says that small numbers may have great effects, and we can add to this that small changes can lead to unexpected outcomes as we saw in Section 4.1.1 on chaos theory. Summing up, the answer to the question "Is Evolution Compatible with the Nature of God?" is yes. It can also be argued that evolution goes some way in explaining suffering.

A related question is: "If evolution is true then what effect does this have on Christian theology?" I am not a theologian so my comments on this will be brief. Before we begin, however, it is clear from this chapter that the current theory of evolution, namely neo-darwinism, has a lot of problems (including those mentioned by evolutionists) so that a revised theory is needed, if one exists. However, it may be that God used evolution to bring life to our planet and, if this is the case, we would need to reinterpret the first few chapters of Genesis; six different interpretations are mentioned in Section 8.9.1 that are compatible with evolution. Finlay asks the question:

> Is it feasible that a lineage of ape-like creatures progressively losing its ability to make vitamin C, its hair, and its sense of smell, and sustaining the random invasion of myriad retrotransposons, could be ancestral to *Homo Divinus*?[82]

and answers:"God works through apparently insignificant and particular players to achieve unimaginably grand ends." He also says: "God consistently creates what is new and magnificent from what is old and flawed."[83]

---

[82]Finlay (2003: 38).
[83]Finlay (2008: 84)

10.7 CONCLUSION

We can conclude that in the formation of proteins and complex biological molecules, chance as it is understood today is currently not supported as the mechanism. We don't have a naturalistic model to explain the origin of the specified genetic information needed to build a living cell and make life inevitable. Also we need to explain such things as the origin of the genetic code, the origin of multicellular life, the origin of sexuality, the gaps in the fossil record, the explosion of life in the Cambrian era, the development of complex organ systems, and the development of irreducibly complex molecular machines. Where does neo-Darwinism fit into all of this? Not too well at present! How do we get from matter and energy to biological function without an input of information from intelligence? This is a key problem.

We find then that evolution, rather than being a haphazard process, appears to have direction and seems to run according to an underlying plan or design, but overlaid with some natural variation. We could call it natural selection with a purpose! Since information implies intelligence, we could infer a designer, irrespective of whether we believed in evolution or not. Those who do not accept the idea of a designer behind the scenes would argue that we don't yet know the laws that lead to self-organization and that appealing to a designer is a "God-of-the-gaps" idea. Of course one can always appeal to future science as eventually providing answers ("Science-of-the-gaps"), which is possible, and there is no way to dispute this. In fact it can be used as an "answer" to anything not understood or apparently outside of "natural law."

It can also be argued that if the laws of nature are from God, then filling in the gaps leads us closer to understanding God's world. In the end, however, it comes back to current evidence and what is the best explanation! In discussing such questions I am not saying that God specifically intervened to produce each intricate system, which is what people seem to regard as the usual intelligent design approach, but I am saying that something more than chance is operating here. If evolution is a true theory, then it appears that there is a final cause—evolution has a direction that suggests a designer is involved. A completely materialistic view of evolution does not provide us with all the answers as Neil Broom, mechanical engineer, sets out very clearly in his book, showing that evolution appears to be purposeful.[84]

Summing up and looking back over this and previous chapters we are faced with the following problems: [85]

(1.) Nothing produces everything (the big bang).

(2.) Reality occurs when we observe it (quantum theory).

---

[84] Broom (2010).
[85] Some of these are from Strobel (2004: 346).

(3.) Randomness produces fine-tuning

(4.) Chaos produces information.

(2.) Non-life produces life.

(5.) Unconsciousness produces consciousness.

(6.) Non-reason produces reason.

(7.) What happened before the Cambrian explosion?

(8.) How did we get from microevolution to macroevolution?

(9.) What is the origin of modern humans?

Clearly science has a long way to go, and as we answer some questions, others turn up. We have an amazing world and the question that arises is an existential one, namely am I alone in this universe.

CHAPTER 11

HOW DO WE GET TO KNOW GOD?

11.1 INTRODUCTION

If the reader does not believe in the existence of God, then this chapter might be irrelevant. However given the evidence presented in this book, it would seem reasonable to keep reading even if just out of curiosity! Initially I was tempted not to write this chapter as I did not want to come across as preaching. However I felt I needed to at least share something with the reader what has meant so much to me. The chapter presupposes four questions:

(1.) What sort of God are we talking about?

(2.) What does it mean to "know"?

(3.) Is God knowable?

(4.) If God is knowable how do we go about knowing God?

In this introductory section we shall look very briefly at these questions through the eyes of the great monotheistic religions, Judaism, Islam, and Christianity. They each have their own way of knowing God, and although there are some common features, there are also some vast differences in both understanding and practice. Some differences have already been mentioned in Section 6.5 and in this chapter I shall first make a few comparative comments below, but will then focus on the Christian approach.

I shall not consider Eastern religions like Hinduism and Buddhism as their god or gods seem to be unknowable in any personal way.

11.2 WHAT IS GOD LIKE?

Art through the ages has portrayed God in a variety of ways that often give false pictures. Philosopher Keith Ward[1] discusses this point in detail, referring to one false idea of God as being a mind that is human in character but immortal, invisible, and transcendent, transcending everything in the physical universe, and being separate from us—out there somewhere. However, the God of Christianity (and of Judaism and Islam) is described as being infinite, that is not being limited by anything else. This means that

> we cannot think of God as a being, even as a large being, who exists in addition to the universe—because if he were outside the universe, he would be limited by it and excluded from it.[2]

The paradox is that God is more than the universe yet is somehow involved with it and upholding it. We just cannot imagine God. That is why in the Old Testament images of God were prohibited, as any image would be bound to be finite and therefore wrong. Ward goes on to say

> that the whole of space and time is the finite image of the infinite God. The whole of the universe expresses, in finite ways, a being which cannot itself be described.

Clearly such an infinite God would lack nothing. As noted in Section 5.3, this idea of God answers the question of whether there is more than one god, that is polytheism versus monotheism. If there were several gods then each god would have some characteristic that would make them different from the others and would therefore be deficient in some aspects of their existence. Such a god would not be an infinite being.

God in Judaism is most often known as Yahweh (YHWH) or Jehovah, which means "I am that I am", or "I will be that I will be" or "the self-existing one", thus indicating the eternal nature of God. Jews believe that even though God is beyond human understanding and imagination, they can have an individual and personal relationship with God because of the covenant relationship that the Jewish nation has with God. In exchange they need to keep God's laws and seek holiness in their daily lives, and thus set a standard for the rest of the world. It is a family and community religion with much of their religious customs such as the Sabbath meal being centered around the home.

---

[1] Ward 1984: 2).
[2] Ward (1984: 3).

God in Islam is known as Allah, the Arabic name for "one God", and traditionally there are 99 other names of God each describing a particular attribute of God. The most familiar and frequent names are "the merciful" and "the compassionate", and in Islam Allah is identified as the same God of Israel who covenanted with Abraham.[3] We find that all three religions describe God's characteristics as all knowing (omniscient), all powerful (omnipotent), and unlimited by space and time (omnipresent).[4] The three religions also agree that God can somehow be known in a personal way and that He reveals Himself through his prophets and through scriptures given to us by people of God.

As mentioned in Section 6.6.5, Muslims believe that the Qur'an is the verbatim word of God revealed through the angel Gabriel over a period of 23 years beginning in 610 AD. They also believe that the Qur'an was exactly memorized, written down, and copied by the Sahaba, Muhammad's companions (disciples, scribes, and family members). Their testimony to Muhammad's life has been accepted by later Islamic scholars as being accurately conveyed through a chain of trusted narrators. They do acknowledge other scriptures such as the Jewish Torah (and encourage Muslims to read it), but believe it has been corrupted. The Psalms of David are recognized except for some written later that are regarded as not divinely inspired. They believe that an original Gospel was divinely revealed to Jesus, but the canonical gospels as we know them give a corrupted version that is not divinely revealed, though they do contain some of Jesus' teaching and details about his life.

For Muslims, the Qur'an is God's authentic and final revelation and Muhammad is the last of a line of prophets including Abraham, Moses, and Jesus. There is also a collection of other writings, called *hadith*, that are believed to be sayings of Muhammad. The choice of which of these writings are used depends for example on whether they are relevant to Shia Islam or Sunni Islam, two sects of Islam, with about 90% being Sunnis.[5]

In the Bible there are many images of God, some male, some female, and others neutral. For example the following are mainly male images: God is described as a father, a husband, a bridegroom, a king, a judge, a shepherd, a builder, and a potter, each demonstrating some aspect of God. We also have female images where God is described as a woman in labour, a nursing mother, and a midwife. Then we have various animal images such as a dove, a hen, an eagle, and a lamb. God is also described as water, rock, fire, wind, light, shield, stronghold, and fortress, each with particular characteristics. Some of these images the Bible describes as God actually using of Himself. As a mathematician, one image that I really like to use is that of God the

---

[3] See Qur'an (2009: Surah 2:137, 3:4–5,4:137, 5:47, 29:47, 6:92); depending on the translation the verse number may need to be reduced by one.
[4] For a discussion of the difference between Islam and Christianity see http://answering-islam.org/Nehls/tt1/tt5.html, accessed August, 2015.
[5] There are also a large number of sub-sects; see the internet, for example http://www.real-islam.org/73_8.htm, accessed August, 2015.

mathematician, especially when we see how mathematics underlies science The Lord said to Job:

> "Where were you when I laid the earth's foundation? Tell me, if you understand. Who marked off its dimensions? Surely you know! Who stretched a measuring line across it'?"[6]

11.3 WHAT DOES IT MEAN TO KNOW GOD?

Here we are not talking about knowing *of* God but whether we can know the "heart of God" in some sort of intimate way. As my focus is on the Bible, we read three times in the New Testament[7] that we can become children of God with God described as our heavenly father who we can call by the more intimate term *Abba* (an Aramaic term that is best translated as "Daddy").[8] This term came to be used by adults with a meaning such as "dear father." Some would say that such language is sexist, and it is true that Jewish and early Christian societies were very patriarchal. In defense it should be noted that fatherhood and sonship convey three ideas: first, God is a loving God who loves us like a parent (including the use of discipline), a concept that seems to be unique to Christianity; second, fatherhood involves protection of and authority over children; and third, the word son rather than being just a sexist word conveys the idea of inheritance that we have from God.

11.4 IS GOD KNOWABLE?

In Psalm 8:3–4 the psalmist ponders about why God should be interested in us, given God's grandeur in creating the moon and the stars. Isaiah 55:8–9 says that God's thoughts and ways are not our thoughts and ways, and as the heavens are higher than the earth so are God's thoughts and ways higher than our thoughts and ways. Does this mean that God is too distant? No. In fact verse 6 says that we should seek the Lord while he may be found and call upon him while he near. When Paul was in Athens he told the men in the middle of the Areopagus that he saw that they were very religious as they had many idols including one addressed to the unknown God (in case they missed one out!). He said that the unknown God is the Creator, the Lord of Heaven and earth, and the giver of life and breath. God does not live in man-made shrines, but quoting from one of their local poets he said that God is not far from each of us as in him we live and move and have our being. Hebrews 11:6 I believe sums it up well when it says that we need faith to please God and if we draw near to God we must believe that He exists and that He rewards

---

[6] Job 38:4–5.
[7] Mark 14:36; Romans 8:15; Galatians 4:6.
[8] Nowhere in the entire devotional literature of ancient Judaism is *abba* a way of addressing God.

those who seek Him. Yes, we can come to know something of God but we need to make an effort and step out in faith.

11.4.1 The Role of God's Creation

In the Old Testament, Psalm 19:1 says that the heavens declare the glory of God and the earth declares his handiwork, while in Psalm 8:3–9 David looks at the heavens that God has established and asks the question of why God is mindful of humankind. He then goes on to say that God has crowned humankind with glory and honor and given them dominion over creation. With regard to the New Testament, we have previously noted that the question of knowing God from external observation is considered in the first chapter of the book of Romans in the New Testament. In verses 19 and 21 of chapter 1 we are told that God's eternal power and deity are plainly evident through the wonders of creation. This is usually referred to as God's *general revelation*. Therefore since God's creation is one key to knowing about God, some would say that we can know God by contemplating the natural world; we must become one with nature somehow. This approach has two problems.

(1.) It doesn't work very well! Although God offers eternal life to all who seek him in well-doing,[9] people tend not honor God but suppress the truth in unrighteousness, ignoring God the creator.[10] They may finally end up worshipping the creation (e.g., idols) rather than the creator, with the idols of today being more subtle (e.g., self, money, power).

(2.) We cannot know what an artist is really like from the work he or she has created. I would not like people to judge me on the basis of the things I have tried to make! However, the awesomeness of the universe at least tells us about God's greatness and his power of creation. It would seem that God has given evidence of Himself sufficiently clear for those with an open heart, but sufficiently vague so as not to compel those whose hearts are closed, as God has given us free will. However, we don't know whether God is personal and whether or not He is a God of love or a God of justice simply from contemplating creation. We might even think that God is capricious or cruel because of suffering in the world! In fact if this world is "fallen" in some mysterious way as we discussed in Chapter 8, then nature may give some a false picture of God. Attributes of God like love and justice would have to be revealed to us from some other source.

---

[9] Romans 2:7.
[10] Romans 1:21.

11.4.2 God's Revelation

If God exists, it is not unreasonable that He would want to communicate with us. We therefore cannot rule out logically a more direct revelation from God so that we could know more about God and have some kind of a relationship with Him. Christians believe that Jesus is God's supreme revelation, the human face of God, who entered into our world and experienced rejection and suffering, and taught us about forgiveness, mercy, and righteousness. He then confirmed who he was through his resurrection and his conquering of death. What about the Bible? It is also part of God's revelation as it describes how God dealt with his chosen people the Jews from whom would come the Messiah. The Old Testament describes God as being a unique, immaterial, uncaused, and timeless creator[11] who is also personal, rational, powerful, loving, and omnipresent.[12] As discussed in Chapter 6, almost everything we know about Jesus comes from the gospels and, since Jesus claimed that Yahweh the God of the Old Testament was his father, the Bible is an important part of the total revelation. Unlike the Qur'an in Islam, Christians do not worship the Bible, but revere it and treat it as a word inspired by God. It gives us a glimpse of God and it also has a role in helping us to be more moral people and understand our role in this world.

The challenge then is which is the right revelation, as every religion claims special revelation or enlightenment. How do we choose? I have already discussed the question of religious pluralism in Section 6.5.1 and noted that all religions are generally exclusive. Also the idea of pluralism runs through almost all aspects of our society, and not just religion. Name any topic and you will get a plethora of views, for example the nature of reality given quantum theory. If we accept Christianity it does not mean that we simply reject all other religions as worthless and devoid of any insight, as there are some ideas not contrary to Christianity. For example there are many things that Islam teaches about God that Christians would agree with. For example, God is creator and sustainer of the world; He is just, all powerful, all knowing, will judge all people, and will have a general resurrection of the dead. Islam emphasizes prayer and alms giving. However there are some things incompatible with Christianity such as the denial of the Trinity and that Jesus was the son of God, and the denial of Christ's crucifixion and his resurrection, a central tenant of Christianity. I emphasize that Christianity is different from other religions as it is not just a code of ethics or a list of rules to follow, but a relationship with God through his Spirit.

We can also know something about God from internal observation. In Section 6.7 we referred to Romans 2:15 that says that the demands of God's moral law are implanted on everyone's heart. We have a faculty called our conscience that helps us to make moral decisions, if we listen to it. In that

---

[11] Deuteronomy 4:35; John 4:24; Psalm 92:2, 102:25.
[12] Genesis 17:1; Psalm 104:24; Nahum 1:3; Psalm 33:5; 1 Kings 8:27.

section we also argued that the existence of a universal moral law suggests the existence of a moral lawgiver who must be perfectly good, and goodness implies personality. We can also argue from Genesis 1:26 that since we are made in the image of God, God must be rational, and has some attributes of a person. This does not mean that we make God in our own image, but we somehow share in a small way some of God's attributes, for example knowing something of love, goodness, and justice. However, in spite of this internal revelation, there are those who deliberately flout the internal moral law.[13] Before discussing how we might come to know God and reach out to God in some way we briefly consider the negative side of rejecting any belief in God.

11.4.3 Life Without God

If a God who promises the possibility of eternal life does not exist, then life is ultimately meaningless and purposeless, and we have to face the fact that one day we will not exist and neither will our universe eventually as we are winding down. What is more, there will be no moral accountability without immortality. We can fool ourselves into pretending that life has meaning and significance, but in the end, in the words of Ecclesiastes in the Old testament, "all is vanity." The writer of this book is referred to as the teacher or preacher who may have been Solomon, and it is surprisingly modern in its cynical outlook on life in spite of its great age. I will give you just some examples from the book.

(1.) *Futility of life.* You spend your life working and laboring, and what have you to show for it? Generations come and go but the world stays the same (Ecclesiastes1:1-4). In the end you leave it all to someone else and we don't know what he will do with it or whether he is wise or a fool (Ecclesiastes 2:18–19).

(2.) *Futility of wisdom.* The wiser you are the more worries you have because you cannot fix things or straighten out what is crooked. What he is saying is that knowledge for its own sake does not bring happiness (Ecclesiastes 1:16–18). Much study makes you weary (Ecclesiastes 12:12).

(3.) *Futility of pleasure for pleasure's sake.* The writer went on a spending spree having wine, women, and song, and not denying himself any pleasure. It was selfish pleasure and it did not bring any satisfaction (Ecclesiastes 2:1–11).

(4.) *Futility of human endeavor.* We are soon forgotten irrespective of whether we are wise or foolish (Ecclesiastes 2:15–16).

---

[13] Romans 1:32.

(5.) *Futility of labor and possessions.* The love of money and wealth does not bring satisfaction. (Ecclesiastes 5:10). We have to leave it all behind as we cannot take it with us (Ecclesiastes 5:15).

There are other discouraging comments as well, and the reader might be surprised at finding such depressing material in the Bible. However, the Bible tells it as it is and brings home to us what life is sometimes like if we believe it is meaningless. There are of course some more positive ideas in Ecclesiastes and I now list some by way of contrast.

(1.) Enjoyment comes from God (Ecclesiastes 2:24–25).

(2.) God gives wisdom to those who please him (Ecclesiastes 2:26).

(3.) God has put eternity in our minds (Ecclesiastes 3:11).

(4.) Whatever God does endures for ever (Ecclesiastes 4:14).

(5.) Although a wicked man can commit much crime and still live a long time, it won't go well with him. It will go much better for God-fearing men (Ecclesiastes 8:12–13).

(6.) The righteous, the wise, and their deeds are in God's hands (Ecclesiastes 9:1).

(7.) The preacher concludes with the admonition to fear God and keep his commandments, which is our whole duty (Ecclesiastes 12:13). This attitude brings positive benefits described briefly throughout the book.

We all struggle with the idea of death because, as the writer of Ecclesiastes reminds us above, nothing in this world whether it is wealth, fame, or power can finally satisfy us. He says that the best does not last and is never enough for us as God has put eternity in our minds. Psychiatrist Viktor Frankl, in writing about the survival of prisoners including himself in a concentration camp said:

> It is a peculiarity of man that he can only live by looking to the future. . . . And this is his salvation in the most difficult moments of his existence, although he sometimes had to force his mind to the task.[14]

For me, having a focus beyond the grave enables me to press on in spite of life's difficulties, and I have had my share. We now return to our question (4.) at the beginning of the chapter of how can we know God and how to somehow reach out to God.

---

[14]Frankl (1992: 81).

11.5 REACHING OUT

I believe that God has created a vacuum within us that can only He can fill. Having made some preliminary comments with regard to knowing about God through nature, our moral consciousness, and biblical revelations I now want to suggest some steps that a seeker might follow in seeking to reach out to God. We are all different in our spiritual walk so I hope that the reader may find some of the following suggestions helpful.

11.5.1 Contemplation

The Bible tells us that God will reward us if we earnestly seek Him.[15] However, if we are first of all uncertain about God's existence we can at least pray a prayer something like "Oh God, I want to get to know you but I am uncertain about your existence. Please forgive me for my doubt and reveal yourself in the depths of my being." A good way to start is to choose a quiet place and contemplate our amazing universe with its vast distances, its multitude of galaxies, and the incomprehensibility of the nature of our universe, or perhaps even go for a quiet walk, say on a beach or in some bush. We can also look out at a picturesque scene and marvel at its beauty (why is it beautiful?), realizing that this didn't happen by accident; we have a great God who has created it all and brought it into being by his power, and has given us the ability to appreciate it.

Music may help us tap into the right hemisphere of our brain and help us to reach out beyond ourselves to God. It can also be helpful to relax through diaphragm breathing as it allows us to focus our thoughts better. Those of us in the West seem to have forgotten how to meditate. In the end, however, believing that God exists requires a step of faith. Reason and arguments may take us much of the way to believing in God, but the final step is an act of faith. Some may find themselves in a battle of faith as I believe there are spiritual forces in this world that would try and stop us from taking that step.

11.5.2 Steps to Communication

Once we have accepted by faith that God exists, we need to know something about this God, especially if we have been brought up in ignorance of God and having no knowledge of any spiritual dimension; something which is happening all too frequently these days. A good place to find out about God is in the gospels of the Bible, as Jesus came into this world to tell us about the nature of God. Reading about his life and teaching can be inspiring, and revisiting the arguments in this book for his resurrection can be helpful. As previously mentioned, the main monotheistic religions all claim that God can be known

---

[15] Hebrews 11:6.

personally and we find this idea particularly throughout the New Testament where God is described intimately using the metaphor of a Father (or even "Daddy") who cares for his children.

We read in the Bible that God is a righteous and holy God, and we have argued in this book that the morality that we are aware of must come from a God who is a perfectly just moral being and, although long suffering, does not tolerate sinfulness. We therefore need to acknowledge that we don't always get things right. A common meaning for the word sin in both Greek and Hebrew is the missing of a standard, mark, or goal. I remember once having a go at archery; no matter how good my intentions were or how hard I concentrated, I could not hit the bulls-eye. You know as well as I do that we cannot even live up to our own internal standards and thereby miss our own target, let alone live up to God's standards described by Jesus (quoting from the Old Testament) as loving God with our heart, soul, and mind, and our neighbors as ourselves.[16] Talking about neighbors, which means anybody we rub shoulders with, we don't always get it right with them. We may even have family members that we don't get on with, a common occurrence. Much of my counseling these days is about broken relationships and trying to encourage healing and reconciliation.

Approaching God

We may have neglected God for a long time or not even given him a thought. How then should we approach God? If I was to meet a royal couple, would I wear my gardening clothes on that occasion? No, I would want to dress to the best of my ability and have my best manners so that I would at least be both presentable and acceptable to the royal couple; I don't have to be perfect. It would be my way of honoring the couple. When we approach God, we are approaching the awesome creator of the universe, which is no trivial matter and demands respect for who God is. We may think our good deeds will make us acceptable, but we read that our righteous acts are like filthy rags in God's sight,[17] which are even worse than gardening clothes! We need to recognize that God is holy and we are not, and we have certainly broken say the ten commandments at some stage in our lives whether by having other "gods" in our lives (e.g., focusing too much on material possessions), lying (or telling half-truths or withholding truth), stealing (a person's possessions or reputation, or tax evasion), dishonoring parents, coveting what other people have, and so forth.

We may sometimes feel we are quite moral and upright, yet we tend to delude ourselves as we are by no means perfect, and regularly make mistakes such as even not doing something when we know we should do it— sin by omission. Let's face it, we are all unrighteous and miss the mark.[18] How then

---

[16] Matthew 22:37–40.
[17] Isaiah 64:6
[18] Romans 3:10–12, 23.

does a sinful person approach a holy God, especially as the Bible tells us to seek the Lord with all our heart and soul.[19]

Acknowledging Our Sinfulness

The first thing we need to do is acknowledge that we are sinful and own up to our inappropriate behavior, misdeeds, and wrong thoughts. If someone does something wrong against us, we cry out for justice. If we wrong someone else, they will want justice and possibly revenge in terms of "an eye for an eye and a tooth for a tooth."[20] The concepts of justice, consequences, punishment for misdeeds, and possibly retribution seem to be part of who we are; God given if you like. God emphasized these ideas with the Jewish nation by setting up a sacrificial system that was costly for them. It was not God's first choice as a method for getting across the importance of purity and forgiveness as He is described as being tired of burnt offerings and the rituals for the atonement of sin. He wanted his people to live righteously instead of having to use such visual reminders.[21]

The Jews were a primitive people in many ways and needed concrete examples to bring home the seriousness of sin in God's sight. As with the Israelites, we also come under God's judgment because of our many transgressions against God's moral law, even the law within us. However, instead of having to offer our own sacrifices, Jesus fulfilled the Old Testament practice by being our sacrifice on a Roman cross, effectively taking our deserved place. God wants us to accept what Jesus has done and believe that Jesus is the son of God, being one of the three persons making up the one God.

Confession

The second thing we need to do is confess our sins to God and believe that Jesus died for us. God will then forgive us because of the sacrifice of Jesus, and we are promised that we will then be with God after we die—our "salvation."[22] Where Christianity is different from other religions is that we do not have to try and earn our salvation by good works; we cannot anyway. There is no way we can be good enough to merit anything from a holy and righteous God; it has to be a free gift by God's grace. As Ephesians 2:8–9 says, we are saved by God's grace (God's unmerited favor) through faith and not by works; salvation is a free gift of God.

When God is first in our lives his Spirit comes into our lives and brings about a transformation called by a variety of names or metaphors such as being born again (a spiritual rebirth), being converted (meaning turned around and going in the opposite direction), or being raised from the dead (resuscitated from being spiritually dead).

---

[19] Deuteronomy 4:29.
[20] Exodus 21:23–25; Leviticus 24:19
[21] See Leviticus in Section 5.2.2 for biblical references.
[22] John 3:16.

Some psychologists have tried to tell us that conversion is nothing but brain washing. While it is true that brainwashing can take place with some cults, genuine conversion does not satisfy the criteria for brainwashing. When it happens to us we will know that it is genuine as God's joy will flood our hearts. So great can be the transformation that the very last person I could ever imagine changing in the past has become a Christian and has completely changed.

Prayer

The third thing that will help our spiritual walk is to pray. One way is to simply relax and allow our minds to be open to God's inner prompting. A helpful verse is Psalm 46:10 that says "be still and know that I am God." It conveys the following ideas.

(1.) *The first phrase is "Be still."* Ever tried to talk to someone who is distracted? It goes on in families! It is even more important to avoid distraction in communicating with God as our minds tend to fill up with activities for the day. Jesus used to depart to a lonely place to pray. Association is important. We need a haven free of external distractions such as a special place or room that we go to, or even to just a particular chair, or shifting a chair to a particular place. When we are in our haven it is for the express purpose of being still. The continual association with such a place helps us to relax more quickly. We don't have to try and do anything but just "be."

We need to practice being still, letting go, and relaxing. In life we use both hemispheres of our brains with the left hemisphere tending to be the source of analytical and logical thinking while the right hemisphere generally deals with the artistic, the intuitive, and the visionary. In Western society we tend to be a bit one-sided with our left side predominating. There are times however when we need to set aside our intellect and open our minds in a free flowing way using our right hemisphere to open up ourselves to God. We don't throw out our reasoning processes but simply step aside from them for the moment.

(2.) *The second phrase is "and know that I am God".* If we have just read a passage from the Bible we can meditate on this to begin with. This can help us to focus on God and open our minds to his vision for us. He might speak inwardly and tell us that we have to get a few things right with others. As we ask him for forgiveness we can ask him to bring to mind those things that are hindering our communication with him and our Christian walk. We cannot undo the past and our past mistakes, but God can help us bring healing into our past if we let Him. Jesus is the express image of the Father so we could imagine Jesus in some casual situation, or picture some incident in the New Testament such as him calming the storm. If we are praying for somebody we can imagine bringing that person to Jesus and seeing what happens.

(3.) *We need to listen for God's voice.* It will be sensed as a spontaneous thought, idea, feeling or vision. We get our minds out of the way and let the images come. We replace thinking by vision. How do we recognize that the thoughts are from God and not simply from self or even a satanic influence that can happen if our minds are relaxed? If the thoughts are not from God they tend to have negative characteristics like being accusing, or contrary to the Bible. Also any image that arises will be afraid to be tested, and its fruit may be fear, compulsion, anxiety, or confusion. Some suggestions for discerning that the thoughts are from God follow:

(a) They can come easily as though God is speaking first person.

(b) They are often light and gentle, and easily cut off by any exertion of self.

(c) They will have an unusual content to them in that they will somehow be better and often different from our own thoughts.

(d) They will cause a special reaction within our beings (e.g., a sense of excitement, conviction, faith, awe, peace).

(e) When embraced, they will come with the strength to carry them out, as well as a joy in doing so.

As time goes on we shall get better at learning to listen. Some people find it helpful to write down the thoughts or images in a journal that they can examine later. It may be like learning to walk all over again: we will fall over from time to time. Journalling is quite biblical as the Psalms, the prophetic books, and Revelation were written this way. For example, the prophet Habakkuk (see 2:1, 2) went to a quiet place and waited for God to speak; he then wrote down what he received. We can of course pray to God at any time. When I am counseling and wonder how to deal with a problem that a counseling client has raised I send a quick "arrow" prayer to God for guidance. It is surprising what comes to mind then.

Community

Although some people like solitude, we are generally a community oriented animal and like to interact with others, even if it is in silence. We can also know God through the Christian church as God uses others to speak to us and support us, so it is important to attend a church or join with a group of people in which the gospel is faithfully preached. You have no doubt heard the comment that the church is full of hypocrites and, given the behavior of some prominent Christians latched on to by the media, I am not surprised at such a reaction. However, I believe that such a criticism misses the point. In the sight of a holy God we are all sinners and fall short of God's standard, irrespective of any good works we might do. None of us is worthy to approach a holy and just God. However, if we believe in what Christ has done for us on the cross in opening up access to God and if we give our allegiance to him, God extends his grace to us and freely forgives and accepts us, imperfections and all. Therefore we don't have to be perfect and completely clean up our act before we can become a Christian, though change happens afterwards when God's Spirit enters our lives.

Jesus was criticized when he socialized with the hated tax collectors and sinners, and he responded by saying that he came to call sinners not the righteous, as those who are well do not need a physician. In fact Jesus was very critical of the self-righteous person. He told the parable of two men who went to the temple to pray,[23] which was aimed at those who trusted in their own righteousness and despised others. One was a Pharisee and the other a hated tax collector. The Pharisee was thankful he was not like other men, especially the tax collector, and boasted that he never committed any major sins, fasting twice a week, and giving tithes (gifts) from his earnings. The tax collector, however, was not so ostentatious, but beat his chest and asked God to be merciful to him a sinner. Jesus said the tax collector was more accepted than the Pharisee and added that whoever exalts himself will be humbled, while the person who is humble will be exalted. Jesus also criticized legalism and hypocrisy in his Sermon on the Mount[24] and in other scriptures;[25] we see the same condemnation on hypocrisy in the Old Testament.[26] What is interesting is that such criticisms are from within Christianity itself, not from outsiders, and include remedies to correct such behavior.

We find that all sorts of people attend church, often broken people with various kinds of hangups. If they are Christians they will be at various stages in getting to know God better, getting rid of the baggage they bring with them, and getting their lives back on track. They are still imperfect and will make mistakes. Someone once said "If you find a perfect church don't join it as it won't be perfect any longer!"

Being Christian does not exempt us from problems and from suffering because of the world we live in, as we saw in Chapter 8. However, we understand that God through the person of Jesus knows about suffering and what it is like, and is able to get alongside us and walk with us. We know that our lives have a purpose, that we are not alone in this universe, and that ultimately our lives and our future are in God's hands no matter whatever the outcome. We can trust God that our future after death is secure.

[23] Luke 18:9–14.
[24] Matthew 6:1–18.
[25] Matthew 21:31.
[26] Isaiah 58:2–7.

11.6 CONCLUSION

I hope the above did not sound too much like a sermon, but as I said at the beginning I wanted to pass on something that has meant so much to me. We are all on a pilgrimage of one sort or another irrespective of our beliefs, and it is a question of where we are heading.

REFERENCES

Adler, M. J. (1980). *How to Think About God: A Guide for the 20th-Century Pagan.* New York: Macmillan.

Ali, F. A. and Das, S. (2015). Cosmology from quantum potential. *Physics Letters B*, **741**:276–279.

Allen, C. and Trestman, M. (2014). Animal consciousness. In E. N. Zalta (Ed.), *The Stanford Encyclopedia of Philosophy.* Standford University. (See http://plato.stanford.edu/archives/sum2014/entries/consciousness-animal/, accessed August, 2015.)

Alberts, B. (1998). The cell as a collection of protein machines: Preparing the next generation of molecular biologists. *Cell*, **92** (3):291–294.

Alexander, D. R. (2012). Creation and evolution. In J. B. Stump, Alan G. Padgett (Eds), *The Blackwell Companion to Science and Christianity*, pp. 231–245. Chichester: Wiley-Blackwell.

Aquinas, Thomas. (1920). *The Summa Theologica of Thomas Aquinas*, Second and Revised edit. Translated by Fathers of the English Dominican Province. Online Edition (2008) by Kevin Knight.

Archer, G. L., Jr. (1994). *A Survey of Old Testament Introduction*, 3rd edit. Chicago: Moody Press.

Ayoub, G. (1996). On the design of the vertebrate retina. *Origins & Designs*, **17** (1). (See http://www.arn.org/docs/odesign/od171/retina171.htm, accessed August, 2015.)

Baháu'lláh. (1976). *Gleanings from the Writings of Baháu'lláh*, 2nd rev. edit. Wilmette: Bahá Publishing Trust.

Baker, M. C. (2011). Brains and Souls: Grammar and Speaking. In M. C. Baker and S. Goetz (Eds), *The Soul Hypothesis: Investigations into the Existence of the Soul*. New York: Continuum.

Baker, M. C. and Goetz , S. (2011). *The Soul Hypothesis: Investigations into the Existence of the Soul*. New York: Continuum.

Ball, P. (2009). *Nature's Patterns: A Tapestry im Three parts*. Vol.1, *Shapes*; vol. 2, *Branches*: and vol. 3, *Flow*. Oxford: Oxford University Press.

Barclay, W. (1958). *The Letters to the Galatians and Ephesians*, 2nd edit.. Edinburgh: Saint Andrew Press.

Barnes, L. A. (2012). The fine-tuning of the universe for intelligent life. *Publications of the Astronomical Society of Australia*, **29** (4):529–564. (See also http://arxiv.org/PS_cache/arxiv/pdf/1112/1112.4647v1.pdf, accessed August 2015, for a slightly expanded version.)

Barnes, L. A. (2013). What to Read: The Fine-Tuning of the Universe for Intelligent life. (See http://letterstonature.wordpress.com/2013/09/10/what-to-read-the-fine-tuning-of-the-universe-for-intelligent-life/, accessed August, 2015.)

Barrow, J. D. (1990). *The World Within the World*. Oxford: Oxford University Press.

Barrow, J. D. and Tipler, F. J. (1986). *The Anthropic Cosmological Principle*. Oxford: Oxford University Press.

Bartholomew, D. J. (1988). Probability, statistics and theology. *Journal of the Royal Statistical Society, Series A*, **151** (1):137-178. (See also the discussion by Wiztum, Rips, and Rosenberg, pp. 173–174.)

Baumeister, R. F. and Tierney, J. (2011). *Willpower: Rediscovering the Greatest Human Strength*. London: Penguin group.

Beauregard, M. and O'Leary, D. (2007). *The Spiritual Brain: A Neuroscientist's Case for the Existence of the Soul*. New York: HarperCollins.

Beauregard, M. and Paquette, V. (2006). Neural correlates of a mystical experience in Carmelite nuns. *Neuroscience Letters*, **405** (3):186–109.

Behe, M. J. (1996). *Darwin's Black Box: The Biochemical Challenge to Evolution*. New York: Free Press. (Second edition, 2006.)

Behe, M. J. (2007). *The Edge of Evolution: The Search for the Limits of Darwinism*. New York: Free Press.

Beilby J. K. and Eddy, P. R. (Eds) (2001). *Divine Foreknowledge: Four Views*. Downers Grove, Ill: InterVarsity Press.

Benson, H. (1996). *Timeless Healing: The Power and Biology of Belief*. New York: Scribner.

Benton, N. J. (1995). Diversification and extinction in the history of life. *Science*, **268**, no. 5207, 7 April, 52–58.

Berry, R. J. (1988). *God and Evolution*. Seven Oaks, Great Britain: Hodder & Stoughton.

Berry, R. J. and Jeeves, M. (2008). The nature of human nature. *Science and Christian Belief*, **20**:3–47.

Best, J. B. (1963). Protopsychology. *Scientific American*, **208**:54–62.

Blomberg, C. L. (2007). *The Historical Reliability of the Gospels*, 2nd Edit. Downers Grove, IL: IVP Academic.

Blomberg, C. L. (2011). *The Historical Reliability of John's Gospel: Issues and Commentary*. Downers Grove, IL: IVP Academic.

Borde, C. L. and Vilenkin, A. (1994). Eternal inflation and the initial singularity. *Physical Review Letters*, **72** (21):3305–3308.

Borde, A., Guth, A. H., and Vilenkin, A. (2003). Inflationary spacetimes are incomplete in past directions. *Physical Review Letters* **90** (15):15301, 1–4.

Bostrom, N. (2002). *Anthropic Bias: Observation Selection Effects in Science and Philosophy*. New York: Routledge.

Bott, M. and Sarfati, J. (1995). What's wrong with Bishop Spong? Laymen Rethink the Scholarship of John Shelby Spong. *Apologia* **4** (1):3–27. (For a 2007 update see http://creation.com/whats-wrong-with-bishop-spong, accessed August, 2015.)

Boyd, G. A. (2001). The open-theism view. In J. K. Beilby and P. R. Eddy (Eds), *Divine Foreknowledge: Four Views*. Downers Grove, Ill.: InterVarsity Press.

Broom, N. (2010). *Life's X Factor: The Missing Link in Materialism's Science of Living Things*. Wellington, New Zealand: Steele Roberts.

Brown, D. E. (1991). *Human Universals*. New York: McGraw-Hill.

Budziszewski, J. (2011). *What We Can't Not Know: A Guide*, 2nd edit. San Francisco: Ignatius Press.

Burns, J. M. and Swerdlow, R. H. (2003). Right orbitofrontal tumor with pedophilia symptom and constructional apraxia sign. *Archives of Neurology*, **60**:437–440.

Carbone, L., and many others. (2014). Gibbon genome and the fast karyotype evolution of small apes. *Nature*, **513** (7517):195–208.

Carlson, R. F. (Ed.) (2000). *Science and Christianity: Four Views*. InterVarsity Press: Ill.

Carroll, S. B. (2005a). *Endless Forms Most Beautiful: The New Science of Evo Devo*. New York: W.W. Norton.

Carroll, S. B. (2005b). Evolution at two levels: on genes and form. *PLoS Biology*, **3** (7):1159–1166.

Carson, D. A. (1981). *Divine Sovereignty and Human Responsibility: Biblical Perspectives in Tension*, 2nd edit. Louisville, KY: John Knox Press, Baker Books.

Cartwright, N. (1983). *How the Laws of Physics Lie*. Oxford: Oxford University Press.

Chalmers, D. (1997). *The Conscious Mind: In Search of a Fundamental Theory*. Oxford, UK: Oxford University Press.

Chalmers, D. (2014). Panpsychism and panprotopsychism. In T. Alter and Y. Nagasawa (Eds), *Russellian Monism*, pp. 246–276. New York: Oxford University Press. (See also http://www.amherstlecture.org/chalmers2013/, accessed August, 2015.)

Chapman, D. G. (1961). Statistical problems in the dynamics of exploited fish populations. *Proceedings of the 4th Berkeley Symposium, 1960*, vol. 4, 153–168.

Charlesworth, J. H. (Ed.) (1992). *Jesus and the Dead Sea Scrolls*. New York: Doubleday.

Collins, R. (2003). Evidence for fine tuning. In N. A. Manson (Ed.), *God and Design: The Teleological Argument and Modern Science*, pp. 178–199. London: Routledge.

Collins, R. (2008). Modern physics and the energy conservation objection to mind-body dualism. *The American Philosophical Quarterly*, **45** (1):31–42. (For an abridged version, http://www.newdualism.org/papers/R.Collins/EC-PEC.htm, accessed August, 2015.)

Collins, R. (2009). The teleological argument: An exploration of the fine-tuning of the cosmos. In W. L. Craig and J. P. Moreland (Eds), *Blackwell Companion in Natural Theology*, pp. 202–281. Chichester: Wiley-Blackwell.

Collins, R. (2011). A scientific case for the soul. In M. C. Baker and S. Goetz (Eds), *The Soul Hypothesis: Investigations into the Existence of the Soul*, pp. 222–246. New York: Continuum International.

Connelly, D. (1997). *What the Bible Says About Miracles*. Downers Grove, IL.: InterVarsity Press. (Available online at http://www.ccel.us/miracles.toc.html, accessed August, 2015.)

Conway-Morris, S. (2003). *Life's Solution: Inevitable Humans in a Lonely Universe*. New York: Cambridge University Press.

Copan, P. and Flannagan, M. (2014). *Did God Really Command Genocide?*. Grand Rapids, MI: Baker Books.

Coyne, J. A. (2005). Switching on evolution—how does evo-devo explain the huge diversity of life on Earth? *Nature*, **435** (7045):1029–1030.

Craig, W. L. (1980). The bodily resurrection of Jesus. In R. T. France and D. Wenham (Eds), *Gospel Perspectives I*, pp. 47–74. Sheffield, England: JSOT Press. (See also http://www.leaderu.com./offices/billcraig/docs/bodily.html, accessed August, 2015.)

Craig, W. L. (1984). The guard at the tomb. *New Testament Studies*, **30**:273–281. See also http://www.leaderu.com./offices/billcraig/docs/guard.html, accessed August, 2015.

Craig, W. L. (1985). The historicity of the empty tomb of Jesus. *New Testament Studies*, **31**:39–67. (http://www.leaderu.com./offices/billcraig/docs/tomb2.html, accessed August, 2015.)

Craig, W. L. (1986). The problem of miracles: A historical and philosophical perspective. In D. Wenham and C. Blomberg (Eds), *Gospel Perspectives*, pp. 9–40. JSOT Press: Sheffield, U.K.

Craig, W. L. (1991). Talbott's universalism. *Religious Studies*, **27**:297–308. (See http://www.reasonablefaith.org/talbotts-universalism, accessed August, 2015.)

Craig, W. L. (1992). The disciples' inspection of the empty tomb (Luke 24,12.24; John 20, 1–10). In A. Dernaux (Ed.), *John and the Synoptics*, pp. 614–619. Bibliotheca Ephemeridum Theologicarum Lovaniensium 101. University Press: Louvain. (See also http://www.leaderu.com./offices/billcraig/docs/tomb1.html, accessed August, 2015.)

Craig, W. L. (1998a). Creation, providence, and miracle. In B. Davies (Ed.), *Philosophy of Religion*, pp. 136–162. Georgetown University Press, Washington, D.C. (See http://www.leaderu.com/offices/billcraig/docs/creation-providence.html, accessed August, 2015.)

Craig, W. L. (1998b). Rediscovering the historical Jesus: The presuppositions and presumptions of the Jesus seminar. *Faith and Mission*, **15**:3–15.

Craig, W. L. (1998c). Divine timelessness and personhood. *International Journal for Philosophy of Religion*, **43**:109-124. (See also http://www.reasonablefaith.org/divine-timelessness-and-personhood, accessed August, 2015.)

Craig, W. L. (2000). Timelessness and omnitemporality. *Philosophia Christi*, Series 2, **2** (1):29–33. (See also http://www.leaderu.com/offices/billcraig/docs/omnitemporality.html, accessed August, 2015.)

Craig, W. L. (2001a). Reply to Fales: On the empty tomb of Jesus. *Philosophia Christi* **3**:67–76. (See http://www.leaderu.com./offices/billcraig/docs/fales.html, accessed August, 2015.)

Craig, W. L. (2001b). The middle-knowledge view. In J. K. Beilby and P. R. Eddy (Eds), *Divine Foreknowledge: Four Views*. Downers Grove, Ill: InterVarsity Press.

Craig, W. L. (2010). *On Guard: Defending Your Faith with Reason and Precision*. Colorado Springs, CO: David C. Cook.

Craig, W. L. (2012). God and abstract objects. In J. B. Stump and A. G. Padgett (Eds), *The Blackwell Companion to Science and Christianity*, pp. 441–452. Chichester: Wiley-Blackwell.

Crockett, W. V., Walvoord, J. F., Hayes, Z. J., and Pinnock, C. H. (1992). *Four Views of Hell*. Grand Rapids, Michigan: Zondervan.

Custance, A. C. (1980). *The Mysterious Matter of Mind*. Probe Ministries (Texas) and Zondervan. (Second online edition 2001 at http://www.custance.org/Library/MIND/index.html, August, 2015.)

Damasio, A. R. (1999). How the brain creates the mind. *Scientific American*, **281** (6):112–117. (See http://www.ucd.ie/artspgs/langmind/braintomind.pdf, accessed August, 2015.)

Davies, P. C. W. (1984). *Superforce: The Search for a Grand Unified Theory of Nature*. New York: Simon and Schuster.

Davies, P. C. W. (1987). *The Cosmic Blueprint*. Penguin Books: London.

Davies, P. C. W. (2003). How bio-friendly is the universe? *International Journal of Astrobiology*, **2** (2):115–120.

Davies, P. C. W. (2006). *The Goldilocks Enigma: Why is the Universe Just Right for Life?*. New York: Allen Lane.

Dawkins, R. (1996). *The Blind Watchmaker: Why the Evidence of Evolution Reveals a Universe without Design*. New York: W. W. Norton.

Dawkins, R. (1995). *River Out of Eden: A Darwinian View of Life*. New York: Basic Books.

Dawkins, R. (2006). *The God Delusion*. London: Bantam press.

Deemer, W. L. and Votaw, D. V. Jr. (1955). Estimation of parameters of truncated or censored exponential distributions. *Annals of Mathematical Statistics*, **26**, 498–504.

d'Espagnat, B. (1979). The quantum theory and reality. *Scientific American*, **241** (5):158–181.

Denham J., Marques, F. Z. , O'Brien, B. J., and Charchar, F. J. (2014). Exercise: putting action into our epigenome. *Sports Medicine*, **44** (2):189–209.

Denton, M. (1999). The inverted retina: Maladaptation or pre-adaptation? *Origins & Designs*. **19** (2). (http://arn.org/docs/odesign/od192/invertedretina192.htm, accessed August, 2015.)

de Waal, F. B. M. (1996). *Good Natured: The Origins of Right and Wrong in Humans and Other Animals*. Cambridge: Harvard University Press.

de Waal, F. B. M. (2005). *Our Inner Ape*. New York: Riverhead, Penguin.

de Waal, F. B. M. (2009). *The Age of Empathy: Nature's Lessons for a Kinder Society*. New York: Harmony, Random House.

Doidge, N. (2007). *The Brain That Changes Itself*. New York: Penguin Books.

D'Souza, D. (2007). *What's So Great About Christianity*. Washington DC: Regnery Publishing.

Dubay, T. (1999). *The Evidential Power of Beauty*. San Francisco: Ignatius.

Durant, W. (1994). *Caesar and Christ (The Story of Civilization Vol.3)*. MJF books: Also available as an e-book, Simon and Schuster (2011).

Einstein, A. (1954). *Ideas and Opinions*. New York: Random House.

Ekman, P. (1980). Biological and cultural contributions to body and facial movement in the expressions of emotions. In A. O. Rorty (Ed.), *Explaining Emotions*. Berkeley: University of California Press.

Ekman, P. (1993). An argument for basic emotions. *Cognition and Emotions*, **6**:169–200.

Ekman, P. (1999). Basic emotions. In T. Dalgleish and M. Power (Eds), *Handbook of Cognition and Emotion*, chapter 3. Sussex, U.K.: John Wiley.

Ellis, G. F. R. (2011). Does the multiverse really exist? *Scientific American* **305** (2):38–43. (See http://www.scientificamerican.com/article/does-the-multiverse-really-exist/, accessed August, 2015.)

Erwin, D. H. and Valentine, J. W. (2013). *The Cambrian Explosion: The Construction of Animal Biodiversity*. Greenwood Village, CO: Roberts and Company.

Feser, E. (2006). *Philosophy of Mind: A Beginners Guide*. Oxford: One World. (e-book edition, 2011)

Feser, E. (2008). *The Last Superstition: A Refutation of the New Atheism*. South Bend, IN: St. Augustine's Press.

Feser, E. (2009). *Aquinas: A Beginner's Guide*. Oxford: One World.

Finlay, G. (2003). Homo divinus: The ape that bears God's image. *Science and Christian Belief*, **15** (1):1–96.

Finlay, G. (2004a). Just glorified apes? In G. Finlay (Ed.), *A Seamless Web: Science and Faith*. Auckland, New Zealand: TELOS Publications.

Finlay, G. (2004b). *God's Books Genetics and Genesis*. Auckland, New Zealand: TELOS Publications.

Finlay, G. (2008). Evolution as created history. *Science and Christian Belief*, **20** (1):67–90.

Finlay, G. (2013). *Human Evolution: Genes, Genealogies and Phylogenies*. Cambridge, UK: Cambridge University Press.

Fischer, J. M., Pereboom, D., and Vargas, M. (2007). *Four Views on Free Will*. Oxford: Blackwell.

Foote, M. (1997). Sampling, taxonomic description, and our evolving knowledge of morphology disparity. *Paleobiology*, **23**:181–206.

Flannagan, M. (2011). Stoning adulterers. *Christian Research Journal*, **34** (6). (See http://www.equip.org/article/stoning-adulterers/#christian-books-3, accessed in August, 2105.)

Frankl, V. E. (1969). Discussion of J. R. Smythies's paper, "Some Aspects of Consciousness" in *Beyond Reductionism*, A. Koestler and J. R. Smythies (Eds). London: Hutchinson Publishing Group.

Frankl, V. E. (1992). *Man's Search for Meaning: An Introduction to Logotherapy*, 4th edit. Boston: Beacon Press. (This book has been republished a number of times.)

Frankl, V. E. (2000). *Man's Search for Ultimate Meaning*. Basic Books.

Funk, R. W. (1985). The Issue of Jesus. *Foundations and Facets Forum* **1**.

Funk, R. M., Hoover, R.W., and the Jesus Seminar. (1993). *Introduction to the Five Gospels*. New York: Macmillan.

Gallup, G. G. Jr., Anderson, J. R., and Shillito, D. J. (2002). The Mirror Test. In M. Bekoff, C. Allen, and G. Burghardt (Eds), *The Cognitive Animal*, pp. 325–334. Cambridge, MA: MIT Press.

Gee, H. (1999). *In Search of Deep Time: Beyond the Fossil Record to a New History of Life*. New York: The Free Press.

Geisler, N. L. (1999). *Baker Encyclopedia of Christian Apologetics*. Grand Rapids, MI: Baker Books.

Geisler, N. L. and Howe, T. (1992). *When Critics Ask*. Grand Rapids, MI: Baker Books.

Geisler, N. L. and Howe, T. (1992a). *The Big Book of Bible Difficulties*. Grand Rapids, MI: Baker Books.

Geisler, N. L. and Nix, W. E. (2012). *From God to Us: How We Got our Bible*. Chicago: Moody Publishers.

Gilbert, S. F. (2000). *Developmental Biology*. Sunderland, MA: Sinauer Associates.

Gisin, N. (2013). Are there quantum effects coming from outside spacetime? Non-locality, free will and "no many-worlds". In A. Suarez and P. Adams (Eds), *Is Science Compatible with Free Will? Exploring Free Will and Consciousness in the Light of Quantum Physics and Neuroscience*, chapter 3. New York: Springer.

Glass, D. H. (2012). *Atheism's New Clothes: Exploring and Exposing the Claims of the New Atheists*. Nottingham, England: Inter-Varsity Press.

Glueck, N. (1959). *Rivers in the Desert*. New York: Farrar, Strous, and Cudahy.

Goetz, S. (2011). Making things happen: souls in action. In M. C. Baker and S. Goetz (Eds), *The Soul Hypothesis: Investigations into the Existence of the Soul*, pp. 222–246. New York: Continuum International.

Gokhman, R., Lavi, E., Prüfer, K., Fraga, M. F., Rinacho, J. A., Kelso, J., Pääbo, S., Meshorer, E., and Carmetl, L. (2014). Reconstructing the DNA methylation maps of the Neandertal and the Denisovan. *Science*, **344** (6183):523–527.

Gould, S. J. (1980). Is a new and general theory of evolution emerging? *Paleobiology*, **6**:119–130.

Greene, B. (2003). *The Elegant Universe: Superstrings, Hidden Dimensions, and the Quest for the Ultimate Theory*. New York: W.W. Norton.

Griffin, D. R. (1976). *The Question of Animal Awareness: Evolutionary Continuity of Mental Experience.* New York: Rockefeller University Press.

Griffin, D. R. (1984). *Animal Thinking.* Cambridge, MA: Harvard University Press.

Griffin, D. R. (1992). *Animal Minds.* Chicago: University of Chicago Press.

Habermas, G. (1996). *The Historical Jesus: Ancient Evidence for the Life of Christ.* Joplin, MO: College Press.

Haggard, P. (2008). Human volition: towards a neuroscience of will. *Nature Reviews Neeuroscience,* **9** (12):934–946.

Haldane, J. B. S. (1927). *Possible Worlds: And Other Essays.* London: Chatto and Windus. Has been reprinted with additions in 2002 by Transaction Publishers, New Brunswick, NJ, and further reprinted several times.

Hamer, D. H. (2004). *The God Gene: How Faith is Hardwired into our Genes.* New York: Doubleday.

Hardy, A. C. (1965). *The Living Stream; a restatement of Evolution theory and its Relation to the Spirit of Man.* London: Collins.

Harris, S. B. (2006). *Letter to a Christian Nation.* New York: Knof Publishing Group.

Hasel, G. F. (1994). The "days" of creation in Genesis 1: Literal "days" or figurative "periods/epochs" of time? *Origins,* **21** (1):5–38.
(See also http://www.ldolphin.org/haseldays.html, accessed August, 2105.)

Hauser, M. D. (2006). *Moral Minds: How Nature Designed Our Universal Sense of Right and Wrong.* New York: Harper Collins.

Hawking S. (1988). *A Brief History of Time.* New York: Bantam Books.

Hawking, S. W. and Mlodinow, L. (2005). *A Briefer History of Time.* New York: Bantam Books.

Heilbron, J. (1999). *The Sun in the Church: Cathedrals as Solar Observatories.* Cambridge: Harvard University Press.

Helm, P. (2001). The Augustinian-Calvinist view. In J. K. Beilby and P. R. Eddy (Eds), *Divine Foreknowledge: Four Views.* Downers Grove, Ill: InterVarsity Press.

Hemer, C. J. (1989). The book of Acts in the setting of Hellenistic history. In C. H. Gempf (Ed.), *Wissenschaftliche Untersuchungen zum Neuen Testament 49.* Tübingen: J. C. B. Mohr.

Hick, J. (1963). *Philosophy of Religion.* Englewood Cliffs, NJ: Prentice-Hall.

Hick, J. (1974). *Evil and the God of Love.* London: Fontana. (Second edition, 1977, with several reissues with new prefaces.)

Hilbert, D. (1964). On the infinite. In P. Benacerraf and H. Putnam (Eds), *Philosophy of Mathematics* 2nd edit. London: Cambridge University Press.

Hoggard-Creegan, N. (2013). *Animal Suffering and the Problem of Evil.* New York: Oxford University Press.

Holder, R. (2013). *Big Bang, Big God: A Universe Designed for Life?.* London: Lyon Books.

Holland, P. W. H. (2013). Evolution of homeobox genes. *Wiley Interdisciplinary Reviews: Developmental Biology,* **2**: 31–45. (doi: 10.1002/wdev.78).

Hordijk, W. Hein, J., and Steel, M. (2010). Autocatalytic sets and the origin of life. *Entropy,* **12**:1733–1742.

Hoyle, F. (1981). The universe: Some past and present reflections. *Engineering & Science*, November:8–12.
(See also http://calteches.library.caltech.edu/527/2/Hoyle.pdf, accessed August, 2015.)

Hoyle, F. (1983). *The Intelligent Universe*. London: Michael Joseph.

Hume, David. (1779). *Dialogues Concerning Natural Religion*. Various publishing companies; see the internet.

Hunt, D. (2001). The simple foreknowledge view. In J. K. Beilby and P. R. Eddy (Eds), *Divine Foreknowledge: Four Views*. Downers Grove, Ill.: InterVarsity Press.

James, W. (1950). *The Principles of Psychology*. Reprint of 1890 text. New York: Dover. (See https://ebooks.adelaide.edu.au/j/james/william/principles/, accessed August, 2015.)

James, W. (2001). *Psychology: Briefer Course*. Reprint of 1892 text. New York: Dover.

Jarvis, M. (2008). *BigBang Christianity*. Wellington, South Africa: Fact and Faith Publications.

Jastrow, R. (1984). *God and the Astronomers*. Warner Books. (New and expanded edition, 2000, W. W. Norton & Co.)

Jeffrey, G. R. (1996). *The Signature of God*. Toronto: Frontier Research Publications.

Jeffrey, G. R. (1997). *The Handwriting of God*. Toronto: Frontier Research Publications.

Jennings, H. S. (1917). *Behavior of the Lower Organisms*. Columbia University Biological Series 10. New York: Columbia University Press.

Johnson, L. T. (1996). *The Real Jesus*. San Francisco: Harper.

Johnson, P. (1995). *Paul Johnson in New Zealand*. Wellington, NZ: New Zealand Business Roundtable.

Johnstone, P. and Mandryk, J. (2001). *Operation World: 21st Century Edition*. Harrisonburg, VA: R. R.. Donnelley and Sons.

Jones, A.Z. and Robbins, D. (2009). *String Theory for Dummies*. New York: Wiley.

Kane, R. H. (2005). *A Contemporary Introduction to Free Will*. New York: Oxford University Press.

Kant, I. (1788). *Critique of Practical Reason*. Translation published, for example, by Cambridge University Press, Cambridge, in 1997, and reprinted several times.

Kaufmann, S. A. (1995). *At Home in the Universe, The Search for Laws of Self-Organization and Complexity*. New York: Oxford University Press.

Keener, C. S. (2011). *Miracles: The Credibility of the New Testament Accounts*, (2 volumes). Grand Rapids, MI: Baker Academic.

Keller, T. (2008). *The Reason for God: Belief in an Age of Scepticism*. London: Hodder & Stoughton.

Kelly, E. F., Kelly, E. W., Crabtree, A., Gauld, A., Grosso, M., and Greyson, B. (2007). *Irreducible Mind: Toward a Psychology for the 21st Century*. Lanham, MD: Rowman and Littlefield.

Kenyon, F. (1940). *The Bible and Archaeology*. New York: Harper.

Kitchen, K. A. (2003). *On the Reliability of the Old Testament.* Grand Rapids, MI: Eerdmans.

Koenig, H. G., King, D. E., and Carson, V. B. (2012). *Handbook of Religion and Health,* 2nd edit. New York: Oxford University Press.

Kreeft, P. (1986). *Making Sense out of Suffering.* Ann Arbor, MI: Servant Books.

Küng, H. (1984). *On Being a Christian.* New York: Doubleday.

Kurtzweil. R. (2002). The evolution of the mind in the twenty-first century. In J. W. Richards (Ed.), *Are We Spiritual Machines?* Seattle, WA: Discovery Institute.

Lashley, K. S. (1950). In search of the engram. *Society of Experimental Biology Symposium,* **4**:454–482.

Leslie, J. (1989). *Universes.* New York: Routledge.

Levin, J. S. (1994). Religion and health: is there an association, is it valid, and is it causal? *Social Science Medicine,* **38**:1475–1482.

Lewin, R. (1980). Is your brain really necessary? *SCIENCE,* **210** (4475):1232-1234. (See also http://www.rifters.com/real/articles/Science_No-Brain.pdf, accessed August, 2015.)

Lewis, C. S. (1967). *Christian Reflections.* Grand Rapids, MI: Wm. B. Eerdmans.

Lewis, C. S. (2001a). *Mere Christianity.* New York: HarperCollins. (Originally published in 1952. Reprinted many times)

Lewis, C. S. (2001b). *The Problem of Pain.* New York: HarperCollins. (Originally published in 1940.)

Lewis, C. S. (2001c). *Miracles: A Preliminary Study.* New York: HarperCollins. (Originally published in 1947 and revised in 1960.)

Lewis, C. S. (2001d). *God in the Dock: Essays on Theology and Ethics.* Grand Rapids, MI: Wm. B. Eerdmans. (Originally published in 1970.)

Libet, B. (2004). *Mind Time: The Temporal Factor in Consciousness.* Cambridge, Mass: Harvard University Press.

Licona, M. R. (2010) *The Resurrection of Jesus: A New Historiographical Approach.* Downers Grove: InterVarsity.

Lindberg, D. C. and Numbers, R. L. (1986). *God and Nature: Historical Essays on the Encounter between Christianity and Science.* Berkeley and Los Angeles: University of California Press.

Livio, M. (2002). *The Golden Ratio: The Story of Phi, the World's Most Astonishing Number.* New York: Broadway Books.

Long, J. and Perry, P. (2010). *Evidence of the Afterlife: The Science of Near-Death Experiences.* New York: HarperCollins.

Lorenz, E. N. (1972). Predictability: Does the flap of a butterflys wings in Brazil set off a tornado in Texas? Presented before the American Association for the Advancement of Science, December 29.

Lovejoy, A. O. (1962). *The Great Chain of Being.* Cambridge, Mass: Harvard University Press.

Lubenow, M. L. (1992). *Bones of Contention.* Grand Rapids, MI: Baker.

Mackay, D. M. (1974). *The Clockwork Image: A Christian Perspective On Science.* Downers Grove, IL: InterVarsity Press.

Malin, S. (2001). *Nature Loves to Hide: Quantum Physics and Reality, a Western Perspective.* Oxford; New York: Oxford University Press.

Margulis, L. (2001). The conscious cell. *Annals of the New York Academy of Sciences*, **929**:55–70.

Marks, M. (2003). Cognitive therapy for OCD. In R. G. Menzies and P. de Silva (Eds), *Obsessive Compulsive Disorder: Theory, Research, and Treatment*, pp. 275–290. Chichester, UK: Wiley.

Mather, J. A. (2008) Cephalopod consciousness: Behavioral evidence. *Consciousness and Cognition*, **17**:37–48.

Mayr, E. (1988). *Towards a New Philosophy of Biology: Observations of an Evolutionist.* Cambridge: Harvard University Press.

Mbiti, J. S. (1990). *African Religions and Philosophy*, 2nd edit. Heinemann: Blackwell companion.

McDonald, J. F. (1983). The molecular basis of of adaption: A critical review of relevant ideas and observations. *Annual Review of Ecology and Systematics*, **14**:77–102.

McGrath, A. E. (1997). *An Introduction to Christianity.* Cambridge, MA: Blackwell.

McGrath, A. E. (2002). *Glimpsing the Face of God: The Search for Meaning in the Universe.* Grand Rapids: Eerdmans.

McGrath, A. (2004). *The Twilight of Atheism: The Rise and Fall of Disbelief in the Modern World.* Oxford University Press.

McGrath, A. E. (2005). *Dawkins's God: Genes, Memes, and the Meaning of Life.* Oxford, UK: Blackwell.

McGrew, T. and McGrew, L. (2009). The argument from miracles: A cumulative case for the resurrection of Jesus of Nazareth. In W. L. Craig and J. P. Moreland (Eds), *The Blackwell Companion to Natural Theology*, pp. 593–662. Blackwell. (See also http://www.lydiamcgrew.com/Resurrectionarticlesinglefile.pdf, accessed August, 2015.)

McKenna, M., and Coates, D. Justin. (2015) Compatibilism. In E. N. Zalta (Ed.), *The Stanford Encyclopedia of Philosophy.* Stanford University Press. (See http://plato.stanford.edu/entries/compatibilism/, accessed August, 2105.)

McRay, J. (1991). *Archaeology and the New Testament.* Grand Rapids: Baker.

Megidish, E., Halevy, A., Shacham, T., Dvir, T., Dovrat, L., and Eisenberg, H. S. (2013). Entanglement swapping between photons that have never coexisted. *Physical Review Letters*, **110** (21):210403.

Melendez-Hevia, E., Wadell, T. G., and Cascante, M. (1996). The puzzle of the Krebs citric acid cycle: assembling the pieces of chemically feasible reactions, and opportunism in the design of metabolic pathways during evolution. *Journal of Molecular Evolution*, **43** (3):293–303.

Meyer, S. C. (1999). Evidence for design in Physics and Biology. In M. J. Behe, W. A. Dembski, and S. C. Meyer (Eds), *Science and Evidence for Design in the Universe* pp. 53–111. San Francisco: Ignatius.

Meyer, S. C. (2012). Signature in the cell. In J. B. Stump and Alan G. Padgett (Eds), *The Blackwell Companion to Science and Christianity*, pp. 270–282. Chichester: Wiley-Blackwell.

Meyer, S. C. (2013). *Darwin's Doubt: The Explosive Origin of Animal Life and the Case for Intelligent Design.* New York: Harper-Collins.

Mills, G. C. and Kenyon, D. (1996). The RNA world. *Origins and Design*, **17** (1). (See http://www.arn.org/docs/odesign/od171/rnaworld171.htm, accessed August, 2015.)

Misler, C. (1999). *Cosmic Codes*. Coeur d'Alene, ID: Koinonia House.

Mithani, A. T. and Vilenkin, A. (2012). Collapse of simple harmonic universe. *Journal of Cosmology and Astroparticle Physics*, **2012** (01):028.

Mohrhoff, U. (2007). Book Review of "The spiritual brain: A neuroscientist's case for the existence of the soul." *AntiMatters* **1** (2):165–174. (Also available online at http://anti-matters.org/articles/33/public/33-28-1-PB.pdf, accessed August, 2015.)

Moody, R. A., Jr. (2001). *Life After Life*, 2nd edit. San Francisco: Harper.

Moody, R. A., Jr., and Perry, P. (2011). *Glimpses of Eternity: Sharing a Loved One's Passage from this Life to the Next*. Franklin, TN: Ideals publications.

Moore, J. N. and Cuffey, R. J. (1972). Paleontologic evidence and organic evolution. *Journal of the American Scientific Affiliation*, **24**:167–174.

Moreland, J. P. and Craig, W. L. (2003). *Philosophical Foundations of a Christian Worldview*. Downers Grove: IVP.

Morgan, R. J. (2011). *Angels: True Stories*. Nashville, TN: Thomas Nelson.

Morison, F. (1930). *Who Moved the Stone?* London: Faber and Faber. This book has been reprinted many times (e.g., 2006).

Murray, M. (2008). *Nature Red in Tooth and Claw: Theism and the Problem of Animal Suffering*. Oxford Scholarship Online.

Myers, D. G. and Reeves, M. A. (2003). *Psychology Through the Eyes of Faith*, revised edit. New York: HarperCollins.

Nagel, A. H. M. (1997). Are Plants Conscious? *Journal of Consciousness Studies*, **4** (3):215–230.

Nagel, T. (1997). *The Last Word*. New York: Oxford University Press.

Nelson, J. M. (2009). *Psychology, Religion, and Spirituality*. New York: Springer.

Nelson, K. (2010). *The Spiritual Doorway in the Brain: A Neurologist's Search for the God Experience*. Dutton: New York.

Nietzsche, F. (1968). *The Will to Power*. New York: Vintage Books.

North, G. (1994). *Leviticus: An Economic Commentary*. Tyler, TX: Institute for Christian Economics,

Norris, P. and Inglehart, R. (2004). *Sacred and Secular: Religion and Politics Worldwide*. New York: Cambridge University Press. (See also http://www.hks.harvard.edu/fs/pnorris/, accessed August, 2015)

O'Malley, J. W., Bailey, G. A., Harris, S. J., and Kennedy, T. F. (1999). *The Jesuits: Cultures, Sciences, and the Arts, 1540–1773*. Toronto: University of Toronto Press.

Overman, D. L. (2009). *A Case for the Existence of God*. Lanham, MD: Rowman & Littlefield.

Payne, J. B. (1973). *Encyclopedia of Biblical Prophecy*. New York: Harper and Row.

Pearce, E. K. V. (1969). *Who was Adam?*. Exeter, Great Britain: Paternoster.

Penfield, W. (1975). *The Mystery of the Mind*. Princeton: Princeton University Press.

Penrose, R. (1989). *The Emperor's New Mind*. New York: Oxford University Press.

Penrose, R. (2010). *Cycles of Time: An Extraordinary New View of the Universe*. London: Random House.

Pepperberg, I. M. (1999). *The Alex Studies: Cognitive and Communicative Abilities of Grey Parrots*. Cambridge, MA: Harvard University Press.

Perry, M. J. (2007). *Toward a Theory of Human Rights: Religion, Law, Courts*. New York: Cambridge University Press.

Plantinga, A. (1977). *God, Freedom, and Evil*. Grand Rapids, MI: Eerdmans. Reprinted 2001.

Plantinga, A. (1982). Tooley and evil: A reply. *Australasian Journal of Philosophy*, **60** (1):66–75.

Plantinga, A. (1993). A Christian life partly lived. In K. J. Clark (Ed.), *Philosophers Who Believe: The Spiritual Journeys of 11 Leading Thinkers*. Downers Grove, IL: InterVarsity Press.

Plantinga, A. (2000). *Warranted Christian Belief*. Oxford: Oxford University Press.

Plantinga, A. (2011). *Where the Conflict Really Lies: Science, Religion, & Naturalism*. New York: Oxford University Press.

Plantinga, A. (2012). The evolutionary argument against naturalism. In J. B. Stump and A. G Padgett (Eds), *The Blackwell Companion to Science and Christianity*, pp. 103–115. New York: Blackwell.

Polkinghorne, J. C. (1991). *Reason and Reality: The Relationship between Science and Theology*. London: SPCK.

Polkinghorne, J. C. (1998a). *Belief in God in an Age of Science*. New Haven: Yale University Press.

Polkinghorne, J. C. (1998b). *Science and Theology: An Introduction*. Minneapolis: Fortress.

Polkinghorne, J. C. (2002). *The God of Hope and the End of the World*. New Haven: Yale University Press.

Polkinghorne, J. C. (2005). *Science and Providence: God's Interaction with the World*, 2nd edit. West Consho-hocken, PA: Templeton Foundation.

Polkinghorne, J. C. (Ed.) (2011). *Meaning in Mathematics*. Oxford; New York: Oxford University Press.

Polkinghorne, J. C. (2012). Reflections of a bottom up thinker. In F. Watts and C. C. Knight (Eds), *God and the Scientist: Exploring the Work of John Polkinghorne*, Chapter 1. Farnham, England: Ashgate.

Polanyi, M. (1967). Life transcending physics and chemistry. *Chemical and Engineering News*, **21**:54–66.

Popper, K. R. (1963). *Conjectures and Refutations: The Growth of Scientific Knowledge*. London: Routledge. (Reprinted in 2004 by Routledge.)

Popper, K. R. and Eccles, J. C. (1977). *The Self and Its Brain*. New York: Springer-Verlag.

Prigogine, I. and Nicholis, G. (1977). *Self-organization in Nonequilibrium Systems: From Dissipative Structures to Order Through Fluctuations*. New York: Wiley.

Prigogine, I., Nicolis, G., and Babloyantz, A. (1972). Thermodynamics of Evolution. *Physics Today*, **23**:23 – 31 .

Prince, D. (1993). *The Spirit-Filled Believer's Handbook*. Orlando, FL: Creation House.

Prothero, D. (2007). *Evolution: What the Fossils Say and Why It Matters*. Columbia University Press: New York.

Qur'an, The Holy. (2009). Translated by Maulawī Sher 'Alī. Tilford, Surrey, UK: Islam International Publications.

Rambsel, Y. (1996). *The Hebrew Factor*. Shippensburg, PA: Companion Press.

Rambsel, Y. (1997). *His Name is Jesus*. Toronto: Frontier Research.

Ramsay, W. M. (1915). *The Bearing Of Recent Discovery On The Trustworthiness Of The New Testament*. London: Hodder and Stoughton.

Rana, F. (2008). *The Cell's Design: How Chemistry Reveals the Creator's Artistry*. Grand Rapids, MI: Baker Books.

Rees, M. (2000). *Just Six Numbers: The Deep Forces that Shape the Universe*. New York: Basic Books.

Ring, K. and Cooper, S. (1999). *Mindsight: Near-Death and Out-of-Body Experiences in the Blind*. Palo Alto, CA: William James Center for Consciousness Studies at the Institute of Transpersonal Psychology.

Robinson, M. D. (2001). *Eternity and Freedom: A Critical Analysis of Divine Timelessness as a Solution to the Foreknowledge/Free Will Debate*. New York: University Press of America.

Ross, H. (1998). *Big Bang Model Refined by Fire*. Pasadena, CA: Reasons To Believe.

Rota, M. (2010). A Problem for Hasker: Freedom with respect to the present, hard facts, and theological incompatibilism. *Faith and Philosophy*, **27** (3):287–305.

Ruse, M. (2001). *Can a Darwinian be a Christian?: The Relationship between Science and Religion*. Cambridge: Oxford University Press.

Ruse, M. (2012). Darwinism and Atheism. In J. B. Stump, A. G. Padgett (Eds), *The Blackwell Companion to Science and Christianity*, pp. 246–257. Chichester: Wiley-Blackwell.

Ryan, F. (2009). *Virolution*. London: HarperCollins.

Sabom, M. B. (1998). *Light and Death: One Doctor's Fascinating Account of Near-Death Experiences*. Zondervan Publishing House: Grand Rapids, Michigan. (See also http://www.near-death.com/experiences/evidence01.html for an interesting example; accessed August, 2015.)

Sanders, E. P. (1993). *The Historical Figure of Jesus*. London: Allen Lane.

Schonfield, H. J. (1967). *The Passover Plot: New Light on the History of Jesus*. New York: Bantam.

Schurger, A., Sitt, J. D., and Dehaene, S. (2012). An accumulator model for spontaneous neural activity prior to self-initiated movement. *Proceedings of the National Academy of Science*, **109** (42):2904–2913.

Schwartz, J. (1997). *Brain Lock: Free Yourself from Obsessive-Compulsive Behavior*. New York: Regan Books.

Schwartz, J. (2011). *You Are Not Your Brain: The 4-Step Solution for Changing Bad Habits, Ending Unhealthy Thinking, and Taking Control of Your Life*. New York: Penguin Group.

Schwartz, J. and Begley, S. (2002). *The Mind and the Brain: Neuroplasticity and the Power of Mental Force.* New York: Regan Books.

Seber, G. A. F. (2013). *Counseling Issues: A Handbook for Counselors and Psychotherapists.* Bloomington, IN: Xlibris.

Sherwin-White, A. N. (1963). *Roman Society and Roman Law in the New Testament.* Oxford: Clarendon Press.

Smith, L. A. (2007). *Chaos: A Very Short Introduction.* Oxford: Oxford University Press.

Smith, Q. (2001). The metaphilosophy of naturalism. *Philo,* **4** (2).

Smolin, L. (1997). *The Life of the Cosmos.* New York: Oxford University Press.

Snowdon, D. (2001). *Aging with Grace: What the Nun Study Teaches Us About Leading Longer, Healthier, and More Meaningful Lives.* New York: Bantam.

Spencer, N. (2014). *Atheists: The Origen of the Species.* New York: Bloomsbury.

Spinney, L. (2008). Back to their roots. *New Scientist,* **194** (2608):48–51.

Stapp, H. P. (2009). *Mind, Matter and Quantum Mechanics,* 3rd edit. Heidelberg: Springer-Verlag.

Stark, R. (2003). *For the Glory of God: How Monotheism Led to Reformations, Science, Witch-Hunts, and the End of Slavery.* Princeton: Princeton University Press.

Stenger, V. J. (2011).*The Fallacy of Fine-Tuning: Why the Universe is Not Designed for Us.* New York: Prometheus Books.

Stenhouse, J. (2004). Science, Religion, and History. In G. Finlay (Ed.), *A Seamless Web: Science and Faith.* Auckland, New Zealand: TELOS Publications.

Stewart, I. (1996). *From Here to Infinity.* Oxford: Oxford University press.

Stewart, I. (1997). *Does God Play Dice?* London: Penguin Books.

Stott, J. R. (1984). *Understanding the Bible.* London: Scripture Union.

Strobel, L. (1998). *The Case for Christ: A Journalist's Personal Investigation of the Evidence of Jesus.* Grand Rapids, MI: Zondervan.

Strobel, L. (2000). *The Case for Faith: A Journalist Investigates the Toughest Objections to Christianity.* Grand Rapids, MI: Zondervan.

Strobel L. (2004). *The Case for a Creator: A Journalist Investigates Scientific Evidence That Points Toward God.* Grand Rapids, MI: Zondervan.

Strobel L. (2008). *Finding the Real Jesus: A Guide for Curious Christians and Skeptical Seekers.* Grand Rapids, MI: Zondervan.

Swenson, R. (2000). Spontaneous order, autocatakinetic closure, and the development of space-time. *Annals New York Academy of Sciences,* **901**:311–319.

Swinburne, R. (2004). *The Existence of God.* Oxford: Oxford University Press. (Also available on line in 2007).

Talbott, T. (1990a). The doctrine of everlasting punishment. *Faith and Philosophy* **7**:19–42. (See also
http://www.reasonablefaith.org/talbotts-universalism#ixzz2MM4itzC9, accessed August, 2105.)

Talbott, T. (1990b). Providence, freedom, and human destiny. *Religious Studies,* **26**:227–245. (See also
http://www.reasonablefaith.org/talbotts-universalism#ixzz2MM55Ecbz, accessed August, 2015.)

Talbott, T. (1999). *The Inescapable Love of God*. Parkland, Florida: Universal.

Taylor, J. (2009). *Not a Chimp: The Hunt to Find the Genes that Make Us Human*. New York: Oxford University Press.

Tegmark, M. (2008). The mathematical universe. *Foundations of Physics*, **38**:101–150.

Thaxton, C. B., Bradley, W. L., and Olsen, R. L. (1984). *The Mystery of Life's Origin: Reassessing Current Theories*. Dallas, TX: Lewis and Stanley. (Second printing, 1992, available publicly online.)

Theobald, D. L. (2012). 29+ evidences for macroevolution: the scientific case for common descent. *The Talk. Origins Archive*. Version 2.89, at http://www.talkorigins.org/faqs/comdesc/, accessed August, 2105.

Thomas, A. (2002). Is your mind your brain? *Richmond Journal of Philosophy* **1** (1):23–26.

Toynbee, A. J. (1971). *Surviving the Future*. New York: Oxford University Press.

Trevena, J., and Miller, J. (2010). Brain preparation before a voluntary action: Evidence against unconscious movement initiation. *Consciousness and Cognition*, **19**:447–456.

Turek, F. and Geisler, N. L. (2004). *I Don't Have Enough Faith to Be an Atheist*. Wheaton, IL: Crossway books.

Udermann, B. E. (2000). The effect of spirituality on health and healing: A critical review for athletic trainers. *Journal of Athletic Training*, **35** (2):194–197.

van Lommel, P., van Wees, R., Meyers, V., and Elfferich I. (2001). Near-death experience in survivors of cardiac arrest: a prospective study in the Netherlands. *Lancet*, **358**:2039–2045. (See also http://profezie3m.altervista.org/archivio/TheLancet_NDE.htm, accessed August, 2015.)

Vilenkin, A. (2006). *Many Worlds in One: The Search for Other Universes*. New York: Hill and Wang.

Wallis, C. (1996). Faith and healing. *TIME*, **147** (26) June 24, 58–63.

Walton, J. H. (2009). *The Lost World of Genesis One*. Downers Grove, IL: InterVarsity Press.

Ward, K. (1982). *Rational Theology and the Creativity of God*. New York: Pilgrim.

Ward, K. (1984). *The Living God*. London: SPCK.

Ward, K. (2008a). *Why There Almost Certainly Is a God: Doubting Dawkins*. Oxford, UK: Lion Hudson.

Ward, K. (2008b). *The Big Questions in Science and Religion*. West Conshohocken, PA: Templeton Foundation.

Weinberg, S. (1994). *Dreams of a Final Theory: The Search for the Fundamental Laws of Nature*. New York: Vintage Books.

White, J. E. (2006). *Contemporary Moral Problems*, 8th edit. Belmont, CA: Wadsworth.

Whitefield, R. (2003). *Genesis One and the Age of the Earth: Comparing Biblical Hebrew with English Translation*. Publisher R. Whitefield (See a brief version at http://www.creationingenesis.com/Genesis_One_and_the_Age_of_the_Earth.pdf, accessed August, 2015.)

Wigner, E. P. (1960). The unreasonable effectiveness of mathematics in the physical sciences. *Communications on Pure and Applied Mathematics*, **13** (1):1–14.

Wilson, J. Q. (1993). *The Moral Sense*. New York: Free Press.

Wiseman, D. J. (Ed.) (1977). *Clues to Creation in Genesis*. London: Marshall, Morgan, and Scott.

Witztum, D, Rips, E., and Rosenberg, Y. (1994). Equidistant letter sequences in the book of Genesis. *Statistical Science*, **9** (3):429–438.

Woese, C. R. (2004). A new biology for a new century. *Microbiology and Molecular Biology Reviews*, **68** (2):173–186.

Wolterstoff, N. (2008). *Justice: Rights and Wrongs*. Princeton: Princeton University Press.

Wood, B. G. (1990). Did the Israelites Conquer Jericho? A New Look at the Archaeological Evidence. *Biblical Archaeological Review*, **16** (2):44–58. (See also http://www.biblearchaeology.org/post/2008/05/Did-the-Israelites-Conquer-Jericho-A-New-Look-at-the-Archaeological-Evidence.aspx#Article, accessed August, 2105.)

Woodberry, R. D. (2012). The missionary roots of liberal democracy. *American Political Review*, **106** (2):244–274.

Wright, C. J. H. (2004). *Old Testament Ethics for the People of God*. Leicester: IVP.

Wright, N. T. (1998). Christian origins and the resurrection of Jesus: The resurrection of Jesus as a historical problem. *Seewanee Theological Review*, **41** (2):107–123.. (Also available online at http://www.ntwrightpage.com/Wright_Historical_Problem.htm, accessed August, 2015.)

Wright, N. T. (2003). *The Resurrection of the Son of God*. Minneapolis, MN: Fortress.

Wright, N. T. (2006). *Simply Christian: Why Christianity Makes Sense*. New York: HarperCollins.

Yancey, (1990). *Where is God When it Hurts?* Grand Rapids, MI: Zondervan.

Zacharias, R. (2000). *Jesus Among Other Gods: The Absolute Claims of the Christian Message*. Nashville: Thomas Nelson.

Zimmer, C. (2004). A review of "The God Gene: How Faith is Hard-wired into Our Genes." *Scientific American* **29** (4):110–111.
(See also an online version at http://carlzimmer.com/articles/2004.php under "Faith-Boosting Genes.")

INDEX

A

AA program, 16
Abstract ideas, 96
Abstract mathematics, 77
Abstract objects, 23
Adam, 148, 278
  farmer in Turkey, 340
African religions, 147
Afterlife, 228
Agnosticism, 18
Altenberg 16, 335
Altruism, 62, 255
Amalekites, 172
Amorites, 173
Analogous features, 319
Ancestry of humans, 325
Angels, 88, 101, 280
Animal consciousness
  quorum sensing, 88
Animal suffering, 268, 281–282
Annihilationism, 233
Ant colony, 25
Anthropic principle
  strong version, 49
  weak version, 49
Antimatter, 43
Antiparticles, 38, 43
Antony Flew, 48
Ants, 91
Apocrypha, 132, 145
Apocryphal gospels, 164
Aquinas five ways, 33
Aquinas
  fifth way, 55
Archaeology, 146, 227
  New Testament, 156
  Old Testament, 140
Aristotle's four causes, 33
Aristotle, 303–304
Ark, 149
Assumptions of science, 2
Athanasius's Easter letter, 165
Atheism
  atrocities and, 300
  numbers of atheists, 72
Autocatalysis, 343

B

Babylonian creation story, 147

384 INDEX

Bahai, 223
Bara, 273
Beauty
  in nature, 39
  of mathematics, 39
  theology of, 39
Bell inequalities, 113
Best possible world, 260
Bhagavad-Gita, 224
Bible and Science, 178
Biblical ark, 149
Biblical contradictions, 167
Biblical creation, 53
Big Bang model, 27
Big Bang
  evidence for, 28
  light elements, 28
  Microwave background for, 28
Binding problem, 89
Black holes, 38, 47
  and new universes, 52
Body mind and spirit, 96
Bohr's correspondence principle, 115
Boundary conditions, 40, 42
Brahman, 220
Brain damage, 83, 95
Brain lock, 95
Brain
  quantum mechanics and, 114
Buddha, 158, 224, 226, 228
Buddhism, 220, 283, 286
Butterfly effect, 108, 256

C

Canaanites, 173
Carbon properties, 48
Cartesian dualism, 90, 110
Cell, 334
Chaos theory, 105, 256
  Mandelbrot set and, 108
  butterfly effect and, 108
  nonlinear processes, 106
  strange attractor, 106
Chaos
  fractals and, 108
Chemical evolution, 330
Child exposure, 308
Christian activism, 310
Christian aid, 300
Christianity
  equality of all people, 311
  health benefits of, 301
  marriage and, 307
  masters and slaves, 309
  parents and children, 308
  past violence and, 299

  science flourishing under, 302
  social contributions of, 310
  universal, 226
  universities and, 305
Church fathers, 151
Cilium, 345
Classical mechanics
  failure of, 110
Classical physics, 111
Closed system, 29, 31
Collapse of quantum states, 111
Collective intelligence, 91
Color vision, 328
Compatibilism, 118
Complex number, 32
Consciousness, 87
  animals and, 88
Conservation of energy, 29
Contemplation, 359
Contingency, 36
  radical, 37
  superficial, 37
Correlation and causation, 80
Cosmic rebound, 31
Cosmological argument, 34
Cosmological constant, 32
Counterfactual, 128
COX8H pseudogene, 327
Creation myths, 53

D

Daniel
  book of, 143
Dark energy, 20
Dark matter, 20
Dead Sea scrolls, 135
  Daniel and, 143
  Essenes and, 187
  Isaiah, 143
  Isaiah and, 135
Deism, 18
Demon possession, 100
Demonic influences, 100
Dependent variables, 86
Depression model, 2
Design
  appearance of, 54
Determinism, 118
  Calvinism and, 118
  theistic, 118
Deterministic process, 105
Deutero-Isaiah theory, 143
Deuterocanonical books, 145
Deuteronomy, 142
Dinosaurs, 348
Divine foreknowledge, 127

DNA code, 57
DNA
  structure, 59
Double-slit experiment, 110
Drugs and spirituality, 82
Dualism, 90

E

Earth's properties, 47
Ebla tablets, 147
Ecclesiastes, 290, 357
Efficient cause, 33
Ego states, 119
Eightfold Path, 229
Elephant story, 5
Elisha, 174
Emergent properties, 86, 89
Entanglement, 112
Entropy, 29
Epigenetics, 59, 263, 320, 322, 326, 338
Epilepsy patients, 90
Equidistant letter sequence, 150
ERVs, 328
Essenes, 135, 187
Euthyphro's dilemma, 66
Eve, 278
Evidence and probability, 68
Evidence for evolution, 318
Evil
  argument for God's existence, 255
  definition of, 99
  God and, 134
  New Testament and, 280
  not created, 255
  Old Testament and, 280
  one day destroyed, 262
Evo-devo, 315
Evolution
  analogous features, 319
  animal vestigial organs and, 318
  biological, 314
  cell and, 334
  change methods and, 320
  chemical, 330
  computer simulation and, 347
  conservatism and, 316
  convergent, 319
  evidence for, 318
  God and, 347
  homology features, 319
  human, 324, 340
  human vestigial organs and, 321
  of everything, 316
  of proteins, 331
  poor design and, 321
  prebiotic, 330

  right conclusions and, 23
  RNA and, 333
  the Fall and, 275
  theistic, 313
  thought processes and, 341
  transitional forms and, 338
  truth and, 317, 342
Evolutionary changes, 320
Evolutionary developmental biology, 60, 315
Exclusive religions, 219
Exodus, 139
Extinction events, 347

F

Faith, 16, 19
Faith healing, 101
Fallen world, 148
Falsifiability in science, 11
Falsification of science, 11
Fatalism, 118
Fibonacci sequence, 6
Fideism, 16
Final cause, 33, 333
Fine-structure constants, 44
Fine tuning, 44
Flagellum, 345
Flood account, 148
Food laws, 179
Fossils
  animal, 336
  Cambrian explosion and, 336
  humanoid, 340
Four noble truths, 229
FOXP2 gene, 326
Free will, 235, 262
  definition of, 103
Fundamental constants, 40–41
Futility, 357

G

Gödel, 12
Galileo, 303
Gambling, 119
Garden of Eden, 148, 278
Gas laws, 241
Gas model, 2
Gene conservatism, 61
Gene switches, 59–60
General revelation, 355
Genesis, 138, 148
  early chapters of, 147
  views of, 271
Genetic code, 57
Genetic inversion, 326
Genetic parasites, 329

God-of-the-gaps, 55
God outside of time, 35
God
  and time, 22
  art and, 352
  communicate with, 359
  community and, 363
  confession to, 361
  definition of, 15
  evil and, 134
  evolution and, 347
  first cause, 34
  free creator, 249
  gene for, 81
  images of, 353
  infinite, 352
  knowable, 351
  life without, 357
  middle knowledge, 128
  more than one, 38
  natural knowledge, 128
  nature of, 38, 226
  providence of, 238, 291
  reaching out to, 358
  role of creation, 355
  simplicity of, 56
  sovereignty and accountability, 124
  sovereignty of, 124, 133, 280
  unfair?, 232
  voice of, 362
  wired for, 81
Golden proportion, 7
Good Samaritan, 195
Goodness, 99
Gospel consistency, 159
Gospel of Mary, 164
Gospel of Thomas, 164
Group mind, 25
Group theory, 43
Guilt, 63

H

Heat death, 30
Heaven, 233
Hebrew language, 150
Hegel, 316
Heisenberg's Uncertainty Principle, 109
Hell, 232, 235
Hilbert, 24
Hinduism, 220, 283
Holy Spirit transformation, 361
Homologies, 319, 339
Hox genes, 61, 315, 339
Hubble Space Telescope, 26
Human accountability, 125
Human biblical models, 97
Human evolution, 324, 340
Human rights, 63
Hume, 240, 255
  miracles and, 239
Husband and wife, 306

I

Image of God, 2, 17, 97
Impartial world, 265
Indeterminism, 117–118
Inerrancy of the Bible, 131
Infinity, 24, 35
Information, 59
Initial male and female, 147
Intrusive thoughts, 122–123
Irreducibly complex system, 344
Isaiah
  book of, 143
Islam, 219, 227
Isolated system, 29

J

James' letter, 155
JEDP theory, 137
Jesus' genealogies, 163
Jesus' genealogy, 148
Jesus seminar, 188
Jesus
  authority of, 197
  abilities of, 192
  claims of, 195
  death of, 202, 295
  ethics of, 185
  evidence for resurrection, 204
  existence of, 184
  hostile witnesses, 251
  miracle birth of, 200
  miracle worker, 199
  Muslim argument and, 198
  Paul's early writing, 203
  portraits of, 186
  prophecies for, 188
  without sin, 193
Jewish sacrificial system, 140
Jews and gentiles, 306
Job, 134, 260, 280, 289
Joseph of Arimathea, 205
Joshua, 142
Judaism, 219
Judges, 142

K

Kālam cosmological argument, 35
Karma, 286
Kepler's law, 11

Krishna, 220, 224

L

Latin vulgate, 145
Laws of nature, 2
Legends
   biblical, 132
   Moses and, 139
   New Testament, 159, 156, 152
Leviticus, 139
Libertarianism, 118
Libet's experiments, 120
Life-permitting universe, 47
Life without God, 357
LINE elements, 329
Linking laws, 85
Locality, 116
Long-Term Evolution Experiment (LTEE), 319
Lorenz, 105, 108
LRT elements, 329
Luke's history, 153
Luke
   archaeology and, 153

M

Macroevolution, 320
Marx, 316
Mary Magdalene, 211
Masoretic text, 135–136
Masters and slaves, 309
Material cause, 33
Materialism, 4, 72
   non-reductive, 77
   reductive, 77
Mathematics
   abstract, 9, 24
   for an idealized world, 9
   in nature, 6
   inconsistent, 12
   limits of, 12
   model for reality, 24
   pendulum experiment, 10
   reality and, 8
   role of, 38
   undecidable problems, 12
Maximum entropy production, 30
Mercury's orbit, 11
Midrash, 218
Middle knowledge, 128, 238
Mind-brain system irreducible, 91
Mind and brain, 84, 91
Miracles
   biblical, 243
   categories of, 249
   definition of, 237
   Hume and, 239
   providence and, 238
   signs wonders and powers, 248
   Spinoza and, 239
Missionaries, 301, 303
Mithraism, 222
Moabite stone, 146
Molinism, 128
Monism, 90
Monotheism
   polytheism and, 147
Moral argument, 61
Moral law giver, 65
Morality, 256
   evolution and, 62
   in animals, 64
   law giver, 65
   natural law and, 64
   relative, 66
   relativism, 62
   universal, 61
Muhammad, 158, 224, 229, 353
Multiverses, 31, 50–51, 53, 114
   and personal explanation, 5
   selection bias, 51
Muslims, 226, 230
Mutations, 314
Mystery religions, 222
Mythology, 275
Nabonidus Chronicle, 144

N

Natural disasters, 265
Natural selection
   insufficient, 315
Natural theology, 71
Naturalism, 4
Near-death experiences, 92
   brain dead, 93
   blindness and, 93
   shared, 94
Neurodynamics
   stochastic, 118
Neurosplasticity, 83
Neurotheology, 79
New Age, 221, 283
New atheists, 16
New Caledonian crows, 75
New Testament, 151
   authorship of, 151
   accuracy of gospels, 158
   archaeology and, 156
   canon of, 164
   dating of, 152
   disagreements in, 163
   gospel consistency, 159

James' letter, 155
Luke's gospel, 153
manuscripts, 157
miracles in, 244
Paul's letters, 154
Peter's letters, 155
sacrificial system, 141
slavery and, 309
Newton, 11, 304
Non-conservation of energy, 91
Non-hearers, 230
Non-locality, 113
Nothing but, 74
Nuclear forces
strong, 44
weak, 44
Numbers, 141

O

Obsessive Compulsive Disorder (OCD), 94, 124
Old Testament
archaeology and, 146
canon, 144
codes and, 150, 191
David, 290
Deuteronomy, 142
Ecclesiastes, 290
Elijah, 288
Exodus, 139
food laws, 179
Genesis, 138
hiegene laws, 179
internal evidence for, 136
Isaiah, 143
Jeremiah, 288
Job, 289
Joseph, 287
Joshua, 142
Judges, 142
Leviticus, 139
manuscripts for, 135
miracles in, 244
Numbers, 141
sacrificial system, 140
unpleasant stories, 171
OldTestament
Daniel, 143
Omnipotence, 259
Oparin-Haldane hypothesis, 331
Open theism, 125
Oral tradition, 158, 204
Original atmosphere, 330
Orphan genes, 339

P

Pain
positive aspects of, 257
Panentheism, 18
Panpsychism, 90
Pantheism, 18, 220, 286
Parable of the wheat and weeds, 270
Parents and children, 308
Patria potestas, 308
Paul's conversion, 204
Paul's letters
dating, 154
Paul's conversion, 154
PAX-6 gene, 316
Pentateuch, 136
JEDP theory and, 137
Personal explanation, 4, 305
Peter's letters, 155
Photon, 110, 112, 114
Photosynthesis, 48
Phylogenic tree, 320
Physicalism, 4
Planck length, 32
Planck time, 32, 44
Plato, 24
Plenitude, 269
Pluralism, 223, 356
Polanyi, 56
Polygamy, 174
Poor design
animals and, 321
humans and, 323
Popper, 11
Prayer, 102, 242, 362
Predictability, 117
Privation, 254
Probability waves, 109
Probability
Bayes' factor, 71
conditional independence and, 213
independence and, 212
nature of, 68
of God's existence, 69
of resurrection, 212
subjective, 69
Process theology, 18
Prophecy, 165, 188, 205–206
Protein evolution, 331
Pseudogenes, 325
Pseudorandom numbers, 104
Purgatory, 234

Q

Qualia, 84
Quantum chaos, 115, 117
Quantum gravity theory, 21
Quantum mechanics, 110

Bell inequalities and, 113
brain and, 114
entanglement, 112
non-locality, 112
non-locality and, 113
uncertainty and, 117
Quarks, 43
Quorum sensing, 88
Qur'an, 219, 224–225, 227–228, 353

R

Randomness, 104
Ransom for penalities, 171
Rape, 174
Readiness potential, 120
Reduced Planck's constant, 109
Reincarnation, 220
  suffering and, 286
Relativism, 223
Relativity, 5
Relativity theory, 116
  twin paradox, 116
Relativity
  general, 11
  special, 116
Religions are exclusive, 219
Religious experiences, 67
Rest on day 7, 274
Resurrection
  evidence for, 204
  some alternative views of, 217
Retroviruses, 328
RNA world hypothesis, 334
Royal Society, 303

S

Salvation, 228
Salvation history, 132
Samaritan Pentateuch, 136
Samaritans, 225
Satan, 280, 285
Schrödinger wave function, 111
Science-of-the-gaps, 55
Science-worship, 19
Science and the Bible, 178
Science
  assumptions of, 2
  Christian input to, 302
  falsifiability of, 11
  God-like status, 13
  laws of, 240
  legacy of, 3
  philosophy and, 3
  popularizers of, 5
  provisional acceptance of, 11
  specialization in, 5

Scriptural inspiration, 131, 180
Second chance, 234
Second law of thermodynamics, 29, 266, 342
Secular humanism, 18
Secularization hypothesis, 78
Self-consciousness, 87, 90
Self-defeating statements, 3
Self-determinism, 117
Self-recognition, 88
Septuagint, 136, 145, 163
Simple foreknowledge, 127
Sin, 227, 360
SINE elements, 329
Singularity, 27
Slavery
  abolishing of, 309
  New Testament and, 309
  Old Testament and, 176
Sodom and Gomorrah, 169
Solipsism, 12
Soul, 86, 89–91
Spinoza, 239
Spiritual battleground, 280
Spiritual death, 88
Spiritual discernment, 16
Spiritual evolution, 278
Spiritual forces, 99
Spiritual laws, 242
Spirituality, 78
  and religion, 78
  drugs and, 82
Spong, 218
Stochastic neurodynamics, 118
String theory, 21
Subatomic particles, 42
Suffering
  all in the mind, 287
  animals and, 281
  character building, 292
  coping with, 295
  disciplinary process, 292
  eternity and, 293
  God's judgment, 284
  other religions and, 283
  positive aspects of, 291
  psychological component of, 268
  refining process, 292
  reincarnation and, 286
  satanic attack, 285
  sickness and, 253
  sins of parents and, 284
  some practical aspects of, 296
  transformed, 295
Surah, 219

Symbiosis, 315

T

Teleological argument, 38
Theistic evolution, 313
Theory of everything, 21
Thermodynamics
  first law, 29
  second law, 29, 342
Thought processes, 341
Three-day motif, 158
Tolerance, 230
Torah, 136, 219
Traditional religions
  rise of, 78
Transcendental self, 87
Transitional forms, 338
Trinity, 198
Twin paradox, 116

U

Universalism, 234
Universe
  beginning of, 26
  center of, 23
  compactified dimensions for, 21
  dimensions of, 21
  expanding, 20
  flat, 20
  infinite past of, 31
  inflation model for, 52
  like a watch, 55
  origin of, 26
  parallel, 52
  size of, 20
Unmoved Mover, 36
Unpleasant Biblical stories, 171
Unpredictability, 104
Untestable hypothesis, 12
Uranus orbit, 11

V

Vacuum
  not nothing, 27
  quantum, 28
Vedas, 220
Vertebrate eye, 323
Vestigial organs
  animals and, 318
  humans and, 321
Virtual particles, 28

W

Watch example, 29, 56, 76
Water's properties, 48
Wave function, 111
Willpower, 115
Wistar conference, 333
World restoration, 283
Worry circuit, 95

Y

Yom, 271

Z

Zoroaster, 158

www.ingramcontent.com/pod-product-compliance
Lightning Source LLC
Chambersburg PA
CBHW071946220426
43662CB00009B/1023